Computable Structure Theory

In mathematics, we know there are some concepts—objects, constructions, structures, proofs—that are more complex and difficult to describe than others. Computable structure theory quantifies and studies the complexity of mathematical structures, structures such as graphs, groups, and orderings.

Written by a contemporary expert in the subject, this is the first full monograph on computable structure theory in 20 years. Aimed at graduate students and researchers in mathematical logic, it brings new results of the author together with many older results that were previously scattered across the literature and presents them all in a coherent framework, making it easier for the reader to learn the main results and techniques in the area for application in their own research. This volume focuses on countable structures whose complexity can be measured within arithmetic; a forthcoming second volume will study structures beyond arithmetic.

ANTONIO MONTALBÁN is Professor of Mathematics at the University of California, Berkeley.

PERSPECTIVES IN LOGIC

The *Perspectives in Logic* series publishes substantial, high-quality books whose central theme lies in any area or aspect of logic. Books that present new material not now available in book form are particularly welcome. The series ranges from introductory texts suitable for beginning graduate courses to specialized monographs at the frontiers of research. Each book offers an illuminating perspective for its intended audience.

The series has its origins in the old *Perspectives in Mathematical Logic* series edited by the Ω-Group for "Mathematische Logik" of the Heidelberger Akademie der Wissenschaften, whose beginnings date back to the 1960s. The Association for Symbolic Logic has assumed editorial responsibility for the series and changed its name to reflect its interest in books that span the full range of disciplines in which logic plays an important role.

More information, including a list of the books in the series, can be found at www.aslonline.org/books/perspectives-in-logic/

Computable Structure Theory

Within the Arithmetic

ANTONIO MONTALBÁN

University of California, Berkeley

ASSOCIATION FOR SYMBOLIC LOGIC

CAMBRIDGE
UNIVERSITY PRESS

CAMBRIDGE
UNIVERSITY PRESS

University Printing House, Cambridge CB2 8BS, United Kingdom

One Liberty Plaza, 20th Floor, New York, NY 10006, USA

477 Williamstown Road, Port Melbourne, VIC 3207, Australia

314–321, 3rd Floor, Plot 3, Splendor Forum, Jasola District Centre, New Delhi – 110025, India

79 Anson Road, #06–04/06, Singapore 079906

Cambridge University Press is part of the University of Cambridge.

It furthers the University's mission by disseminating knowledge in the pursuit of education, learning, and research at the highest international levels of excellence.

www.cambridge.org
Information on this title: www.cambridge.org/9781108423298
DOI: 10.1017/9781108525749

Association for Symbolic Logic
Richard A. Shore, Publisher
Department of Mathematics, Cornell University, Ithaca, NY 14853
http://aslonline.org

First published 2021

A catalogue record for this publication is available from the British Library.

ISBN 978-1-108-42329-8 Hardback

Dedicated to my family back home:
Manu, Facu, Mariana, Javier, Patricia, and Nino.

CONTENTS

PREFACE

We all know that in mathematics there are proofs that are more difficult than others, constructions that are more complicated than others,
and objects that are harder to describe than others. The objective of
computable mathematics is to study this complexity, to measure it, and
to find out where it comes from. Among the many aspects of mathematical practice, this book concentrates on the complexity of structures.
By *structures*, we mean objects like rings, graphs, or linear orderings,
which consist of a domain on which we have relations, functions, and
constants.

Computable structure theory studies the interplay between complexity
and structure. By *complexity*, we mean descriptional or computational
complexity, in the sense of how difficult it is to describe or compute a
certain object. By *structure*, we refer to algebraic or structural properties
of mathematical structures. The setting of computable structure theory is
that of infinite countable structures and thus, within the whole hierarchy
of complexity levels developed by logicians, the appropriate tools come
from computability theory: Turing degrees, the arithmetic hierarchy, the
hyperarithmetic hierarchy, etc. These structures are like the ones studied
in model theory, and we will use a few basic tools from there too. The
intention is not, however, to effectivize model theory, and our motivations
are very different that those of model theory. Our motivations come
from questions of the following sort: Are there syntactical properties that
explain why certain objects (like structures, relations, or isomorphisms)
are easier or harder to compute or to describe?

The objective of this book is to describe some of the main ideas and
techniques used in the field. Most of these ideas are old, but for many
of them, the style of the presentation is not. Over the last few years,
the author has developed new frameworks for dealing with these old
ideas—for instance, for forcing, r.i.c.e. relations, jumps, Scott ranks, and
back-and-forth types. One of the objectives of the book is to present these
frameworks in a concise and self-contained form.

The modern state of the field, and also the author's view of the subject,
has been influenced greatly by the monograph by Ash and Knight [AK00]

published in 2000. There is, of course, some intersection between that book and this one. But even within that intersection, the approach is different.

The intended readers are graduate students and researchers working on mathematical logic. Basic background in computability and logic, as is covered in standard undergraduate courses in logic and computability, is assumed. The objective of this book is to describe some of the main ideas and techniques of the field so that graduate students and researchers can use it for their own research.

This book is part I of a monograph that actually consists of two parts: *within the arithmetic* and *beyond the arithmetic*.

Part I, Within the arithmetic, is about the part of the theory that can be developed below a single Turing jump. The first chapters introduce what the author sees as the basic tools to develop the theory: ω-presentations, relations, and $\exists$-atomic structures, as treated by the author in [Mon09, Mon12, Mon13c, Mon16a]. Many of the topics covered in Part I (like Scott sentences, 1-generics, the method of true stages, categoricity, etc.) will then be generalized through the transfinite in part II. Σ-small classes, covered in the last chapter, have been a recurrent topic in the author's work, as they touch on many aspects of the theory and help to explain previously observed behaviors (cf. [HM12, HM14, Mon10, Mon13b]).

Part II, Beyond the arithmetic, moves into the realm of the hyperarithmetic and the infinitary languages. To fully analyze the complexity of a structure, staying within the arithmetic is not enough. The hyperarithmetic hierarchy goes far enough to capture the complexity levels of relations in almost all structures, though we will see there are some structures whose complexity goes just beyond. The first half of Part II develops the basic theory of infinitary logic, Π^1_1 sets, and the hyperarithmetic hierarchy. In the second half, the main chapters are those on forcing and the α-priority method. The exposition of forcing is only aesthetically new (similar to that in [HTMM]). The presentation of Ash's α-priority method will be more than just aesthetically different. It will use the method of α-true stages developed in [Mon14b]. We also draw connections with descriptive set theory, and some of the more recent work from [Mon13a, Mon15, MM18]. The chapter on comparability of classes treats old topics like Borel reducibility, but also newer topics like effective reducibility of classes of computable structures (cf. [FF09, FFH+12, Mon16b]) and the connections between functors and interpretability (cf. [HTMMM, HTMM]). Here is the tentative list of chapters of part II [MonP2]:

Acknowledgements. Many people helped in different ways throughout the long process that was writing this book, and I'm grateful to them all, though it'd be impossible to name them all. Many people sent me comments and typos along the years. First, I'd like to thank James Walsh for proof reading the whole book. Julia Knight, Peter Cholak, Richard A. Shore, and Barbara Csima ran seminars in Notre Dame, Cornell, and Waterloo following earlier drafts. They then sent me typos and comments, and got their students to send me typos too—that was extremely useful. In particular, Julia Knight gave me lots of useful feedback. So did my students Matthew Harrison-Trainor and Noah Schweber, and also Asher Kach, Jonny Stephenson, and Dino Rossegger.

I learned the subject mostly from Julia Knight and from my Ph.D. advisor Richard A. Shore. I also learned a lot form Ted Slaman, Rod Downey, and Denis Hirschfeldt. I owe them all a great debt.

My work was partially supported by NSF grants DMS-1363310 and DMS-1700361, by the Packard fellowship, and by the Simons fellowship 561299.

NOTATION AND CONVENTIONS

The intention of this section is to refresh the basic concepts of computability theory and structures and set up the basic notation we use throughout the book. If the reader has not seen basic computability theory before, this section would be too fast an introduction and we recommend starting with other textbooks like Cutland [Cut80], Cooper [Coo04], Enderton [End11], or Soare [Soa16].

The computable functions. A function is *computable* if there a purely mechanical process to calculate its values. In today's language, we would say that $f : \mathbb{N} \to \mathbb{N}$ is computable if there is a computer program that, on input n, outputs $f(n)$. This might appear to be too informal a definition, but the Turing–Church thesis tells us that it does not matter which method of computation you choose, you always get the same class of functions from $\mathbb{N}$ to $\mathbb{N}$. The reader may choose to keep in mind whichever definition of computability feels intuitively more comfortable, be it Turing machines, μ-recursive functions, lambda calculus, register machines, Pascal, Basic, C++, Java, Haskell, or Python.[1] We will not use any particular definition of computability, and instead, every time we need to define a computable function, we will just describe the algorithm in English and let the reader convince himself or herself that it can be written in the programing language he or she has in mind.

The choice of $\mathbb{N}$ as the domain and image for the computable functions is not as restrictive as it may sound. Every hereditarily finite object[2] can be encoded by just a single natural number. Even if formally we define computable functions as having domain $\mathbb{N}$, we think of them as using any kind of finitary object as inputs or outputs. This should not be surprising. It is what computers do when they encode everything you see on the screen using finite binary strings, or equivalently, natural numbers written in binary. For instance, we can encode pairs of natural numbers by a single number using the *Cantor pairing function* $\langle x, y \rangle \mapsto ((x+y)(x+y+1)/2+y)$,

---

[1] For the reader with a computer science background, let us remark that we do not impose any time or space bound on our computations—computations just need to halt and return an answer after a finitely many steps using a finite amount of memory.

[2] A hereditarily finite object consist of a finite set or tuple of hereditarily finite objects.

which is a bijection from $\mathbb{N}^2$ to $\mathbb{N}$ whose inverse is easily computable too. One can then encode triples by using pairs of pairs, and then encode n-tuples, and then tuples of arbitrary size, and then tuples of tuples, etc. In the same way, we can consider standard effective bijections between $\mathbb{N}$ and various other sets like $\mathbb{Z}$, $\mathbb{Q}$, V_ω, $\mathcal{L}_{\omega,\omega}$, etc. Given any finite object a, we use Quine's notation $\ulcorner a \urcorner$ to denote the number coding a. Which method of coding we use is immaterial for us so long as the method is sufficiently effective. We will just assume these methods exist and hope the reader can figure out how to define them.

Let

$$\Phi_0, \Phi_1, \Phi_2, \Phi_3, \ldots$$

be an enumeration of the computer programs ordered in some effective way, say lexicographically. Given n, we write $\Phi_e(n)$ for the output of the eth program on input n. Each program Φ_e calculates the values of a *partial computable function* $\mathbb{N} \rightharpoonup \mathbb{N}$. Let us remark that, on some inputs, $\Phi_e(n)$ may run forever and never halt with an answer, in which case $\Phi_e(n)$ is undefined. If Φ_e returns an answer for all n, Φ_e is said to be *total*— even if total, these functions are still included within the class of partial computable functions. The *computable functions* are the total functions among the partial computable ones. We write $\Phi_e(n)\downarrow$ to mean that this computation *converges*, that is, that it halts after a finite number of steps; and we write $\Phi_e(n)\uparrow$ to mean that it *diverges*, i.e., it never returns an answer. Computers, as Turing machines, run on a step-by-step basis. We use $\Phi_{e,s}(n)$ to denote the output of $\Phi_e(n)$ after s steps of computation, which can be either not converging yet $(\Phi_{e,s}(n)\uparrow)$ or converging to a number $(\Phi_{e,s}(n)\downarrow = m)$. Notice that, given e, s, n, we can decide whether $\Phi_{e,s}(n)$ converges or not, computably: All we have to do is run $\Phi_e(n)$ for s steps. If f and g are partial functions, we write $f(n) = g(m)$ to mean that either both $f(n)$ and $g(m)$ are undefined, or both are defined and have the same value. We write $f = g$ if $f(n) = g(n)$ for all n. If $f(n) = \Phi_e(n)$ for all n, we say that e is an *index* for f. The *Padding Lemma* states that every partial computable function has infinitely many indices—just add dummy instructions at the end of a program, getting essentially the same program, but with a different index.

In his famous 1936 paper, Turing showed there is a partial computable function $U\colon \mathbb{N}^2 \to \mathbb{N}$ that encodes all other computable functions in the sense that, for every e, n,

$$U(e, n) = \Phi_e(n).$$

This function U is said to be a *universal partial computable function*. It does essentially what computers do nowadays: You give them an index for a program and an input, and they run it for you. We will not use U explicitly throughout the book, but we will constantly use the fact that we can computably list all programs and start running them one at the time, using U implicitly.

We identify subsets of $\mathbb{N}$ with their characteristic functions in $2^\mathbb{N}$, and we will move from one viewpoint to the other without even mentioning it. For instance, a set $A \subseteq \mathbb{N}$ is said to be *computable* if its characteristic function is.

An *enumeration* of a set A is nothing more than an onto function $g: \mathbb{N} \to A$. A set A is *computably enumerable* (*c.e.*) if it has an enumeration that is computable. The empty set is computably enumerable too. Equivalently, a set is computably enumerable if it is the domain of a partial computable function.[3] We denote

$$W_e = \{n \in \mathbb{N} : \Phi_e(n)\!\downarrow\} \quad \text{and} \quad W_{e,s} = \{n \in \mathbb{N} : \Phi_{e,s}(n)\!\downarrow\}.$$

As a convention, we assume that $W_{e,s}$ is finite, and furthermore, that only on inputs less than s can Φ_e converge in less than s steps. One way to make sense of this is that numbers larger than s should take more than s steps to even be read from the input tape. We sometimes use Lachlan's notation: $W_e[s]$ instead of $W_{e,s}$. In general, if a is an object built during a construction and whose value might change along the stages of the construction, we use $a[s]$ to denote its value at stage s. A set is *co-c.e.* if its complement is c.e.

Recall that a set is computable if and only if it and its complement are computably enumerable.

The *recursion theorem* gives us one of the most general ways of using recursion when defining computable functions. It states that for every computable function $f: \mathbb{N}^2 \to \mathbb{N}$ there is an index $e \in \mathbb{N}$ such that $f(e,n) = \varphi_e(n)$ for all $n \in \mathbb{N}$. Thus, we can think of $f(e,\cdot) = \varphi_e(\cdot)$ as a function of n which uses its own index, namely e, as a parameter during its own computation, and in particular is allowed to call and run itself.[4] An equivalent formulation of this theorem is that, for every computable function $h: \mathbb{N} \to \mathbb{N}$, there is an e such that $W_{h(e)} = W_e$.

Sets and strings. The natural numbers are $\mathbb{N} = \{0,1,2,\ldots\}$. For $n \in \mathbb{N}$, we sometimes use n to denote the set $\{0,\ldots,n-1\}$. For instance, $2^\mathbb{N}$ is the set of functions from $\mathbb{N}$ to $\{0,1\}$, which we will sometimes refer to as *infinite binary sequences* or *infinite binary strings*. For any set X, we use $X^{<\mathbb{N}}$ to denote the set of finite tuples of elements from X, which we call *strings* when $X = 2$ or $X = \mathbb{N}$. For $\sigma \in X^{<\mathbb{N}}$ and $\tau \in X^{\leq\mathbb{N}}$, we use $\sigma^\smallfrown\tau$ to denote the concatenation of these sequences. Similarly, for $x \in X, \sigma^\smallfrown x$ is obtained by appending x to σ. We will often omit the $^\smallfrown$ symbol and just

---

[3]If $A = range(g)$, then A is the domain of the partial function that, on input m, outputs the first n with $g(n) = m$ if it exists.

[4]To prove the recursion theorem, for each i, let $g(i)$ be an index for the partial computable function $\varphi_{g(i)}(n) = f(\varphi_i(i),n)$. Let e_0 be an index for the total computable function g, and let $e = g(e_0)$. Then $\varphi_e(n) = \varphi_{g(e_0)} = f(\varphi_{e_0}(e_0),n) = f(g(e_0),n) = f(e,n)$.

write $\sigma\tau$ and σx. We use $\sigma \subseteq \tau$ to denote that σ is an initial segment of τ, that is, that $|\sigma| \le |\tau|$ and $\sigma(n) = \tau(n)$ for all $n < |\sigma|$. This notation is consistent with the subset notation if we think of a string σ as its graph $\{\langle i, \sigma(i)\rangle : i < |\sigma|\}$. We use $\langle\rangle$ to denote the empty tuple. If Y is a subset of the domain of a function f, we use $f \restriction Y$ for the restriction of f to Y. Given $f \in X^{\le\mathbb{N}}$ and $n \in \mathbb{N}$, we use $f \restriction n$ to denote the initial segment of f of length n. We use $f \Vert n$ for the initial segment of length $n + 1$. For a tuple $\bar{n} = \langle n_0, \ldots, n_k\rangle \in \mathbb{N}^{<\mathbb{N}}$, we use $f \restriction \bar{n}$ for the tuple $\langle f(n_0), \ldots, f(n_k)\rangle$. Given a nested sequence of strings $\sigma_0 \subseteq \sigma_1 \subseteq \cdots$, we let $\bigcup_{i\in\mathbb{N}} \sigma_i$ be the possibly infinite string $f \in X^{\le\mathbb{N}}$ such that $f(n) = m$ if $\sigma_i(n) = m$ for some i.

Given $f, g \in X^{\mathbb{N}}$, we use $f \oplus g$ for the function $(f \oplus g)(2n) = f(n)$ and $(f \oplus g)(2n + 1) = g(n)$. We can extend this to ω-*sums* and define $\bigoplus_{n\in\mathbb{N}} f_n$ to be the function defined by $(\bigoplus_{n\in\mathbb{N}} f_n)(\langle m, k\rangle) = f_m(k)$. Conversely, we define $f^{[n]}$ to be the nth column of f, that is, $f^{[n]}(m) = f(\langle n, m\rangle)$. All these definitions work for sets if we think in terms of their characteristic functions. So, for instance, we can encode countably many sets $\{A_n : n \in \mathbb{N}\}$ with one set $A = \{\langle n, m\rangle : m \in A_n\}$.

For a set $A \subseteq \mathbb{N}$, the complement of A with respect to $\mathbb{N}$ is denoted by A^c.

A *tree* on a set X is a subset T of $X^{<\mathbb{N}}$ that is closed downward, i.e., if $\sigma \in T$ and $\tau \subseteq \sigma$, then $\tau \in T$ too. A *path* through a tree T is a function $f \in X^{\mathbb{N}}$ such that $f \restriction n \in T$ for all $n \in \mathbb{N}$. We use $[T]$ to denote the set of all paths through T. A tree is *well-founded* if it has no paths.

Reducibilities. There are various ways to compare the complexity of sets of natural numbers. Depending on the context or application, some may be more appropriate than others.

Many-one reducibility. Given sets $A, B \subseteq \mathbb{N}$, we say that A is *many-one reducible* (or *m-reducible*) to B, and write $A \le_m B$, if there is a computable function $f : \mathbb{N} \to \mathbb{N}$ such that $n \in A \iff f(n) \in B$ for all $n \in \mathbb{N}$. One should think of this reducibility as saying that all the information in A can be decoded from B. Notice that the classes of computable sets and of c.e. sets are both closed downwards under $\le_m$. A set B is said to be *c.e. complete* if it is c.e. and, for every other c.e. set A, $A \le_m B$.

Two sets are *m-equivalent* if they are *m*-reducible to each other, denoted $A \equiv_m B$. This is an equivalence relation, and the equivalence classes are called *m-degrees*

There are, of course, various other ways to formalize the idea of one set encoding the information from another set. Many-one reducibility is somewhat restrictive in various ways: (1) to figure out if $n \in A$, one is allowed to ask only one question of the form "$m \in B$?"; (2) the answer to "$n \in A$?" has to be the same as the answer to "$f(n) \in B$?". Turing reducibility is much more flexible.

One-one reducibility. 1-*reducibility* is like m-reducibility but requiring the reduction to be one-to-one. The equivalence induced by it, 1-*equivalence*, is one of the strongest notions of equivalence between sets in computability theory—a computability theorist would view sets that are 1-equivalent as being the same. Myhill's theorem states that two sets of natural numbers are 1-equivalent, i.e., each is 1-reducible to the other, if and only if there is a computable bijection of $\mathbb{N}$ that matches one set with the other.

Turing reducibility. Given a function $f : \mathbb{N} \to \mathbb{N}$, we say that a partial function $g : \mathbb{N} \to \mathbb{N}$ is *partial f-computable* if it can be computed by a program that is allowed to use the function f as a primitive function during its computation; that is, the program can ask questions about the value of $f(n)$ for different n's and use the answers to make decisions while the program is running. The function f is called the *oracle* of this computation. If g and f are total, we write $g \leq_T f$ and say that g is *Turing reducible* to f, that f *computes* g, or that g is f-*computable*. The class of partial f-computable functions can be enumerated the same way as the class of the partial computable functions. Programs that are allowed to query an oracle are called *Turing operators* or *computable operators*. We list them as $\Phi_0, \Phi_1, \dots,$ and we write $\Phi_e^f(n)$ for the output of the eth Turing operator on input n when it uses f as oracle. Notice that Φ_e represents a fixed program that can be used with different oracles. When the oracle is the empty set, we may write Φ_e for $\Phi_e^{\emptyset}$ matching the previous notation.

As we already mentioned, for a fixed input n, if $\Phi_e^f(n)$ converges, it does so after a finite number of steps s. As a convention, let us assume that in just s steps, it is only possible to read the first s entries from the oracle. Thus, if σ is a finite substring of f of length greater than s, we could calculate $\Phi_e^{\sigma}(n)$ without ever noticing that the oracle is not an infinite string.

Convention: For $\sigma \in \mathbb{N}^{<\mathbb{N}}$, $\Phi_e^{\sigma}(n)$ is shorthand for $\Phi_{e,|\sigma|}^{\sigma}(n)$, which runs for at most $|\sigma|$ stages.

Notice that given e, σ, n, it is computable to decide if $\Phi_e^{\sigma}(n){\downarrow}$.

As the class of partial computable functions, the class of partial X-computable functions contains the basic functions; is closed under composition, recursion, and minimization; can be listed in such a way that we have a universal partial X-computable function (that satisfies the s-m-n theorem). In practice, with very few exceptions, those are the only properties we use of computable functions. This is why almost everything we can prove about computable functions, we can also prove about X-computable functions. This translation is called *relativization*. All notions whose definition are based on the notion of partial computable function can be relativized by using the notion of partial X-computable function instead. For instance, the notion of c.e. set can be relativized to that of c.e. in X or X-c.e. set:

These are the sets which are the images of X-computable functions (or empty), or, equivalently, the domains of partial X-computable functions. We use W_e^X to denote the domain of Φ_e^X.

When two functions are Turing reducible to each other, we say that they are *Turing equivalent*, which we denote by $\equiv_T$. This is an equivalence relation, and the equivalence classes are called *Turing degrees*.

Computable operators can be encoded by computable subsets of $\mathbb{N}^{<\mathbb{N}} \times \mathbb{N} \times \mathbb{N}$. Given $\Phi \subseteq \mathbb{N}^{<\mathbb{N}} \times \mathbb{N} \times \mathbb{N}$, $\sigma \in \mathbb{N}^{<\mathbb{N}}$, n, m, we write $\Phi^\sigma(n) = m$ as shorthand for $\langle \sigma, n, m \rangle \in \Phi$. Then, given $f \in \mathbb{N}^{\mathbb{N}}$, we let

$$\Phi^f(n) = m \iff (\exists \sigma \subset f)\, \Phi^\sigma(n) = m.$$

We then have that g is computable in f if and only if there is a c.e. subset $\Phi \subseteq \mathbb{N}^{<\mathbb{N}} \times \mathbb{N} \times \mathbb{N}$ such that $\Phi^f(n) = g(n)$ for all $n \in \mathbb{N}$. A standard assumption is that $\langle \sigma, n, m \rangle \in \Phi$ only if $n, m < |\sigma|$.

We can use the same idea to encode c.e. operators by computable subsets of $\mathbb{N}^{<\mathbb{N}} \times \mathbb{N}$. Given $W \subseteq \mathbb{N}^{<\mathbb{N}} \times \mathbb{N}$, $\sigma \in \mathbb{N}^{<\mathbb{N}}$, and $f \in \mathbb{N}^{\mathbb{N}}$, we let

$$W^\sigma = \{n \in \mathbb{N} : \langle \sigma, n \rangle \in W\} \quad \text{and} \quad W^f = \bigcup_{\sigma \subset f} W^\sigma.$$

We then have that X is c.e. in Y if and only if there is a c.e. subset $W \subseteq \mathbb{N}^{<\mathbb{N}} \times \mathbb{N}$ such that $X = W^Y$. A standard assumption is that $\langle \sigma, n \rangle \in W$ only if $n < |\sigma|$.

Enumeration reducibility. Recall that an enumeration of a set A is just an onto function $f : \mathbb{N} \to A$. Given $A, B \subseteq \mathbb{N}$, we say that A is *enumeration reducible* (or *e-reducible*) to B, and write $A \leq_e B$, if every enumeration of B computes an enumeration of A. Selman [Sel71] showed that we can make this reduction uniformly: $A \leq_e B$ if and only if there is a Turing operator Φ such that, for every enumeration f of B, Φ^f is an enumeration of A. (See Theorem 4.2.2.) Another way of defining enumeration reducibility is via *enumeration operators*: An enumeration operator is a c.e. set Θ of pairs that acts as follows: For $B \subseteq \mathbb{N}$, we define

$$\Theta^B = \{n : (\exists D \subseteq_{fin} B)\, \langle \ulcorner D \urcorner, n \rangle \in \Theta\},$$

where $\subseteq_{fin}$ means 'finite subset of'. Selman also showed that $A \leq_e B$ if and only if there is an enumeration operator Θ such that $A = \Theta^B$.

The Turing degrees embed into the enumeration degrees via the map $\iota(A) = A \oplus A^c$. It is not hard to show that $A \leq_T B \iff \iota(A) \leq_e \iota(B)$.

Positive reducibility. We say that A *positively reduces* to B, and write $A \leq_p B$, if there is a computable function $f : \mathbb{N} \to (\mathbb{N}^{<\mathbb{N}})^{<\mathbb{N}}$ such that, for every $n \in \mathbb{N}$, $n \in A$ if and only if there is an $i < |f(n)|$ such that every entry of $f(n)(i)$ is in B (cf. [Joc68]). That is,

$$n \in A \iff \bigvee_{i < |f(n)|} \bigwedge_{j < |f(n)(i)|} f(n)(i)(j) \in B.$$

Notice that $\leq_p$ implies both Turing reducibility and enumeration reducibility, and is implied by many-one reducibility. In particular, the classes of computable sets and of c.e. sets are both closed downwards under $\leq_p$.

The Turing jump. Let K be the domain of the universal partial computable function. That is,

$$K = \{\langle e, n \rangle : \Phi_e(n)\!\downarrow\} = \bigoplus_{e \in \mathbb{N}} W_e.$$

K is called the *halting problem*.[5] It is not hard to see that K is c.e. complete. Using a standard diagonalization argument, one can show that K is not computable.[6] It is common to define K as $\{e : \Phi_e(e)\!\downarrow\}$ instead—the two definitions give 1-equivalent sets. We will use whichever is more convenient in each situation. We will often write $0'$ for K.

We can relativize this definition and, given a set X, define the *Turing jump* of X as

$$X' = \{e \in \mathbb{N} : \Phi_e^X(e)\!\downarrow\}.$$

Relativizing the properties of K, we get that X' is X-c.e.-complete, that $X \leq_T X'$, and that $X' \nleq_T X$. The Turing degree of X' is strictly above that of X—this is why it is called a jump. The jump defines an operation on the Turing degrees. Furthermore, for $X, Y \subseteq \mathbb{N}$, $X \leq_T Y \iff X' \leq_m Y'$.

The double iteration of the Turing jump is denoted X'', and the n-th iteration by $X^{(n)}$.

Vocabularies and languages. Let us quickly review the basics about vocabularies and structures. Our vocabularies will always be countable. Furthermore, except for a few occasions, they will always be computable.

A *vocabulary* τ consists of three sets of symbols $\{R_i : i \in I_R\}$, $\{f_i : i \in I_F\}$, and $\{c_i : i \in I_C\}$; and two functions $a_R : I_R \to \mathbb{N}$ and $a_F : I_F \to \mathbb{N}$. Each of I_R, I_F, and I_C is an initial segment of $\mathbb{N}$. The symbols R_i, f_i, and c_i represent *relations, functions,* and *constants,* respectively. For $i \in I_R$, $a_R(i)$ is the arity of R_i, and for $i \in I_F$, $a_F(i)$ is the arity of f_i.

A vocabulary τ is *computable* if the arity functions a_R and a_F are computable. This only matters when τ is infinite; finite vocabularies are trivially computable.

Given such a vocabulary τ, a *τ-structure* is a tuple

$$\mathcal{M} = (M; \{R_i^\mathcal{M} : i \in I_R\}, \{f_i^\mathcal{M} : i \in I_F\}, \{c_i^\mathcal{M} : i \in I_C\}),$$

where M is just a set called the *domain* of $\mathcal{M}$, and the rest are interpretations of the symbols in τ. That is, $R_i^\mathcal{M} \subset M^{a_R(i)}$, $f_i^\mathcal{M} : M^{a_F(i)} \to M$, and $c_i^\mathcal{M} \in M$. A *structure* is a τ-structure for some τ.

---

[5] The 'K' is for Kleene.

[6] If it were computable, so would be the set $A = \{e : \langle e, e \rangle \notin K\}$. But then $A = W_e$ for some e, and we would have that $e \in A \iff \langle e, e \rangle \notin K \iff e \notin W_e \iff e \notin A$.

Given two τ-structures $\mathcal{A}$ and $\mathcal{B}$, we write $\mathcal{A} \subseteq \mathcal{B}$ to mean that $\mathcal{A}$ is a substructure of $\mathcal{B}$, that is, that $A \subseteq B$, $f_i^{\mathcal{A}} = f_i^{\mathcal{B}} \upharpoonright A^{a_F(i)}$, $R_j^{\mathcal{A}} = R_j^{\mathcal{B}} \upharpoonright A^{a_R(i)}$ and $c_k^{\mathcal{A}} = c_k^{\mathcal{B}}$ for all symbols f_i, R_j and c_k. This notation should not be confused with $A \subseteq B$ which only means that the domain of $\mathcal{A}$ is a subset of the domain of $\mathcal{B}$. If $\mathcal{A}$ is a τ_0-structure and $\mathcal{B}$ a τ_1-structure with $\tau_0 \subseteq \tau_1$,[7] $\mathcal{A} \subseteq \mathcal{B}$ means that $\mathcal{A}$ is a τ_0-substructure of $\mathcal{B} \upharpoonright \tau_0$, where $\mathcal{B} \upharpoonright \tau_0$ is obtained by forgetting the interpretations of the symbols of $\tau_1 \smallsetminus \tau_0$ in $\mathcal{B}$. $\mathcal{B} \upharpoonright \tau_0$ is called the τ_0-*reduct* of $\mathcal{B}$, and $\mathcal{B}$ is said to be an *expansion* of $\mathcal{B} \upharpoonright \tau_0$.

Given a vocabulary τ, we define various languages over it. First, recursively define a τ-*term* to be either a variable x, a constant symbol c_i, or a function symbol applied to other τ-terms, that is, $f_i(t_1, \ldots, t_{a_F(i)})$, where each t_j is a τ-term we have already built. The *atomic τ-formulas* are the ones of the form $R_i(t_1, \ldots, t_{a_R(i)})$ or $t_1 = t_2$, where each t_i is a τ-term. A τ-*literal* is either a τ-atomic formula or a negation of a τ-atomic formula. A *quantifier-free τ-formula* is built out of literals using conjunctions, disjunctions, and implications. If we close the quantifier-free τ-formulas under existential quantification, we get the *existential τ-formulas*, of ∃-*formulas*. Every τ-existential formula is equivalent to one of the form $\exists x_1 \cdots \exists x_k \, \varphi$, where φ is quantifier-free. A *universal τ-formula*, or ∀-*formula*, is one equivalent to $\forall x_1 \cdots \forall x_k \, \varphi$ for some quantifier-free τ-formula φ. An *elementary τ-formula* is built out of quantifier-free formulas using existential and universal quantifiers. We also call these the *finitary first-order formulas*.

Given a τ structure $\mathcal{A}$, and a tuple $\bar{a} \in A^{<\mathbb{N}}$, we write $(\mathcal{A}, \bar{a})$ for the $\tau \cup \bar{c}$-structure where $\bar{c}$ is a new tuple of constant symbols and $\bar{c}^{\mathcal{A}} = \bar{a}$. Given $R \subseteq \mathbb{N} \times A^{<\mathbb{N}}$, we write $(\mathcal{A}, R)$ for the $\tilde{\tau}$ structure where $\tilde{\tau}$ is defined by adding to τ relations symbols $R_{i,j}$ of arity j for $i, j \in \mathbb{N}$, and $R_{i,j}^{\mathcal{A}} = \{\bar{a} \in A^j : \langle i, \bar{a} \rangle \in R\}$.

Orderings. Here are some structures we will use quite often in examples. A *partial order* is a structure over the vocabulary $\{\leq\}$ with one binary relation symbol which is transitive $(x \leq y \, \& \, y \leq z \rightarrow x \leq z)$, reflexive $(x \leq x)$, and anti-symmetric $(x \leq y \, \& \, y \leq x \rightarrow x = y)$. A *linear order* is a partial order where every two elements are comparable $(\forall x, y \, (x \leq y \vee y \leq x))$. We will often add and multiply linear orderings. Given linear orderings $\mathcal{A} = (A; \leq_A)$ and $\mathcal{B} = (B; \leq_B)$, we define $\mathcal{A} + \mathcal{B}$ to be the linear ordering with domain $A \sqcup B$, where the elements of A stand below the elements of B. We define $\mathcal{A} \times \mathcal{B}$ to be the linear ordering with domain $A \times B$ where $\langle a_1, b_1 \rangle \leq_{A \times B} \langle a_2, b_2 \rangle$ if either $b_1 <_B b_2$ or $b_1 = b_2$ and $a_1 \leq_A a_2$—notice we compare the second coordinate first.[8] We will use ω to denote the linear ordering of the natural numbers and $\mathbb{Z}$ and $\mathbb{Q}$ for the orderings of the integers and the rationals. We denote the finite linear

---

[7] By $\tau_0 \subseteq \tau_1$ we mean that every symbol in τ_0 is also in τ_1 and with the same arity.

[8] $\mathcal{A}$ times $\mathcal{B}$ is $\mathcal{A}$ $\mathcal{B}$ times.

ordering with n elements by **n**. We use $\mathcal{A}^*$ to denote the reverse ordering $(A; \geq_A)$ of $\mathcal{A} = (A, \leq_A)$. For $a <_A b \in \mathcal{A}$, we use the notation $\mathcal{A} \restriction (a, b)$ or the notation $(a, b)_{\mathcal{A}}$ to denote the open $\{x \in A : a <_A x <_A b\}$. We also use $\mathcal{A} \restriction a$ to denote the initial segment of $\mathcal{A}$ below a, which we could also denote as $(-\infty, a)_{\mathcal{A}}$.

As mentioned above, a *tree* T is a downward closed subset of $X^{<\mathbb{N}}$. As a structure, a tree can be represented in various ways. One is as a partial order $(T; \subseteq)$ using the ordering on strings. Another is as a graph where each node $\sigma \in T$ other than the root is connected to its parent node $\sigma \restriction |\sigma - 1|$, and there is a constant symbol used for the root of the tree. We will refer to these two types of structures as *trees as orders* and *trees as graphs*.

A partial order where every two elements have a least upper bound $(x \vee y)$ and a greatest lower bound $(x \wedge y)$ is called a *lattice*. A lattice with a top element 1, a bottom element 0, and where $\vee$ and $\wedge$ distribute over each other, and every element x has a *complement* (that is an element x^c such that $x \vee x^c = 1$ and $x \wedge x^c = 0$) is called a *Boolean algebra*. The vocabulary for Boolean algebras is $\{0, 1, \vee, \wedge, \cdot^c\}$, and the ordering can be defined by $x \leq y \iff y = x \vee y$.

The arithmetic hierarchy. Consider the structure $(\mathbb{N}; 0, 1, +, \times, \leq)$. In this vocabulary, the *bounded formulas* are built out of the quantifier-free formulas using bounded quantifiers of the form $\forall x < y$ and $\exists x < y$. A Σ_1^0 formula is one of the form $\exists x \, \varphi$, where φ is bounded; and a Π_1^0 formula is one of the form $\forall x \, \varphi$, where φ is bounded. By coding tuples of numbers by a single natural number, one can show that formulas of the form $\exists x_0 \exists x_1 \ldots \exists x_k \, \varphi$ are equivalent to Σ_1^0 formulas. Post's theorem asserts that a set $A \subseteq \mathbb{N}$ is c.e. if and only if it can be defined by a Σ_1^0 formula. Thus, a set is computable if and only if it is Δ_1^0, that is, if it can be defined both by a Σ_1^0 formula and by a Π_1^0 formula.

By recursion, we define the Σ_{n+1}^0 formulas as those of the form $\exists x \, \varphi$, where φ is Π_n^0; and the Π_{n+1}^0 formulas as those of the form $\forall x \, \varphi$, where φ is Σ_n^0. A set is Δ_n^0 if it can be defined by both a Σ_n^0 formula and a Π_n^0 formula. Again, in the definition of Σ_{n+1}^0 formulas, using one existential quantifier or many makes no difference. What matters is the number of alternations of quantifiers. Post's theorem asserts that a set $A \subseteq \mathbb{N}$ is c.e. in $0^{(n)}$ if and only if it can be defined by a Σ_{n+1}^0 formula. In particular, a set is computable from $0'$ if and only if it is Δ_2^0. The Shoenfield *Limit Lemma* says that a set A is Δ_2^0 if and only if there is a computable function $f : \mathbb{N}^2 \to \mathbb{N}$ such that, for each $n \in \mathbb{N}$, if $n \in A$ then $f(n, s) = 1$ for all sufficiently large s, and if $n \notin A$ then $f(n, s) = 0$ for all sufficiently large s. This can be written as $\chi_A(n) = \lim_{s \to \infty} f(n, s)$, where χ_A is the characteristic function of A and the limit with respect to the discrete topology of $\mathbb{N}$ where a sequence converges if and only if it is eventually constant.

The language of second-order arithmetic is a two-sorted language for the structure $(\mathbb{N}, \mathbb{N}^{\mathbb{N}}; 0, 1, +, \times, \leq)$. The elements of the first sort, called *first-order elements*, are natural numbers. The elements of the second sort, called *second-order elements* or *reals*, are functions $\mathbb{N} \to \mathbb{N}$. The vocabulary consists of the standard vocabulary of arithmetic, $0, 1, +, \times$, $\leq$ which is used on the first-order elements, and an application operation denoted $F(n)$ for a second-order element F and a first-order element n. A formula in this language is said to be *arithmetic* if it has no quantifiers over second-order objects. Among the arithmetic formulas, the hierarchy of Σ_n^0 and Π_n^0 formulas are defined exactly as above. Post's theorem that Σ_1^0 sets are c.e. also applies in this context: For every Σ_1^0 formula $\psi(F, n)$, where n a number variable and F is a function variable, there is c.e. operator W such that $n \in W^F \iff \psi(F, n)$. We can then build the computable tree $T_n = \{\sigma \in \mathbb{N}^{<\mathbb{N}} : n \notin W^\sigma\}$ and we have that $\psi(F, n)$ holds if and only if F is not a path through T_n. A Π_1^0 *class* is a set of the form $\{F \in \mathbb{N}^{\mathbb{N}} : \psi(F)\}$ for some Π_1^0 formula $\psi(F)$. The observation above shows how every Π_1^0 class is of the form $[T]$ for some computable tree $T \subseteq \mathbb{N}^{<\mathbb{N}}$.

Chapter 1

STRUCTURES

Algorithms, Turing machines, and modern computer programs all work with finitary objects, objects that usually can be encoded by finite binary strings or just by natural numbers. For this reason, computability theory concentrates on the study of the complexity of sets of natural numbers. To study the computational properties of a countable mathematical structure, the first approach is to set the domain of the structure to be a subset of the natural numbers and then borrow the tools we already have from computability theory. One issue comes up: There might be many bijections between the domain of a structure and the natural numbers, inducing many different *presentations* of the structure with different computability-theoretic properties. The interplay between properties of presentations (computational properties) and properties of isomorphism types (structural properties) is one of the main themes of computable structure theory.

We start this chapter by introducing various ways of representing structures so that we can analyze their computational complexity. These different types of presentations are essentially equivalent, and the distinctions are purely technical and not deep. However, they will allow us to be precise later. At the end of the chapter we prove Knight's theorem that all non-trivial structures have presentations that code any given set.

1.1. Presentations

All the structures we consider are countable. So, unless otherwise stated, "structure" means "countable structure." Furthermore, we usually assume that the domains of our structures are subsets of $\mathbb{N}$. This will allow us to use everything we already know about computable functions on $\mathbb{N}$.

DEFINITION 1.1.1. An *ω-presentation* is nothing more than a structure whose domain is $\mathbb{N}$.[9] Given a structure $\mathcal{A}$, when we refer to *an ω-presentation of $\mathcal{A}$* or to a *copy of $\mathcal{A}$*, we mean an ω-presentation $\mathcal{M}$ which is

---

[9]The use of the word *presentation* here has nothing to do with its use in group theory. There, a presentation of a group consists a list of generators and a list of relations among them. You might have a group with a computable presentation, meaning that this list of relations is computable, but which has no computable ω-presentation in our sense.

isomorphic to $\mathcal{A}$. An ω-presentation $\mathcal{M}$ is *computable* if all its relations, functions, and constants are uniformly computable; that is, if the set $\tau^{\mathcal{M}}$, defined as

$$\tau^{\mathcal{M}} = \bigoplus_{i \in I_R} R_i^{\mathcal{M}} \oplus \bigoplus_{i \in I_F} F_i^{\mathcal{M}} \oplus \bigoplus_{i \in I_c} \{c_i^{\mathcal{M}}\}, \qquad (1)$$

is computable. Note that via standard coding, we can think of $\tau^{\mathcal{M}}$ as a subset of $\mathbb{N}$.

1.1.1. Atomic diagrams. Another standard way of defining when an ω-presentation is computable is via its atomic diagram. Let $\{\varphi_i^{\text{at}} : i \in \mathbb{N}\}$ be an effective enumeration of all atomic τ-formulas with free variables from the set $\{x_0, x_1, \dots\}$. (An *atomic τ-formula* is one of the form $R(t_1, \dots, t_a)$, where R is either "$=$" or R_j for $j \in I_R$, and each t_i is a term built out of the function, constant, and variable symbols.)

Definition 1.1.2. The *atomic diagram* of an ω-presentation $\mathcal{M}$ is the infinite binary string $D(\mathcal{M}) \in 2^{\mathbb{N}}$ defined by

$$D(\mathcal{M})(i) = \begin{cases} 1 & \text{if } \mathcal{M} \models \varphi_i^{\text{at}}[x_j \mapsto j : j \in \mathbb{N}], \\ 0 & \text{otherwise.} \end{cases}$$

It is not hard to see that $D(\mathcal{M})$ and $\tau^{\mathcal{M}}$ are Turing equivalent. We will often treat the ω-presentation $\mathcal{M}$, the real $\tau^{\mathcal{M}}$, and the real $D(\mathcal{M})$ as the same thing. For instance, we define the *Turing degree of the ω-presentation* $\mathcal{M}$ to be the Turing degree of $D(\mathcal{M})$. When we say that $\mathcal{M}$ *is computable from a set X*, that *a set X is computable from* $\mathcal{M}$, that $\mathcal{M}$ is Δ_2^0, that $\mathcal{M}$ is *arithmetic*, that $\mathcal{M}$ is *low*, etc., we mean $D(\mathcal{M})$ instead of $\mathcal{M}$.

Let us also point out that the quantifier-free diagram, which is defined like the atomic diagram but using a listing of the quantifier-free formulas instead, is Turing equivalent to $D(\mathcal{M})$ too.

1.1.2. An example. Unless it is trivial, a structure will have many different ω-presentations—continuum many actually (see Theorem 1.2.1)—and these different ω-presentations will have different computability theoretic properties. For starters, some of them may be computable while others may not. But even among the computable copies of a single structure one may find different computability theoretic properties.

Consider the linear ordering $\mathcal{A} = (\mathbb{N}; \leq)$, where $\leq$ is the standard ordering on the natural numbers. We can build another ω-presentation $\mathcal{M} = (\mathbb{N}; \leq_M)$ of $\mathcal{A}$ as follows. Let $\{k_i : i \in \mathbb{N}\}$ be a one-to-one computable enumeration of the halting problem $0'$. First, order the even natural numbers in the natural way: $2n \leq_M 2m$ if $n \leq m$. Second, place the odd number $2s + 1$ right in between $2k_s$ and $2k_s + 2$, that is, let $2k_s \leq_M 2s + 1 \leq_M 2k_s + 2$. Using transitivity we can then define $\leq_M$ on all pairs of numbers. Thus $2n <_M 2s + 1$ if and only if $n < k_s$, and

$2s + 1 <_M 2t + 1$ if and only if $k_s < k_t$. (Early codings of sets into ω-presentations of linear orderings appear in [Mar82].)

One can show that $\mathcal{A}$ and $\mathcal{M}$ are two computable ω-presentations of the same structure.[10] However, computationally, they behave quite differently. For instance, the successor function is computable in $\mathcal{A}$ but not in $\mathcal{M}$: In $\mathcal{A}$, $\text{Succ}^{\mathcal{A}}(n) = n + 1$ is clearly computable. On the other hand, in $\mathcal{M}$, $\text{Succ}^{\mathcal{M}}(2n) = 2n + 2$ if and only if there is no odd number placed $\leq_M$-in-between $2n$ and $2n+2$, which occurs if and only if $n \notin 0'$. Therefore, $\text{Succ}^{\mathcal{M}}$ computes $0'$ and $\text{Succ}^{\mathcal{A}}$ does not.

The reason $\mathcal{A}$ and $\mathcal{M}$ can behave differently despite being isomorphic is that they are not *computably isomorphic*: There is no computable isomorphism between them. To see this, note that if there was one, we could use $\text{Succ}^{\mathcal{A}}$ and the isomorphism to compute $\text{Succ}^{\mathcal{M}}$, contradicting that $\text{Succ}^{\mathcal{M}}$ computes $0'$.

1.1.3. Relaxing the domain. In many cases, it will be useful to consider structures whose domain is a subset of $\mathbb{N}$. We call those $(\subseteq\omega)$-*presentations*. If M, the domain of $\mathcal{M}$, is a proper subset of $\mathbb{N}$, we can still define $D(\mathcal{M})$ by letting $D(\mathcal{M})(i) = 0$ if φ_i^{at} mentions a variable x_j with $j \notin M$. In this case, we have

$$D(\mathcal{M}) \equiv_T M \oplus \tau^{\mathcal{M}}.$$

To see that $D(\mathcal{M})$ computes M, notice that, for $j \in \mathbb{N}$, $j \in M \leftrightarrow D(\mathcal{M})(\ulcorner x_j = x_j \urcorner) = 1$, where $\ulcorner \varphi \urcorner$ is the index of the atomic formula φ in the enumeration $\{\varphi_i^{\text{at}} : i \in \mathbb{N}\}$.

The following observation will simplify many of our constructions later on.

Observation 1.1.3. We can always associate to an infinite $(\subseteq\omega)$-presentation $\mathcal{M}$, an isomorphic ω-presentation $\mathcal{A}$: If $M = \{m_0 < m_1 < m_2 < \cdots\} \subseteq \mathbb{N}$, we can use the bijection $i \mapsto m_i \colon \mathbb{N} \to M$ to get a copy $\mathcal{A}$ of $\mathcal{M}$, now with domain $\mathbb{N}$. Since this bijection is computable in M, it is not hard to see that $D(\mathcal{A}) \leq_T D(\mathcal{M})$, and furthermore that $D(\mathcal{A}) \oplus M \equiv_T D(\mathcal{M})$.

One of the advantages of $(\subseteq\omega)$-presentations is that they allow us to present finite structures.

1.1.4. Relational vocabularies. A vocabulary is *relational* if it has no function or constant symbols, and has only relational symbols. Every vocabulary τ can be made into a relational one, $\tilde{\tau}$, by replacing each n-ary function symbol by an $(n + 1)$-ary relation symbol coding the graph of the function, and each constant symbol by a 1-ary relation symbol coding it as a singleton. Depending on the situation, this change in

---

[10]To show that $\mathcal{M}$ is isomorphic to the standard ordering on $\mathbb{N}$, one has to observe that every element of $M = \mathbb{N}$ has finitely many elements $<_M$-below it: $2n$ has at most $2n$, and $2s + 1$ has at most $2k_s$.

vocabulary might be more or less significant. For instance, the class of
quantifier-free definable sets changes, but the class of ∃-definable sets
does not (see Exercise 1.1.4). For most computational properties, this
change is nonessential; for instance, if $\mathcal{M}$ is an ω-presentation of a τ-
structure, and $\widetilde{\mathcal{M}}$ is the associated ω-presentation of $\mathcal{M}$ as a $\widetilde{\tau}$-structure,
then $D(\mathcal{M}) \equiv_T D(\widetilde{\mathcal{M}})$ (as it follows from Exercise 1.1.4). Because of this,
and for the sake of simplicity, we will often restrict ourselves to relational
vocabularies.

EXERCISE 1.1.4. Show that the ∃-*diagram of* $\mathcal{M}$ as a τ-structure is m-
equivalent to its ∃-diagram as a $\widetilde{\tau}$-structure. More concretely, let $\{\varphi_i^\exists :
i \in \mathbb{N}\}$ and $\{\widetilde{\varphi}_i^\exists : i \in \mathbb{N}\}$ be the standard effective enumerations of
the existential τ-formulas and the existential $\widetilde{\tau}$-formulas on the variables
$x_0, x_1, \ldots$. Show that

$$\{i \in \mathbb{N} : \mathcal{M} \models \varphi_i^\exists[x_j \mapsto j : j \in \mathbb{N}]\} \equiv_m$$
$$\{i \in \mathbb{N} : \widetilde{\mathcal{M}} \models \widetilde{\varphi}_i^\exists[x_j \mapsto j : j \in \mathbb{N}]\}.$$

One could also show these sets are $\equiv_1$-equivalent.

1.1.5. Finite structures and approximations. We can represent finite
structures using $(\subseteq\omega)$-presentations. However, when working with infinitely many finite structures at once, we often want to be able to compute
things about them uniformly, for instance the sizes of the structures,
which we could not do from $(\subseteq\omega)$-presentations (see Exercise 1.1.5). For
that reason, we sometimes consider $(\sqsubseteq\omega)$-presentations, which are $(\subseteq\omega)$-
presentations whose domains are initial segments of $\mathbb{N}$. Given a finite
$(\sqsubseteq\omega)$-presentation, we can easily find the first k that is not in the domain
of the structure.

EXERCISE 1.1.5. Show that there exists a computable list $\{\mathcal{M}_n : n \in \mathbb{N}\}$
of $(\subseteq\omega)$-presentations of finite structures whose sizes cannot be computed
uniformly, that is, a list such that the domains and relations of the $\mathcal{M}_n$'s
are uniformly computable, but there is no computable function f such
that $f(n)$ is the size of M_n.

When τ is a finite vocabulary, finite τ-structures can be coded by a finite
amount of information. Suppose $\mathcal{M}$ is a finite τ-structure with domain
$\{0, \ldots, k-1\}$, and τ is a finite relational vocabulary. Then there are only
finitely many atomic τ-formulas on the variables $x_0, \ldots, x_{k-1}$, let us say ℓ_k
of them. Assume the enumeration $\{\varphi_i^{\text{at}} : i \in \mathbb{N}\}$ of the atomic τ-formulas
is such that those ℓ_k formulas come first, and the formulas mentioning
variables beyond x_k come later. Then $D(\mathcal{M})$ is determined by the finite
binary string of length ℓ_k that codes the values of those formulas. We will
often assume $D(\mathcal{M})$ *is* that string.

When dealing with infinite structures, very often we will want to approximate them using finite substructures. We need to take care of two technical

details. First, if τ is an infinite vocabulary, we need to approximate it using finite sub-vocabularies. We assume that all computable vocabularies τ come with an associated effective approximation $\tau_0 \subseteq \tau_1 \subseteq \cdots \subseteq \tau$, where each τ_s is finite and $\tau = \bigcup_s \tau_s$. In general and unless otherwise stated, we let τ_s consist of the first s relation, constant and function symbols in τ, but in some particular cases, we might prefer other approximations. For instance, if τ is already finite, we usually prefer to let $\tau_s = \tau$ for all s. Second, to be able to approximate a τ-structure $\mathcal{M}$ using τ_s-substructures, we need the τ_s-reduct of $\mathcal{M}$ to be *locally finite*, i.e., every finite subset generates a finite substructure. To avoid unnecessary complications, we will just assume τ is relational and, in particular, locally finite. Even if τ is not originally relational, we can make it relational as in Section 1.1.4.

DEFINITION 1.1.6. Given an ω-presentation $\mathcal{M}$, we let $\mathcal{M}_s$ be the finite τ_s-substructure of $\mathcal{M}$ with domain $\{0, \ldots, s-1\}$. We call the sequence $\{\mathcal{M}_s : s \in \mathbb{N}\}$ a *finite approximation* of $\mathcal{M}$. We identify this sequence with the sequence of codes $\{D(\mathcal{M}_s) : s \in \mathbb{N}\} \subseteq 2^{<\mathbb{N}}$, which allows us to consider its computational complexity.

In general, when we refer to a $\tau_{|\cdot|}$-*structure*, we mean a τ_s-structure where s is the size of the structure itself. For instance, the structures $\mathcal{M}_s$ above are all $\tau_{|\cdot|}$-structures.

Observation 1.1.7. Here is a simple, but very important observation we will use throughout the book. For each s, $D(\mathcal{M}_s) = D(\mathcal{M}) \restriction \ell_s$, and hence

$$D(\mathcal{M}_0) \subseteq D(\mathcal{M}_1) \subseteq D(\mathcal{M}_2) \subseteq \cdots \quad \text{and} \quad D(\mathcal{M}) = \bigcup_{s \in \mathbb{N}} D(\mathcal{M}_s).$$

The convention here is that for each s, the τ_s-atomic formulas on the variables $\{x_0, \ldots, x_{s-1}\}$ are listed before the rest; that is, they are $\varphi_0^{\text{at}}, \ldots, \varphi_{\ell_s-1}^{\text{at}}$ for some $\ell_s \in \mathbb{N}$.

Also, let us remark that the inclusion is an inclusion of stings, not of sets, and so is the union, as defined on page xv.

Thus, from a computational viewpoint, having an ω-presentation is equivalent to having a finite approximation of a structure $\mathcal{M}$. This is why, when we are working with an ω-presentation, we often visualize the structure as being given to us little by little.

Observation 1.1.8. Another simple but important observation is that an $\exists$-formula is true of a tuple $\bar{m}$ in $\mathcal{M}$ if and only if it is true in some finite substructure $\mathcal{M}_s$ that contains $\bar{m}$. Thus, if $\exists\text{-}Th(\mathcal{M})$ denotes the set of $\exists$-τ-sentences true of $\mathcal{M}$, and $\exists\text{-}Th(\mathcal{M}_s)$ the set of $\exists$-τ_s-sentences true of $\mathcal{M}_s$, then

$$\exists\text{-}Th(\mathcal{M}) = \bigcup_{s \in \mathbb{N}} \exists\text{-}Th(\mathcal{M}_s),$$

where the union here refers to the union of sets, not sequences.

As a useful technical device, we define the atomic diagram of a finite tuple as the finite binary sequence coding the set of atomic formulas true of the tuple restricted to the smaller vocabulary. Again, we assume that τ is relational.

DEFINITION 1.1.9. Let $\mathcal{M}$ be a τ-structure and let $\bar{a} = \langle a_0, \ldots, a_{s-1} \rangle \in M^s$. We define the *atomic diagram of $\bar{a}$ in $\mathcal{M}$, denoted $D_{\mathcal{M}}(\bar{a})$*, as the string in 2^{ℓ_s} such that

$$D_{\mathcal{M}}(\bar{a})(i) = \begin{cases} 1 & \text{if } \mathcal{M} \models \varphi_i^{\text{at}}[x_j \mapsto a_j, j < s], \\ 0 & \text{otherwise.} \end{cases}$$

So, if $\mathcal{M}$ were an ω-presentation and $a_0, \ldots a_s, \ldots$ were the elements $0, \ldots, s, \cdots \in M = \mathbb{N}$, then $D_{\mathcal{M}}(\langle a_0, \ldots, a_{s-1} \rangle) = D(\mathcal{M}_s)$ as in Definition 1.1.6.

Observation 1.1.10. For every $\sigma \in 2^{<\mathbb{N}}$ and every s with $\ell_s \geq |\sigma|$, there is a quantifier-free τ-formula $\varphi_\sigma^{\text{at}}(x_0, \ldots, x_{s-1})$ such that

$$\mathcal{A} \models \varphi_\sigma^{\text{at}}(\bar{a}) \iff \sigma \subseteq D_{\mathcal{A}}(\bar{a})$$

for every τ-structure $\mathcal{A}$ and tuple $\bar{a} \in A^s$, namely

$$\varphi_\sigma^{at}(\bar{x}) \equiv \left(\bigwedge_{i < |\sigma|, \sigma(i)=1} \varphi_i^{at}(\bar{x}) \right) \wedge \left(\bigwedge_{i < |\sigma|, \sigma(i)=0} \neg\varphi_i^{at}(\bar{x}) \right).$$

1.1.6. Congruence structures. It will often be useful to consider structures where equality is interpreted by an equivalence relation. A *congruence τ-structure* is a structure $\mathcal{M} = (M; =^{\mathcal{M}}, \{R_i^{\mathcal{M}} : i \in I_R\}, \{f_i^{\mathcal{M}} : i \in I_F\}, \{c_i^{\mathcal{M}} : i \in I_C\})$, where $=^{\mathcal{M}}$ is an equivalence relation on M, and the interpretations of all the τ-symbols are invariant under $=^{\mathcal{M}}$ (that is, if $\bar{a} =^{\mathcal{M}} \bar{b}$, then $\bar{a} \in R_i^{\mathcal{M}} \iff \bar{b} \in R_i^{\mathcal{M}}$ and $f_j^{\mathcal{M}}(\bar{a}) =^{\mathcal{M}} f_j(\bar{b})$ for all relations symbols R_i and function symbols f_j). If $M = \mathbb{N}$, we say that $\mathcal{M}$ is a *congruence ω-presentation*. We can then define $D(\mathcal{M})$ exactly as in Definition 1.1.2, using $=^{\mathcal{M}}$ to interpret equality.

Given a congruence τ-structure, one can always take the quotient $\mathcal{M}/=^{\mathcal{M}}$ and get a τ-structure where equality is the standard $\mathbb{N}$-equality. To highlight the difference, we will sometimes use the term *injective ω-presentations* when equality is $\mathbb{N}$-equality.

LEMMA 1.1.11. *Given a congruence ω-presentation $\mathcal{M}$ with infinitely many equivalence classes, the quotient $\mathcal{M}/=^{\mathcal{M}}$ has an injective ω-presentation $\mathcal{A}$ computable from $D(\mathcal{M})$. Furthermore, the natural projection $\mathcal{M} \to \mathcal{A}$ is also computable from $D(\mathcal{M})$.*

PROOF. All we need to do is pick a representative for each $=^{\mathcal{M}}$-equivalence class in a $D(\mathcal{M})$-computable way. Just take the $\mathbb{N}$-least element of each class: Let

$$A = \{a \in M : \forall b \in M \ (b <_{\mathbb{N}} a \Longrightarrow b \neq^{\mathcal{M}} a)\}$$

be the domain of $\mathcal{A}$. Define the functions and relations in the obvious way to get a $(\subseteq\omega)$-presentation of $\mathcal{M}$. To get an ω-presentation, use Observation 1.1.3. □

Therefore, from a computational viewpoint, there is no real difference in considering congruence structures or injective structures.

EXAMPLE 1.1.12. Suppose that $\mathcal{R}$ is a computable ring, and $I \subseteq R$ is a computable ideal. The quotient ring $\mathcal{R}/I$ has a natural congruence ω-presentation where the domain and the operations stay as in $\mathcal{R}$, but the equality relation $=^{\mathcal{R}/I}$ is the equivalence relation induced by I, namely $r =^{\mathcal{R}/I} q \iff r - q \in I$. We can then use the lemma above to get a computable injective ω-presentation of $\mathcal{R}/I$.

EXERCISE 1.1.13. Given a sequence of structures $\{\mathcal{A}_i : i \in \mathbb{N}\}$ and sequence of embeddings $f_{i,i+1} \colon \mathcal{A}_i \hookrightarrow \mathcal{A}_{i+1}$, the *direct limit* of such a sequence is a structure $\mathcal{A}_\infty$ for which there are embeddings $f_{i,\infty} \colon \mathcal{A}_i \to \mathcal{A}_\infty$ that commute with the previous embeddings (i.e. $f_{i,\infty} = f_{i+1,\infty} \circ f_{i,i+1}$ for all $i \in \mathbb{N}$), with the property that there is a increasing sequence of structures $\mathcal{B}_0 \subseteq \mathcal{B}_1 \subseteq \cdots \subseteq \mathcal{B}_\infty$, with $\mathcal{B}_\infty = \bigcup_s \mathcal{B}_s$, that is isomorphic to the original sequence, in the sense that there are isomorphisms $g_i \colon \mathcal{B}_i \to \mathcal{A}_i$ for $i \in \mathbb{N} \cup \{\infty\}$ such that $f_{i,j} \circ g_i = g_j \upharpoonright \mathcal{B}_i$ for all $i < j \in \mathbb{N} \cup \{\infty\}$. Prove that if the sequences $\{\mathcal{A}_i : i \in \mathbb{N}\}$ and $\{f_{i,i+1} : i \in \mathbb{N}\}$ of structures and embeddings are computable, then $\mathcal{A}_\infty$ has a computable copy.

1.1.7. Enumerations. Assume τ is a relational vocabulary. An *enumeration of a structure* $\mathcal{M}$ is just an onto map $g \colon \mathbb{N} \to M$. To each such enumeration we can associate a congruence ω-presentation $g^{-1}(\mathcal{M})$ by taking the *pull-back* of $\mathcal{M}$ through g:

$$g^{-1}(\mathcal{M}) = (\mathbb{N}; \sim, \{R_i^{g^{-1}(\mathcal{M})} : i \in I_R\}),$$

where $a \sim b \iff g(a) = g(b)$ and $R_i^{g^{-1}(\mathcal{M})} = g^{-1}(R_i^{\mathcal{M}}) \subseteq \mathbb{N}^{a(i)}$. The assumption that τ is relational was used here so that the pull-backs of functions and constants are not multi-valued. Let us remark that if g is injective, then $\sim$ becomes $=_{\mathbb{N}}$, and hence $g^{-1}(\mathcal{M})$ is an injective ω-presentation. In this case, the assumption that τ is relational is not important, as we can always pull-back functions and constants through bijections.

It is not hard to see that

$$D(g^{-1}(\mathcal{M})) \leq_T g \oplus D(\mathcal{M}).$$

Furthermore, $D(g^{-1}(\mathcal{M})) \leq_T g \oplus \tau^{\mathcal{M}}$, where $\tau^{\mathcal{M}}$ is as in Definition 1.1.1. As a corollary we get the following lemma.

LEMMA 1.1.14. *Let $\mathcal{A}$ be a computable structure in a relational vocabulary and M be an infinite c.e. subset of A. Then, the substructure $\mathcal{M}$ of $\mathcal{A}$ with domain M has a computable ω-presentation.*

PROOF. Just let g be an injective computable enumeration of $\mathcal{M}$. Then $g^{-1}(\mathcal{M})$ is a computable copy of $\mathcal{M}$. □

Throughout the book, there will be many constructions where we need to build a copy of a given structure with certain properties. In most cases, we will do it by building an enumeration of the structure and then taking the pull-back. The following observation will allow us to approximate the atomic diagram of the pull-back, and we will use it countless times.

Observation 1.1.15. Let g be an enumeration of $\mathcal{M}$. Notice that for every tuple $\bar{a} \in M^{<\mathbb{N}}$,

$$D_{g^{-1}(\mathcal{M})}(\bar{a}) = D_{\mathcal{M}}(g(\bar{a})).$$

For each k, use $g \restriction k$ to denote the tuple $\langle g(0), \ldots, g(k-1) \rangle \in M^k$. Then $D_{g^{-1}(\mathcal{M})}(\langle 0, \ldots, k-1 \rangle) = D_{\mathcal{M}}(g \restriction k)$ and the diagram of the pull-back can be calculated in terms of the diagrams of tuples in $\mathcal{M}$ as follows:

$$D(g^{-1}(\mathcal{M})) = \bigcup_{k \in \mathbb{N}} D_{\mathcal{M}}(g \restriction k).$$

1.2. Presentations that code sets

In this section, we show that the Turing degrees of ω-presentations of a non-trivial structure can be arbitrarily high. Furthermore, we prove a well-known theorem of Julia Knight that states that the set of Turing degrees of the ω-presentations of a structure is upwards closed. This set of Turing degrees is called the *degree spectrum* of the structure, and we will study it in detail in Chapter 5. Knight's theorem applies only to non-trivial structures: A structure $\mathcal{A}$ is *trivial* if there is a finite tuple such that every permutation of the domain fixing that tuple is an automorphism. Notice that these structures are essentially finite in the sense that anything relevant about them happens within that finite tuple.

THEOREM 1.2.1 (Knight [Kni98]). *Suppose that X can compute an ω-presentation of a non-trivial τ-structure $\mathcal{M}$. Then there is an ω-presentation $\mathcal{A}$ of $\mathcal{M}$ of Turing degree X.*

Before proving the theorem, let us remark that if instead of an ω-presentation we wanted a $(\subseteq\omega)$-presentation or a congruence ω-presentation, it would be very easy to code X into either the domain or the equality relation of $\mathcal{A}$: Recall that $\mathcal{D}(\mathcal{A}) = A \oplus (=^{\mathcal{A}}) \oplus \tau^{\mathcal{A}}$. Requiring $\mathcal{A}$ to be an injective ω-presentation forces us to code X into the structural part of $\mathcal{A}$, namely $\tau^{\mathcal{A}}$.

Proof. We will build an X-computable injective enumeration g of $\mathcal{M}$ and let $\mathcal{A} = g^{-1}(\mathcal{M})$. Since g and $\mathcal{M}$ are X-computable, that already gives us $D(\mathcal{A}) \leq_T X$; the actual work comes from ensuring that $D(\mathcal{A}) \geq_T X$. We build g as a limit

$$g = \bigcup_s \bar{p}_s \in M^{\mathbb{N}},$$

where the $\bar{p}_s$ are a nested sequence of injective tuples $\bar{p}_0 \subseteq \bar{p}_1 \subseteq \cdots$ in $M^{<\mathbb{N}}$. Recall from Observation 1.1.15 that we can approximate the atomic diagram of $\mathcal{A}$ by the atomic diagrams of the tuples $\bar{p}_s$:

$$D(\mathcal{A}) = \bigcup_{s \in \mathbb{N}} D_{\mathcal{M}}(\bar{p}_s).$$

Let $\bar{p}_0 = \emptyset$. Suppose now we have already defined $\bar{p}_s$. At stage $s + 1$, we build $\bar{p}_{s+1} \supseteq \bar{p}_s$ with the objective of coding the bit $X(s) \in \{0, 1\}$ into $D(\mathcal{A})$. The idea for coding $X(s)$ is as follows: We would like to find $a, b \in M \setminus \bar{p}_s$ such that $D_{\mathcal{M}}(\bar{p}_s a) \neq D_{\mathcal{M}}(\bar{p}_s b)$. Suppose we find them and $D_{\mathcal{M}}(\bar{p}_s a) <_{lex} D_{\mathcal{M}}(\bar{p}_s b)$, where $\leq_{lex}$ is the lexicographical ordering on strings in $2^{<\mathbb{N}}$. Then, depending on whether $X(s) = 0$ or 1, we can define $\bar{p}_{s+1}$ to be either $\bar{p}_s ab$ or $\bar{p}_s ba$. To decode $X(s)$, all we have to do is compare the binary strings $D_{\mathcal{A}}\langle 0, \ldots, k_s - 1, \ k_s \rangle$ and $D_{\mathcal{A}}\langle 0, \ldots, k_s - 1, \ k_s + 1 \rangle$ lexicographically, where $k_s = |\bar{p}_s|$.

The problem with this idea is that such a and b may not exist, and $D_{\mathcal{M}}(\bar{p}_s a)$ might be the same for all $a \in M$. Since $\mathcal{M}$ is non-trivial, we know there is some bijection of M preserving $\bar{p}_s$ which is not an isomorphism, and hence there exist tuples $\bar{a}$ and $\bar{b} \in (M \setminus \bar{p}_s)^{<\mathbb{N}}$ of the same length with $D_{\mathcal{M}}(\bar{p}_s \bar{a}) \neq D_{\mathcal{M}}(\bar{p}_s \bar{b})$. Furthermore, there exists disjoint such $\bar{a}$ and $\bar{b}$: To see this, take a third tuple disjoint from $\bar{a}$ and $\bar{b}$. Its diagram must be different from that of either $\bar{a}$ or $\bar{b}$ (as those diagrams are different) and we can replace it for $\bar{b}$ or $\bar{a}$ accordingly, two get two disjoint tuples with different diagrams. So we search for such a pair of tuples $\bar{a}, \bar{b}$, say of length h. We also require the pair $\bar{a}, \bar{b}$ to be minimal, in the sense that $D_{\mathcal{M}}(\bar{p}_s a_0, \ldots, a_{i-1}) = D_{\mathcal{M}}(\bar{p}_s b_0, \ldots, b_{i-1})$ for $i < h$; if they are not, truncate them. Suppose $D_{\mathcal{M}}(\bar{p}_s \bar{a}) <_{lex} D_{\mathcal{M}}(\bar{p}_s \bar{b})$ (otherwise replace $\bar{a}$ for $\bar{b}$ in what follows). If $X(s) = 0$, let $\tilde{p}_{s+1} = \bar{p}_s a_0 b_0 a_1 b_1, \ldots, a_{h-1} b_{h-1}$. If $X(s) = 1$, let $\tilde{p}_{s+1} = \bar{p}_s b_0 a_0 b_1 a_1, \ldots, b_{h-1} a_{h-1}$. Finally, to make sure g is onto, we let $\bar{p}_{s+1} = \tilde{p}_{s+1} c$, where c is the $\mathbb{N}$-least element of $M \setminus \tilde{p}_{s+1}$.

To recover X from $D(\mathcal{A})$, we need to also simultaneously recover the sequence of lengths $\{k_s : s \in \mathbb{N}\}$, where $k_s = |\bar{p}_s|$, for which we use the minimality of $\bar{a}$ and $\bar{b}$. Given k_s, we can compute k_{s+1} uniformly in $D(\mathcal{A})$ as follows: k_{s+1} is the least $k > k_s$ such that

$$D_{\mathcal{A}}(0, \ldots, k_s - 1, \ k_s, k_s + 2, k_s + 4, \ldots, k - 3) \neq$$
$$D_{\mathcal{A}}(0, \ldots, k_s - 1, \ k_s + 1, k_s + 3, k_s + 5, \ldots, k - 2).$$

Once we know which of these two binary strings is lexicographically smaller, we can tell if $X(s)$ is 0 or 1: It is 0 if the former one is $<_{lex}$-smaller than the latter one.   □

Notice that for trivial structures, all presentations are isomorphic via computable bijections, and hence all presentations have the same Turing degree. When the vocabulary is finite, all trivial structures are computable.

Chapter 2

RELATIONS

A *relation* is nothing more than a set of tuples from a structure. The study of the complexity and definability of this basic concept is one of the main components of computable structure theory. In model theory, a relation on a structure $\mathcal{A}$ is usually a subset of A^n for some fixed n. Here, we allow ourselves to consider infinitely many relations at once, and hence consider subsets of $A^{<\mathbb{N}}$ and even $\mathbb{N} \times A^{<\mathbb{N}}$ as relations. Thus, while in model theory one is interested in structures of lesser computational complexity than the natural numbers, here we purposely allow our relations to interact with the natural numbers.

Many of the notions of computability on subsets of $\mathbb{N}$ can be extended to such relations on a structure, but the space of relations is usually much richer than the space of subsets of $\mathbb{N}$, and understanding that space allows us to infer properties about the underlying structure. In this chapter we will introduce the analogues of the notions of c.e.ness, Turing reducibility, join, and jump for the space of relations. These tools will be used throughout the book.

From now on, unless otherwise stated, when we are given a structure, we are given an ω-presentation of a structure. Throughout this chapter, $\mathcal{A}$ always denotes an ω-presentation of a τ-structure.

2.1. Relatively intrinsic notions

We start by defining a notion of *c.e.-ness* for relations on a given structure. This will open the door for generalizing other notions of computability theory from subsets of $\mathbb{N}$ to relations on a structure.

2.1.1. R.i.c.e. relations. Let us try to capture what is happening underneath the following examples:

EXAMPLE 2.1.1. Consider $\mathbb{Q}$-vector spaces where the vocabulary contains a constant $\vec{0}$ for the zero vector, a binary operation $+$ for vector addition, and, for each rational $q \in \mathbb{Q}$, a unary operation $q \cdot \_$ for scalar multiplication by q. The field $\mathbb{Q}$ is not part of the domain of the structure, only the vectors are. Over a $\mathbb{Q}$-vector space $\mathcal{V}$, the relation LD $\subseteq V^{<\mathbb{N}}$ of linear dependence[11] is always c.e. in $\mathcal{V}$. To enumerate LD in a $D(\mathcal{V})$-computable way, go through all the possible non-trivial $\mathbb{Q}$-linear combinations $q_0 \cdot v_0 + \cdots + q_k \cdot v_k$ of all possible tuples of vectors $\langle v_0, \ldots, v_k \rangle \in V^{<\mathbb{N}}$, and if you find one that is equal to $\vec{0}$, enumerate $\langle v_0, \ldots, v_k \rangle$ into LD.

EXAMPLE 2.1.2. Over a ring $\mathcal{R}$, the relation that holds of $\langle r_0, \ldots, r_k \rangle \in R^{<\mathbb{N}}$ if the polynomial $r_0 + r_1 x + \cdots + r_k x^k$ has a root is c.e. in $\mathcal{R}$: As in the previous example, search for a root of the polynomial by evaluating the polynomial (which can be done $D(\mathcal{R})$-computably) on all the possible values of $x \in R$, and if you ever find one that makes the polynomial 0, enumerate $\langle r_0, \ldots, r_k \rangle$ into the relation.

DEFINITION 2.1.3. Let $\mathcal{A}$ be a structure. A relation $R \subseteq \mathbb{N} \times A^{<\mathbb{N}}$ is *relatively intrinsically computably enumerable* (*r.i.c.e.*) if, for every copy $(\mathcal{B}, R^{\mathcal{B}})$ of $(\mathcal{A}, R)$, the relation $R^{\mathcal{B}}$ (viewed as a subset of $\mathbb{N}^{<\mathbb{N}}$) is c.e. in $D(\mathcal{B})$.

The relations from Examples 2.1.1 and 2.1.2 are both r.i.c.e. A relation like linear independence, whose complement is r.i.c.e., is said to be *co-r.i.c.e.*

Notice that the notion of being r.i.c.e. is independent of the presentation of $\mathcal{A}$, and depends only on its isomorphism type.

Let us remark that we can view $(\mathcal{A}, R)$ as a structure in the sense we defined on page xix, by thinking of R as an infinite sequence of relations $\langle R_{m,n} : m, n \in \mathbb{N} \rangle$, where $R_{m,n} = \{\vec{r} \in A^n : \langle m, \vec{r} \rangle \in R\}$ is a relation of arity n. The original definitions of r.i.c.e. (cf. [AK00, p. 165] [Mon12, Definition 3.1]) are only on n-ary relations for fixed n, but that is too restrictive for us. The reason we choose to define r.i.c.e. on subsets of $\mathbb{N} \times A^{<\mathbb{N}}$ is that it is a simple enough setting which, at the same time, is fully general. This is the same reason we choose to develop computability theory on sets of natural numbers instead of on the set of hereditarily finite sets: The natural numbers are simpler, and yet every finite object can be encoded by a single natural number. We will get back to this point in Section 2.4.

EXAMPLE 2.1.4. Let $\mathcal{A}$ be a linear ordering $(A; \leq)$. We say that x and $y \in A$ are *adjacent*, and write Adj(x, y), if $x < y$ and there is no element in between them. Notice that the complement of this relation, $\neg$Adj$(a, b) \subseteq A^2$, is c.e. in $D(\mathcal{A})$: At stage s, we are monitoring the first s elements of the ω-presentation of $\mathcal{A}$, and if we see an element appear in between a and b, we enumerate the pair $\langle a, b \rangle$ into $\neg$Adj(a, b). This is also the case for any other ω-presentation of $\mathcal{A}$. Therefore, $\neg$Adj is r.i.c.e. There is something intrinsic about $\neg$Adj that makes it c.e. in whatever ω-presentation we consider. The reason is actually quite explicit: It has an

---

[11]LD is the set of tuples $\langle v_0, \ldots, v_k \rangle \in V^{<\mathbb{N}}$ of vectors that are linearly dependent.

$\exists$-definition, namely

$$\neg\mathsf{Adj}(x, y) \iff x \not< y \vee \exists z \, (x < z < y).$$

There are, however, r.i.c.e. relation that do not have $\exists$-definitions:

EXAMPLE 2.1.5. Consider a linear ordering with the adjacency relation as part of the structure $\mathcal{A} = (A; <, \mathsf{Adj})$. We call these structures *adjacency linear orderings*. On it, consider the set R of pairs of elements from A for which the number of elements in between them is a number that belongs to $0'$. We note that $R \subseteq A^2$ is r.i.c.e.: Given $a, b \in A$, wait to find elements $a_1, \ldots, a_n$ with $\mathsf{Adj}(a, a_1) \wedge \mathsf{Adj}(a_1, a_2) \wedge \cdots \wedge \mathsf{Adj}(a_{n-1}, a_n) \wedge \mathsf{Adj}(a_n, b)$, and if we ever find them, wait to see if n enters $0'$, and if that ever happens, enumerate $\langle a, b \rangle$ into R. The relation R cannot be defined by an $\exists$-formula in the vocabulary $\{\leq, \mathsf{Adj}\}$. But it can be defined by a computable infinite disjunction of them.

EXAMPLE 2.1.6. On the standard computable ω-presentation of the rationals $\mathcal{Q} = (\mathbb{Q}; 0, 1, +, \times)$, a relation $R \subseteq \mathbb{N} \times \mathbb{Q}^{<\mathbb{N}}$ is r.i.c.e. if and only if it is c.e. This is because if $\mathcal{A}$ is a copy of $\mathcal{Q}$, then there is a $D(\mathcal{A})$ computable isomorphism between $\mathcal{A}$ and $\mathcal{Q}$, and hence if R is c.e., $R^{\mathcal{A}}$ is c.e. in $D(\mathcal{A})$.

Observation 2.1.7. For the definition of r.i.c.e., it does not matter whether we use ω-presentations or congruence ($\subseteq\omega$)-presentations. That is, a relation $R \subseteq \mathbb{N} \times A^{<\mathbb{N}}$ is r.i.c.e. as in Definition 2.1.3 if and only if, for every congruence ($\subseteq\omega$)-presentation $(\mathcal{B}, R^{\mathcal{B}})$ of $(\mathcal{A}, R)$, we have that $R^{\mathcal{B}}$ is c.e. in $D(\mathcal{B})$.

2.1.2. R.i. computability. The same way we generalized the notion of c.e.ness to define r.i.c.e. relations, we can extend other standard concepts from computability theory to the space of relations on a structure.

DEFINITION 2.1.8. A relation $R \subseteq \mathbb{N} \times A^{<\mathbb{N}}$ is *relatively intrinsically computable* (*r.i. computable*) if $R^{\mathcal{B}}$ is computable in $D(\mathcal{B})$ whenever $(\mathcal{B}, R^{\mathcal{B}})$ is a copy of $(\mathcal{A}, R)$.

Observe that R is r.i. computable if and only if it is r.i.c.e. and co-r.i.c.e. The reader can imagine how to continue in this line of definitions for other notions of complexity, like *relatively intrinsically* Δ_2^0, *relatively intrinsically arithmetic*, etc. These notions relativize in an obvious way to produce a notion of relative computability:

DEFINITION 2.1.9. Given $R \subseteq \mathbb{N} \times A^{<\mathbb{N}}$ and $Q \subseteq \mathbb{N} \times A^{<\mathbb{N}}$, we say that R *is r.i.c.e. in* Q if R is r.i.c.e. in the structure $(\mathcal{A}, Q)$, that is, if $\mathcal{R}^{\mathcal{B}}$ is c.e. in $D(\mathcal{B}) \oplus Q^{\mathcal{B}}$ for every copy $(\mathcal{B}, R^{\mathcal{B}}, Q^{\mathcal{B}})$ of $(\mathcal{A}, R, Q)$. R *is r.i. computable in* Q, and we write $R \leq_{rT} Q$, if R is r.i. computable in the structure $(\mathcal{A}, Q)$.

The 'rT' stands for "relatively Turing."

EXAMPLE 2.1.10. Let $\mathcal{A} = (A; \leq)$ be a linear ordering, and consider the relation given by the pairs of elements which have at least two elements in between:

$$T = \{\langle a, b \rangle \in A^2 : a < b \wedge \exists c, d(a < c < d < b)\}.$$

Then $T \leq_{rT}$ Adj: Suppose we are given $\langle a, b \rangle \in A^2$ with $a < b$ and we want to decide if $\langle a, b \rangle \in T$ using Adj. If Adj(a, b), we know $\langle a, b \rangle \notin T$. Otherwise, search for c in between a and b, which we know we will find. If Adj(a, c) and Adj(c, b), we know that $\langle a, b \rangle \notin T$. Otherwise, we must have $\langle a, b \rangle \in T$.

On the linear ordering of the natural numbers $\omega = (\mathbb{N}; \leq)$, we also have Adj $\leq_{rT} T$: To decide if a and b are adjacent wait either for an element to appear in between them or for an element $c > b$ with $\neg T(a, c)$. In the former case we know that a and b are not adjacent, while in the latter case we can deduce that they are.

On the other hand, there are linear orderings where Adj $\nleq_{rT} T$. As an example, consider the linear ordering

$$\mathcal{A} = 2\mathbb{Q} + 3 + 2\mathbb{Q} + 3 + 2\mathbb{Q} + 3 + \cdots,$$

where $2\mathbb{Q}$ is built by replacing each element of $\mathbb{Q}$ by a pair of adjacent elements, obtaining densely many copies of 2. To show that Adj $\nleq_{rT} T$, it is enough to build a computable copy $\mathcal{B}$ of $\mathcal{A}$, where $T^{\mathcal{B}}$ is computable, but Adj$^{\mathcal{B}}$ is not. To do this, let us start by fixing a computable ω-presentation $\mathcal{C}$ of the linear ordering $2\mathbb{Q} = 2\mathbb{Q} + 1 + 1 + 2\mathbb{Q} + 1 + 1 + \cdots$, and picking a computable increasing sequence of adjacent pairs $c_{n,0}, c_{n,1}$ for $n \in \mathbb{N}$.

$$\mathcal{C} = 2\mathbb{Q} + \{c_{0,0}\} + \{c_{0,1}\} + 2\mathbb{Q} + \{c_{1,0}\} + \{c_{1,1}\} + 2\mathbb{Q} + \\ \{c_{2,0}\} + \{c_{2,1}\} + \cdots$$

To build the ω-presentation $\mathcal{B}$ of $\mathcal{A}$, we will add an element in between $c_{n,0}$ and $c_{n,1}$ if and only if $n \in 0'$; we can then decode $0'$ from Adj$^{\mathcal{B}}$ by checking if $c_{n,0}$ and $c_{n,1}$ are adjacent in $\mathcal{B}$. More formally, to define $\mathcal{B}$, put a copy of $\mathcal{C}$ on the even numbers in the domain of $\mathcal{B}$, and use the odd numbers to add those "in-between" elements. Let $2s + 1$ be $\leq_{\mathcal{B}}$-between $c_{k_s,0}$ and $c_{k_s,1}$, where $\{k_s : s \in \mathbb{N}\}$ is a computable enumeration of $0'$. Notice that $\leq_{\mathcal{B}}$ is computable. The relation $T^{\mathcal{B}}$ is also computable, as it holds between any two elements of $\mathcal{C}$ which are not in the same 2-block, and holds between $2s + 1$ and any other element, except for $c_{k_s,0}$ and $c_{k_s,1}$. The adjacency relation is not computable because $k \in 0' \iff \neg$Adj$(c_{k,0}, c_{k,1})$ for all $k \in \mathbb{N}$.

EXERCISE 2.1.11. On a linear ordering, let T_n be the n-in-between relation that holds of a pair $\langle a, b \rangle$ if $a < b$ and there are at least n elements in between a and b.

(a) Show that on every linear ordering, $T_{n+1} \leq_{rT} T_n$ for all $n \in \mathbb{N}$.

(b) Show that there is a linear ordering on which $T_{n+1} <_{rT} T_n$ for all n.

2.1.3. A syntactic characterization. R.i.c.e. relations can be characterized in a purely syntactical way using computably infinitary formulas and without referring to the different copies of the structure. We will define computably infinitary formulas in [MonP2]. For now, we define just the class of *computably infinitary Σ_1 formulas* or *Σ_1^c formulas*.

DEFINITION 2.1.12. An *infinitary Σ_1 formula* (denoted Σ_1^{in}) is a countable (finite or infinite) disjunction of $\exists$-formulas over a finite set of free variables. A *computable infinitary Σ_1 formula* (denoted Σ_1^c) is a finite or infinite disjunction of a computable list of $\exists$-formulas over a finite set of free variables.

Thus, a Σ_1^c formula is one of the form

$$\psi(\bar{x}) \equiv \bigvee_{i \in I} \exists \bar{y}_i \varphi_i(\bar{x}, \bar{y}_i),$$

where each φ_i is quantifier-free, I is an initial segment of $\mathbb{N}$, and the Gödel indices $\langle \ulcorner \varphi_i \urcorner : i \in I \rangle$ can be listed computably, i.e. it is a c.e. set of indices. The definition of satisfaction is straightforward: $\mathcal{A} \models \psi(\bar{a})$ if and only if there exist $i \in I$ and $\bar{b} \in A^{|\bar{y}_i|}$ such that $\mathcal{A} \models \varphi_i(\bar{a}, \bar{b})$. Using the effective enumeration $\{W_e : e \in \mathbb{N}\}$ of the c.e. sets, we can enumerate all Σ_1^c formulas as follows: If $\{\varphi_{i,j}^{\exists}(x_1, \ldots, x_j) : i \in \mathbb{N}\}$ is an effective enumeration of the existential τ-formulas with j free variables, we define

$$\varphi_{e,j}^{\Sigma_1^c}(\bar{x}) \equiv \bigvee_{\langle i,j \rangle \in W_e} \varphi_{i,j}^{\exists}(\bar{x})$$

for each $e \in \mathbb{N}$. We then get that $\{\varphi_{e,j}^{\Sigma_1^c} : e \in \mathbb{N}\}$ is an effective enumeration of the Σ_1^c τ-formulas with j free variables. Note that if $\psi(\bar{x})$ is Σ_1^c, then $\{\bar{a} \in A^{|\bar{x}|} : \mathcal{A} \models \psi(\bar{a})\}$ is c.e. in $D(\mathcal{A})$, uniformly in ψ and $\mathcal{A}$. In other words, there is a c.e. operator W such that

$$\langle \ulcorner \psi \urcorner, \bar{a} \rangle \in W^{D(\mathcal{A})} \iff \mathcal{A} \models \psi(\bar{a})$$

for all τ-ω-presentations $\mathcal{A}$, Σ_1^c τ-formulas ψ, and tuples $\bar{a} \in A^{|\bar{x}|}$.

EXAMPLE 2.1.13. In a group $\mathcal{G} = (G; *)$, the set of torsion elements can be described by the Σ_1^c formula:

$$\text{torsion}(x) \equiv \bigvee_{i \in \mathbb{N}} (x * \underbrace{x * x * \cdots * x}_{i \text{ times}} = e),$$

where e is the identity of the group.

EXAMPLE 2.1.14. On a graph $\mathcal{G} = (V; E)$, the relation of being path-connected can be described by the Σ_1^c formula:

$$\text{connected}(x, y) \equiv \bigvee_{i \in \mathbb{N}} \exists z_1, \ldots, z_i \, (xEz_1 \wedge z_1Ez_2 \wedge \cdots \wedge z_iEy).$$

We would like to consider Σ^c_1 definability, not only for n-ary relations, but also for subsets of $A^{<\mathbb{N}}$.

DEFINITION 2.1.15. A relation $R \subset \mathbb{N} \times A^{<\mathbb{N}}$ is Σ^c_1-definable in $\mathcal{A}$ with parameters if there is a tuple $\bar{p} \in A^{<\mathbb{N}}$ and a computable sequence[12] of Σ^c_1 formulas $\psi_{i,j}(x_1, \ldots, x_{|\bar{p}|}, y_1, \ldots, y_j)$, for $i, j \in \mathbb{N}$, such that

$$R = \{\langle i, \bar{b}\rangle \in \mathbb{N} \times A^{<\mathbb{N}} : \mathcal{A} \models \psi_{i,|\bar{b}|}\langle \bar{p}, \bar{b}\rangle\}.$$

The elements in $\bar{p}$ are the *parameters* in the definition of R.

From the observation before Example 2.1.13, it is not hard to see that if $R \subset \mathbb{N} \times A^{<\mathbb{N}}$ is Σ^c_1 definable in $\mathcal{A}$ with parameters, it is r.i.c.e. The next theorem shows that this is a characterization. The theorem was proved for n-ary relations by Ash, Knight, Manasse, and Slaman [AKMS89], and independently by Chisholm [Chi90]. The proof for subsets of $\mathbb{N} \times A^{<\mathbb{N}}$ is no different.

THEOREM 2.1.16. (Ash, Knight, Manasse, and Slaman [AKMS89]; Chisholm [Chi90]). *Let $\mathcal{A}$ be a structure, and $R \subseteq \mathbb{N} \times A^{<\mathbb{N}}$ a relation on it. The following are equivalent*:

(A1) *R is r.i.c.e.*
(A2) *R is Σ^c_1 definable in $\mathcal{A}$ with parameters.*

PROOF. As we mentioned above, (A2) easily implies (A1). We prove the other direction. We will build a copy $\mathcal{B}$ of $\mathcal{A}$ by taking the pull-back of an enumeration $g \colon \mathbb{N} \to A$ that we construct step by step, and we will apply (1) to that copy. We define g as the union of a nested sequence of tuples $\{\bar{p}_s : s \in \mathbb{N}\} \subseteq A^{<\mathbb{N}}$, where $\bar{p}_s$ is defined at stage s. Then we define $\mathcal{B}$ to be the pull-back $g^{-1}(\mathcal{A})$ as in Subsection 1.1.7.[13] Thus, we will have

$$\bar{p}_0 \subseteq \bar{p}_1 \subseteq \cdots \subseteq \bar{p}_s \subseteq \cdots \xrightarrow{s \to \infty} g \quad \text{and} \quad D(\mathcal{B}) = \bigcup_s D_A(\bar{p}_s).$$

Throughout the construction, we try as much as possible to make $R^{\mathcal{B}}$ not c.e. in $D(\mathcal{B})$. But, because of (A1), this attempt will fail somewhere, and we will have that

$$g^{-1}(R) = R^{\mathcal{B}} = W_e^{D(\mathcal{B})}$$

for some $e \in \mathbb{N}$.[14] We will then turn this failure into a Σ^c_1 definition of R.

---

[12] When we say "computable sequence of Σ^c_1 formulas," we of course mean a computable sequence of indices of Σ^c_1 formulas.

[13] Let us observe that the fact that the congruence ω-presentation $\mathcal{B}$ is non-injective is not important here by Observation 2.1.7. Alternatively, we can make g one-to-one by requiring the tuples p_s to be injective.

[14] By $g^{-1}(R)$ we of course mean

$$\{\langle i, \langle j_0, \ldots, j_\ell\rangle\rangle \in \mathbb{N} \times \mathbb{N}^{<\mathbb{N}} : \langle i, \langle g(j_0), \ldots, g(j_\ell)\rangle\rangle \in R\}.$$

We will use $\langle i, j_1, \ldots, j_\ell\rangle$ as shorthand for $\langle i, \langle j_1, \ldots, j_\ell\rangle\rangle$.

Here is the construction of $\mathcal{B}$. Let $\bar{p}_0$ be the empty sequence. At odd stages, we take one step towards making g onto: At stage $s + 1 = 2e + 1$, if the eth element of A is not already in $\bar{p}_s$, we add it to the range of $\bar{p}_{s+1}$ (i.e., we let $\bar{p}_{s+1} = \bar{p}_s{}^\frown e$), and otherwise let $\bar{p}_{s+1} = \bar{p}_s$.

At the even stages, we work towards making $R^{\mathcal{B}}$ not c.e. in $D(\mathcal{B})$: At stage $s + 1 = 2e$, we try to force $W_e^{D(\mathcal{B})} \not\subseteq g^{-1}(R)$ for which we need a tuple $\langle i, j_1, \ldots, j_\ell \rangle \in W_e^{D(\mathcal{B})}$ with $\langle i, g(j_1), \ldots, g(j_\ell) \rangle \notin R$. We do this as follows: Ask if there is an extension $\bar{q}$ of $\bar{p}_s$ in the set

$$Q_e = \{\bar{q} \in A^{<\mathbb{N}} : \exists \ell, i, j_1, \ldots, j_\ell < |\bar{q}|$$
$$\left(\langle i, j_1, \ldots, j_\ell \rangle \in W_e^{D_A(\bar{q})} \text{ and } \langle i, q_{j_1}, \ldots, q_{j_\ell} \rangle \notin R \right) \}.$$

If there is one, we let $\bar{p}_{s+1} = \bar{q}$. If not, we do nothing and let $\bar{p}_{s+1} = \bar{p}_s$. This ends the construction of g and $\mathcal{B}$.

$$
\begin{array}{ccccc}
\mathbb{N} \times B^{<\mathbb{N}} & = & \mathbb{N} \times \mathbb{N}^{<\mathbb{N}} & \xrightarrow{\ \ g\ \ } & \mathbb{N} \times A^{<\mathbb{N}} \\
& & \cup\!\shortmid & & \cup\!\shortmid \\
& & W_e^{D(\mathcal{B})} & & R \\
& & \cup\!\shortmid & & \not\cup\!\shortmid \\
& & \langle i, j_1, \ldots, j_\ell \rangle & \xrightarrow{\ \ \bar{q}\ \ } & \langle i, q_{j_1}, \ldots, q_{j_\ell} \rangle
\end{array}
$$

Notice that if at a stage $s + 1 = 2e$, we succeed in defining $\bar{p}_{s+1} = \bar{q} \in Q_e$, then we succeed in making $W_e^{D(\mathcal{B})} \neq g^{-1}(R)$: This is because we would have $\bar{q} \subseteq g$ and hence that[15]

$$\langle i, j_1, \ldots, j_\ell \rangle \in W_e^{D_A(\bar{q})} \subseteq W_e^{D(\mathcal{B})}$$

$$\text{while } \langle i, g(j_1), \ldots, g(j_\ell) \rangle = \langle i, q_{j_1}, \ldots, q_{j_\ell} \rangle \notin R.$$

However, we cannot succeed at all such stages because $R^{\mathcal{B}} = W_e^{D(\mathcal{B})}$ for some $e \in \mathbb{N}$. Thus, for that particular e, at stage $s + 1 = 2e$, there was no extension of $\bar{p}_s$ in Q_e.

CLAIM 2.1.17. *If $R^{\mathcal{B}} = W_e^{D(\mathcal{B})}$ and there are no extensions of $\bar{p}$ in Q_e, then R is Σ_1^c-definable in $\mathcal{A}$ with parameters $\bar{p}$.*

PROOF OF THE CLAIM. Notice that if we find some $\bar{q} \supseteq \bar{p}$ and a sub-tuple $\langle i, \bar{a} \rangle = \langle i, q_{j_1}, \ldots, q_{j_\ell} \rangle$ such that $\langle i, j_1, \ldots, j_\ell \rangle \in W_e^{D_A(\bar{q})}$, then we must have $\langle i, \bar{a} \rangle \in R$, as otherwise we would get $\bar{q} \in Q_e$. This is the key idea we use to enumerate elements into R.

More formally, we will show that R is equal to the set

$$S = \{ \langle i, q_{j_1}, \ldots, q_{j_\ell} \rangle \in \mathbb{N} \times A^{<\mathbb{N}} : \text{ for some } \bar{q} \in A^{<\mathbb{N}} \text{ and}$$
$$\ell, i, j_1, \ldots, j_\ell < |\bar{q}| \text{ satisfying } \bar{q} \supseteq \bar{p} \text{ and } \langle i, j_1, \ldots, j_\ell \rangle \in W_e^{D_A(\bar{q})} \}.$$

---

[15]Observe that since $D(\mathcal{B}) = \bigcup_s D_A(\bar{p}_s)$, we have that $W_e^{D(\mathcal{B})} = \bigcup_s W_e^{D_A(\bar{p}_s)}$.

If $\langle i, \bar{a} \rangle \in R$, let $j_1, \ldots, j_{|\bar{a}|}$ be indices such that $\bar{a} = \langle g(j_1), \ldots, g(j_{|\bar{a}|}) \rangle$, and we get that $\langle i, \bar{a} \rangle \in S$ witnessed by a long enough segment $\bar{q}$ of g. For the other direction, if $\bar{a} = \langle i, q_{j_1}, \ldots, q_{j_\ell} \rangle \in S$, then we must have $\langle i, \bar{a} \rangle \in R$: Otherwise we would have $\bar{q} \in Q_e$, contradicting the assumption of the claim.

Now that we know that $R = S$, let us show that S is Σ_1^c definable with parameters $\bar{p}$. For every $i \in \mathbb{N}$ and $\bar{a} \in A^{<\mathbb{N}}$,

$$\langle i, \bar{a} \rangle \in S \iff$$

$$\exists \bar{q} \supseteq \bar{p} \bigvee_{j_1, \ldots, j_{|\bar{a}|} < |\bar{q}|} \left(\langle q_{j_1}, \ldots, q_{j_{|\bar{a}|}} \rangle = \bar{a} \ \& \ \langle i, j_1, \ldots, j_{|\bar{a}|} \rangle \in W_e^{D_A(\bar{q})} \right).$$

But "$\langle i, j_1, \ldots, j_{|\bar{a}|} \rangle \in W_e^{D_A(\bar{q})}$" is not a formula in the language. So we need to re-write it as:

$$\langle i, \bar{a} \rangle \in S \iff$$

$$\bigvee_{\sigma \in 2^{<\mathbb{N}}, \ \langle i, j_1, \ldots, j_{|\bar{a}|} \rangle \in W_e^\sigma} \exists \bar{q} \supseteq \bar{p} \left(\langle q_{j_1}, \ldots, q_{j_{|\bar{a}|}} \rangle = \bar{a} \ \& \ \text{"}\sigma \subseteq D_A(\bar{q})\text{"} \right).$$

Recall that, for each $\sigma \in 2^{<\mathbb{N}}$, there is a quantifier-free formula with the meaning "$\sigma \subseteq D_A(\bar{x})$" (Observation 1.1.10). □

Thus, R is Σ_1^c-definable in A with parameters $\bar{p}_s$. □

Let us comment on where the parameters come from. We just showed that: either for every e, every $\bar{p} \in A^{<\mathbb{N}}$ can be extend to a tuple $\bar{q} \in Q_e$, in which case we can satisfy every diagonalization requirement getting that $R^{\mathcal{B}}$ is not c.e. in $D(\mathcal{B})$; or there exists some e and some tuple $\bar{p}$ which cannot be extended in Q_e, in which case R is Σ_1^c definable in A with parameters $\bar{p}$. This tuple $\bar{p}$ *forces* $R^{\mathcal{B}}$ to be equal to $W_e^{D(\mathcal{B})}$ as we will see in Chapter 4.

Very often, we will deal with relations that are Σ_1^c-definable without parameters. These relations are not just r.i.c.e., but uniformly r.i.c.e.:

DEFINITION 2.1.18. A relation $R \subseteq \mathbb{N} \times A^{<\mathbb{N}}$ is *uniformly r.i.c.e.* (denoted u.r.i.c.e.) if there is a c.e. operator W such that $R^{\mathcal{B}} = W^{D(\mathcal{B})}$ for all $(\mathcal{B}, R^{\mathcal{B}}) \cong (A, R)$.

The difference between r.i.c.e. and u.r.i.c.e. relations is just that the former needs parameters in its Σ_1^c definition—parameters that one may not be able to find computably and hence require "non-uniform" information.

COROLLARY 2.1.19. *Let A be a structure and $R \subseteq \mathbb{N} \times A^{<\mathbb{N}}$ a relation on it. The following are equivalent:*

(A1) *R is u.r.i.c.e.*

(A2) *R is Σ_1^c definable in A without parameters.*

PROOF. It is easy to see that (A2) implies (A1). For the other direction, let W_e be the c.e. operator witnessing that R is u.r.i.c.e.. Let Q_e be as in the proof of Theorem 2.1.16. No tuple $\bar{q} \in A^{<\mathbb{N}}$ can be in Q_e because, otherwise, any extension of $\bar{q}$ to an enumeration g of $\mathcal{A}$ would satisfy $W_e^{D(g^{-1}(\mathcal{A}))} \not\subseteq g^{-1}(R)$, contradicting our choice of W_e. The corollary then follows from Claim 2.1.17 where $\bar{p}$ is the empty tuple. $\square$

EXERCISE 2.1.20. Show that a relation $R \subseteq A^k$ is u.r.i.c.e. if and only if there exists a c.e. set $W \subseteq 2^{<\mathbb{N}}$ such that, for $\bar{a} \in A^k$,

$$\bar{a} \in R \iff \bigvee_{\sigma \in W} \exists \bar{q} \in A^{<\mathbb{N}} \left(\sigma \subseteq D_{\mathcal{A}}(\bar{a}\bar{q}) \right).$$

Hint in footnote.[16]

2.1.4. Coding sets of natural numbers. Another feature that is useful when working with subsets of $\mathbb{N} \times A^{<\mathbb{N}}$ is that we can code subsets of $\mathbb{N}$ in an obvious way: We represent $X \subseteq \mathbb{N}$ by $X \times \{\langle\rangle\} \subseteq \mathbb{N} \times A^{<\mathbb{N}}$, where $\langle\rangle$ is the empty tuple. We will sometimes abuse notation and refer to a set $X \subseteq \mathbb{N}$ as if it was a subset of $\mathbb{N} \times A^{<\mathbb{N}}$. For instance, if we say that X is r.i.c.e. in $\mathcal{A}$, we would formally mean that $X \times \{\langle\rangle\}$ is r.i.c.e. in $\mathcal{A}$. Thus, $X \times \{\langle\rangle\}$ is r.i.c.e. in $\mathcal{A}$ if and only if X is c.e. in every ω-presentation of $\mathcal{A}$. If that is the case, we say that X is *c.e.-coded* by $\mathcal{A}$ (cf. [Mon10, Definition 1.8]). If $X \times \{\langle\rangle\}$ r.i. computable in $\mathcal{A}$, or equivalently if X is computable in every ω-presentation of $\mathcal{A}$, we say that X is *computably coded* by $\mathcal{A}$. A characterization of the sets that are c.e.-coded by a given structure was first given by Knight [Kni86, Theorem 1.4']. We get it as a corollary of Theorem 2.1.16. Let us first see an example.

EXAMPLE 2.1.21. Given $X \subseteq \mathbb{N}$, let $\mathcal{G}$ be the group $\bigoplus_{i \in X} \mathbb{Z}_{p_i}$, where p_i is the ith prime number and $\mathbb{Z}_p$ is the cyclic group of size p, namely $\mathbb{Z}/p\mathbb{Z}$. We then have that X is c.e.-coded by $\mathcal{G}$, as $i \in X$ if and only if there is an element of $\mathcal{G}$ of order p_i.

A more general family of examples are the $\exists$-types of tuples from the structure.

DEFINITION 2.1.22. Given $\bar{a} \in A^{<\mathbb{N}}$, we define the $\exists$-*type* of $\bar{a}$ in $\mathcal{A}$ as

$$\exists\text{-}tp_{\mathcal{A}}(\bar{a}) = \{i \in \mathbb{N} : \mathcal{A} \models \varphi^{\exists}_{i,|\bar{a}|}(\bar{a})\}$$

where $\{\varphi^{\exists}_{i,j} : i \in \mathbb{N}\}$ is an effective enumeration of the $\exists$-τ-formulas with j-free variables.

Clearly, for any tuple $\bar{a} \in A^{<\mathbb{N}}$, we can enumerate $\exists\text{-}tp_{\mathcal{A}}(\bar{a})$ from any ω-presentation of $\mathcal{A}$ once we recognize where the tuple $\bar{a}$ is in the ω-presentation (non-uniformly). Knight's theorem essentially says that $\exists$-types

---

[16]Prove it first for atomic formulas, then quantifier free formulas, then $\exists$-formulas, and then Σ^c_1 formulas.

are essentially all that a structure can c.e.-code. To state Knight's results, we need to review *enumeration reducibility*.

DEFINITION 2.1.23. An *enumeration of Y* is an onto function $f : \mathbb{N} \to Y$. A set $X \subseteq \mathbb{N}$ is *e-reducible to $Y \subseteq \mathbb{N}$* if every enumeration of Y computes an enumeration of X. See page xviii in the background section for more on *e*-reducibility.

Suppose we have a set $X \subseteq \mathbb{N}$ that is *e*-reducible to the $\exists$-type of some tuple $\bar{p}$ in $\mathcal{A}$. Then any ω-presentation of $\mathcal{A}$ can enumerate $\exists\text{-}tp_{\mathcal{A}}(\bar{p})$ and hence also X. Thus, X is c.e.-coded by $\mathcal{A}$. Knight showed that these are all the sets $\mathcal{A}$ codes:

COROLLARY 2.1.24. (Knight [Kni86, Theorem 1.4'], see also Ash and Knight [AK00, Theorem 10.17]). *Let $X \subseteq \mathbb{N}$. The following are equivalent*:

(B1) X *is c.e.-coded by $\mathcal{A}$ (i.e., X is c.e. in every copy of $\mathcal{A}$).*
(B2) X *is e-reducible to $\exists\text{-}tp_{\mathcal{A}}(\bar{p})$ for some $\bar{p} \in A^{<\mathbb{N}}$.*

PROOF. We have already mentioned how (B2) implies (B1). We prove the other direction.

As we mentioned before, X is c.e. in every copy of $\mathcal{A}$ if and only if $X \times \{\langle\rangle\}$ is r.i.c.e. in $\mathcal{A}$. By Theorem 2.1.16, we have a Σ_1^c definition of $X \times \{\langle\rangle\}$ over some parameters $\bar{p}$. This means that we have a computable list $\{\psi_n : n \in \mathbb{N}\}$ of Σ_1^c sentences such that $n \in X \iff \mathcal{A} \models \psi_n$. We can then transform this Σ_1^c definition into an enumeration operator Φ that outputs X when $\exists\text{-}tp_{\mathcal{A}}(\bar{p})$ is given as input: The operator Φ enumerates n into $\Phi^{\exists\text{-}tp_{\mathcal{A}}(\bar{p})}$ if (the index of) one of the disjuncts of ψ_n appears in $\exists\text{-}tp_{\mathcal{A}}(\bar{p})$. If the reader wants to be very explicit: if the Σ_1^c definition of $X \times \{\langle\rangle\}$ with parameters $\bar{p}$ is of the form

$$\langle n, \langle\rangle\rangle \in X \times \{\langle\rangle\} \iff \mathcal{A} \models \psi_n \iff \mathcal{A} \models \bigvee_{i : \langle n, i\rangle \in W} \varphi_{i, |\bar{p}|}^{\exists}(\bar{p})$$

for some c.e. set W. Then

$$n \in X \iff \exists i \in \mathbb{N}\big(\langle n, i\rangle \in W \wedge i \in \exists\text{-}tp_{\mathcal{A}}(\bar{p})\big),$$

and hence X is *e*-reducible to $\exists\text{-}tp_{\mathcal{A}}(\bar{p})$.   □

EXERCISE 2.1.25. Let $\mathcal{A}$ be a structure and X a set c.e.-coded by $\mathcal{A}$. Let $\psi(x_0, \ldots, x_{k-1})$ be a Σ_1^{in} formula of the form $\bigvee_{i \in Y} \varphi_{i,k}^{\exists}(\bar{x})$ where Y is *e*-reducible to X. Show that $\{\bar{a} \in A^k : \mathcal{A} \models \psi(\bar{a})\}$ is r.i.c.e.

EXERCISE 2.1.26. We say that $X \subseteq \mathbb{N}$ is *uniformly c.e.-coded* by $\mathcal{A}$ if $X \times \langle\rangle$ is u.r.i.c.e. in $\mathcal{A}$. Show that X is uniformly c.e.-coded by $\mathcal{A}$ if and only if $X \leq_e \exists\text{-}Th(\mathcal{A})$.

2.1.5. Joins. The use of subsets of $\mathbb{N} \times A^{<\mathbb{N}}$ allows us to consider not only natural numbers and all n-tuples simultaneously, but also all finite objects that can be built over A. We will see more on this in Section 2.4. For now, we see how to code many relations using just one.

DEFINITION 2.1.27. Given $R, Q \subseteq \mathbb{N} \times A^{<\mathbb{N}}$, we define $R \oplus Q$ by $\langle m, \bar{b} \rangle \in R \oplus Q$ if either $m = 2n$ and $\langle n, \bar{b} \rangle \in R$, or $m = 2n + 1$ and $\langle n, \bar{b} \rangle \in Q$.

It is not hard to see that $\oplus$ defines a least-upper-bound operation for r.i. computability. That is, R and Q are r.i. computable in $R \oplus Q$, and whenever both R and Q are r.i. computable in a relation $S \subseteq \mathbb{N} \times A^{<\mathbb{N}}$, $R \oplus Q$ is r.i. computable in S too.

We can then take joins of $\mathbb{N}$-sequences of relations in a straightforward way too. We can keep on pushing this idea much further. For instance, given $Q \subseteq (A^{<\mathbb{N}})^2$, we can encode it by a relation $\mathbb{N} \times R \subseteq \mathbb{N} \times A^{<\mathbb{N}}$ as follows: $\langle n, \bar{b} \rangle \in R$ if $\langle \langle b_0, \ldots, b_{n-1} \rangle, \langle b_n, \ldots, b_{|\bar{b}|-1} \rangle \rangle \in Q$. In a similar way, the reader can imagine how to code subsets of $(A^{<\mathbb{N}})^{<\mathbb{N}}$ by subsets of $A^{<\mathbb{N}}$. We will see the most general form of this in Section 2.4.1.

Remark 2.1.28. Given $Q \subseteq \mathbb{N} \times A^{<\mathbb{N}}$, define $R \subseteq A^{<\mathbb{N}}$ as follows: $\bar{b} \in R$ if and only if $|\bar{b}|$ is a number coding a pair $\langle n, m \rangle \in \mathbb{N}^2$ and $\langle n, \bar{b} \upharpoonright m \rangle \in Q$. We then have that Q is r.i.c.e. if and only if R is r.i.c.e. Thus, working in the setting of subsets of $A^{<\mathbb{N}}$ would have been as general as working in the setting of subsets of $\mathbb{N} \times A^{<\mathbb{N}}$.

2.2. Complete relations

So far we have notions of c.e.-ness, computability, and join on the subsets of $\mathbb{N} \times A^{<\mathbb{N}}$. The next step is to get an analogue for the Turing jump.

2.2.1. R.i.c.e. complete relations.

DEFINITION 2.2.1. A relation $R \subseteq \mathbb{N} \times A^{<\mathbb{N}}$ is *complete* in A if every r.i.c.e. relation $Q \subseteq \mathbb{N} \times A^{<\mathbb{N}}$ is r.i. computable in R. R is *r.i.c.e. complete* if it is also r.i.c.e. itself.[17]

If we view $0'$ as a subset of $\mathbb{N} \times A^{<\mathbb{N}}$ as in Section 2.1.4, $0'$ is always r.i.c.e. in A and hence every complete relation must r.i. compute it. For some structures A, $0'$ is r.i.c.e. complete itself, but in most cases, it is not. This is not surprising as $0'$ contains no structural information about A.

EXAMPLE 2.2.2. On a $\mathbb{Q}$-vector space, $LD \oplus 0'$ is r.i.c.e. complete, and LD is not r.i. computable from $0'$ when the space has infinite dimension. Recall that LD is the linear dependence relation (Example 2.1.1).

---

[17]This is the analogue of Turing-complete and not of m-complete.

On a linear ordering, $(\neg\mathsf{Adj}) \oplus 0'$ is r.i.c.e. complete, and $\neg\mathsf{Adj}$ is not r.i. computable from $0'$ unless there are only finitely many adjacencies. We will prove these facts in Lemmas 2.3.2 and 2.3.1.

These examples of complete relations are particularly nice and clean, but we will not always be able to find such simple complete relations. Simple or not, r.i.c.e. complete relations always exist. We consider the analogue of Kleene's predicate K by putting together all Σ_1^c-definable relations. Recall from Section 2.1.3 that $\{\varphi_{i,j}^{\Sigma_1^c} : i \in \mathbb{N}\}$ is an effective enumeration of the Σ_1^c τ-formulas with j free variables.

DEFINITION 2.2.3 (Montalbán [Mon12]). The *Kleene relation* relative to $\mathcal{A}$, $\vec{K}^{\mathcal{A}} \subseteq \mathbb{N} \times A^{<\mathbb{N}}$, is defined by

$$\langle i, \bar{b} \rangle \in \vec{K}^{\mathcal{A}} \iff \mathcal{A} \models \varphi_{i,|\bar{b}|}^{\Sigma_1^c}(\bar{b}).$$

It is clear that $\vec{K}^{\mathcal{A}}$ is r.i.c.e. It follows from Theorem 2.1.16 that, for every r.i.c.e. relation $R \subseteq A^n$, there are $i \in \mathbb{N}$ and $\bar{a} \in A^{<\mathbb{N}}$ such that $R = \{\bar{b} \in A^n : \langle i, \bar{a}^\frown \bar{b} \rangle \in \vec{K}^{\mathcal{A}}\}$. The following lemma shows $\vec{K}^{\mathcal{A}}$ is complete among all r.i.c.e. relations in $\mathbb{N} \times A^{<\mathbb{N}}$.

LEMMA 2.2.4. *For every r.i.c.e. $R \subseteq \mathbb{N} \times A^{<\mathbb{N}}$, there is a tuple $\bar{a}$ and a computable function $f : \mathbb{N} \to \mathbb{N}$ such that*

$$\langle m, \bar{b} \rangle \in R \iff \langle f(m), \bar{a}^\frown \bar{b} \rangle \in \vec{K}^{\mathcal{A}} \quad \text{for all } m \in \mathbb{N} \text{ and } \bar{b} \in A^{<\mathbb{N}}.$$

PROOF. It follows from Theorem 2.1.16 that, for every r.i.c.e. $R \subseteq \mathbb{N} \times A^{<\mathbb{N}}$, there is a tuple $\bar{a}$ such that each column of the form $R \cap \{m\} \times A^n$ is uniformly Σ_1^c definable with parameters $\bar{a}$. More precisely, there is a computable function $m, j \mapsto e_{m,j}$ such that $R \cap (\{m\} \times A^j)$ is definable by the $e_{m,j}$-th Σ_1^c formula using $\bar{a}$ as parameters:

$$\langle m, \bar{b} \rangle \in R \iff \mathcal{A} \models \varphi_{e_{m,|\bar{b}|},|\bar{a}\bar{b}|}^{\Sigma_1^c}(\bar{a}, \bar{b}) \text{ for all } m \in \mathbb{N} \text{ and } \bar{b} \in A^{<\mathbb{N}}.$$

Notice that the right-hand-side is equivalent to $\langle e_{m,|\bar{b}|}, \bar{a}\bar{b} \rangle \in \vec{K}^{\mathcal{A}}$, which is almost what we want; what is left is to remove the dependence of $e_{m,|\bar{b}|}$ on $|\bar{b}|$.

The rest of the proof uses a standard technical argument to define $f(m)$ so that is does not depend on $|\bar{b}|$. Recall that $\varphi_{e,j}^{\Sigma_1^c}(\bar{x})$ was defined as $\bigvee_{\langle i,j \rangle \in W_e} \varphi_{i,j}^{\exists}(\bar{x})$ where $j = |\bar{x}|$. Given m, define $f(m)$ to be the index of a c.e. set, $W_{f(m)}$, such that

$$\langle i, |\bar{a}| + n \rangle \in W_{f(m)} \iff \langle i, |\bar{a}| + n \rangle \in W_{e_{m,n}} \quad \text{for all } i, n \in \mathbb{N}.$$

Then, we get $(\forall m, n \in \mathbb{N})\ \varphi_{f(m),|\bar{a}|+n}^{\Sigma_1^c} \equiv \varphi_{e_{m,n},|\bar{a}|+n}^{\Sigma_1^c}$, and that

$$\langle m, \bar{b} \rangle \in R \iff \langle f(m), \bar{a}\bar{b} \rangle \in \vec{K}^{\mathcal{A}} \quad \text{for all } \bar{b} \in A^{<\mathbb{N}}. \qquad \square$$

Remark 2.2.5. In particular, it follows that, given an enumeration of all tuples in A, we can get an enumeration of all r.i.c.e. subsets of $\mathbb{N} \times A^{<\mathbb{N}}$. Furthermore,

$$\vec{\mathsf{K}}_1^{\mathcal{A}} = \{\langle\langle e, \bar{a}\rangle, \langle i, \bar{b}\rangle\rangle : \Phi_e(i){\downarrow} \ \& \ \langle\Phi_e(i), \bar{a}\bar{b}\rangle \in \vec{\mathsf{K}}^{\mathcal{A}}\}$$
$$\subseteq (\mathbb{N} \times A^{<\mathbb{N}}) \times (\mathbb{N} \times A^{<\mathbb{N}})$$

is a r.i.c.e. relation such that every other r.i.c.e. relation $R \subseteq \mathbb{N} \times A^{<\mathbb{N}}$, is a column of $\vec{\mathsf{K}}_1^{\mathcal{A}}$. That is, $R = \{r : \langle s, r\rangle \in \vec{\mathsf{K}}_1^{\mathcal{A}}\}$ for some $s \in \mathbb{N} \times A^{<\mathbb{N}}$.

In terms of Turing degrees, it is easy to see that $\vec{\mathsf{K}}^{\mathcal{A}} \leq_T D(\mathcal{A})'$ for any ω-presentation $\mathcal{A}$. The reverse reducibility holds in some ω-presentations (Lemma 4.4.4) but not in others:

EXERCISE 2.2.6. Show that any non-trivial structure has an ω-presentation $\mathcal{A}$ with $D(\mathcal{A}) \equiv_T \vec{\mathsf{K}}^{\mathcal{A}}$. Hint in footnote.[18]

By relativizing Kleene's relation, we can define a jump operator on subsets of $\mathbb{N} \times A^{<\mathbb{N}}$.

DEFINITION 2.2.7. Given $Q \in \mathbb{N} \times A^{<\mathbb{N}}$, we define *the jump of Q in $\mathcal{A}$* to be $\vec{\mathsf{K}}^{(\mathcal{A}, Q)}$, that is, Kleene's relation as in Definition 2.2.3 relative to the structure $(\mathcal{A}, Q)$. We denote it by Q'.

Observation 2.2.8. The jump operator on relations is well-defined on $\leq_{rT}$-degrees. Furthermore, if $Q \leq_{rT} R$, then $Q' \leq_{rT} R'$. This is because if $Q \leq_{rT} R$, then $Q^{\mathcal{B}}$ is computable in $D(\mathcal{B}, R^{\mathcal{B}})$ for any ω-presentation $\mathcal{B}$ of $\mathcal{A}$, and hence Q' is c.e. in $D(\mathcal{B}, R^{\mathcal{B}})$, getting that Q' is r.i.c.e. in $(\mathcal{A}, R)$. From the completeness of R', it follows that Q is r.i. computable in $(\mathcal{A}, R')$.

2.2.2. Diagonalization. We now prove that, on the space of subsets of $\mathbb{N} \times A^{<\mathbb{N}}$, the jump operation actually jumps.

THEOREM 2.2.9. *For every structure $\mathcal{A}$, $\vec{\mathsf{K}}^{\mathcal{A}}$ is not r.i. computable in $\mathcal{A}$.*

PROOF. This proof is essentially the same as Kleene's diagonalization argument for showing that $0'$ is not computable (see footnote in page xix), but adapted to this setting. Suppose that $\vec{\mathsf{K}}^{\mathcal{A}}$ is co-r.i.c.e.. We will produce a contradiction by finding a pair that is supposed to be in $\vec{\mathsf{K}}^{\mathcal{A}}$ if and only if it is supposed to be out.

We consider the following relation reminiscent of the complement of the diagonal in Kleene's argument:

$$\vec{R} = \{\langle i, \bar{b}\rangle \in \mathbb{N} \times A^{<\mathbb{N}} : \ \Phi_i(i){\downarrow} \text{ and } \langle\Phi_i(i), \bar{b}\bar{b}\rangle \notin \vec{\mathsf{K}}^{\mathcal{A}}\}.$$

Since we are assuming $\vec{\mathsf{K}}^{\mathcal{A}}$ is co-r.i.c.e., $\vec{R}$ is r.i.c.e.. By the r.i.c.e.-completeness of $\vec{\mathsf{K}}^{\mathcal{A}}$ as in Lemma 2.2.4, we have that there is an index

---

[18]Use Theorem 1.2.1.

$e \in \mathbb{N}$ for a total computable function Φ_e and a tuple $\bar{a} \in A^{<\mathbb{N}}$ such that

$$\langle i, \bar{b} \rangle \in \vec{R} \iff \langle \Phi_e(i), \bar{a}\bar{b} \rangle \in \vec{K}^A \quad \text{for all } \langle i, \bar{b} \rangle \in \mathbb{N} \times A^{<\mathbb{N}}.$$

If we use $\langle e, \bar{a} \rangle$ for $\langle i, \bar{b} \rangle$, we then get the following contradiction:

$$\langle e, \bar{a} \rangle \in \vec{R} \iff \langle \Phi_e(e), \bar{a}\bar{a} \rangle \in \vec{K}^A \iff \langle e, \bar{a} \rangle \notin \vec{R},$$

the latter equivalence coming from the definition of $\vec{R}$ □

COROLLARY 2.2.10. *For every* $Q \in \mathbb{N} \times A^{<\mathbb{N}}$, $Q <^A_{rT} Q'$; *that is, Q is r.i. computable in Q', but Q' is not r.i. computable in Q.*

PROOF. It is easy to see that $Q \leq_{rT} Q'$ because the Σ^c_1 diagram of (A, Q) clearly computes the atomic diagram of (A, Q) in any copy of A. That Q' is not r.i. computable in Q follows from the theorem above applied to the structure (A, Q). □

Historical Remark 2.2.11. The proof of Theorem 2.2.9 given above is from [Mon12], although it is clearly similar to the standard proof of the incomputability of the halting problem. Theorem 2.2.9 had been previously proved for a different, yet equivalent, notion of jump by Vatev in [Vat11]. Vatev's proof, restated in our terms, goes by showing that if $\mathcal{B}$ is a generic copy of A, then $\vec{K}^{\mathcal{B}}$ has degree $D(\mathcal{B})'$ (which, of course, is not computable in $D(\mathcal{B})$), and hence $\vec{K}^A$ is not r.i. computable in A. From a personal communication, Stukachev has another proof which has not been translated into English yet.

2.2.3. Structural versus binary information. As we saw in Section 2.1.4, we can trivially code reals $X \subseteq \mathbb{N}$ with relations $X \times \{\langle\rangle\} \subset \mathbb{N} \times A^{<\mathbb{N}}$. There is no structural information on the relation $X \times \{\langle\rangle\}$. The information content in $X \times \{\langle\rangle\}$ is *purely binary*:

DEFINITION 2.2.12. A relation $R \subseteq \mathbb{N} \times A^{<\mathbb{N}}$ is *purely binary* if there is an $X \in 2^{\mathbb{N}}$ such that R is r.i. computable in (A, X).

EXAMPLE 2.2.13. Let $\mathcal{L} = (L; \leq, \mathrm{Adj})$ be an adjacency linear ordering isomorphic to $\mathbb{Z}$, and let $R \subseteq L^2$ be the set of pairs $\langle a, b \rangle$ for which the number of elements in between a and b is a number in $0'$. R is not r.i. computable, but it is clearly r.i. computable in $0'$. Its information content is purely binary.

In contrast, relations like Adj on a linear ordering contain structural information and no binary information. Relations like the r.i.c.e.-complete relation on a linear ordering, $\vec{K}^{\mathcal{L}} \equiv_{rT} \neg\mathrm{Adj} \oplus 0'$, are a mix of both. In many occasions, one is interested only in structural behavior. In that case, one should consider the *structural* versions of the notions from earlier in this chapter by modding out the binary information: A relation $R \subseteq \mathbb{N} \times A^{<\mathbb{N}}$ is *structurally r.i.c.e.* in A if it is r.i.c.e. in (A, X) for some $X \in 2^{\mathbb{N}}$. R is *structurally r.i. computable* from Q within A if R is r.i. computable in

$(\mathcal{A}, Q, X)$ for some $X \in 2^{\mathbb{N}}$. We sometimes refer to these versions as the *boldface* versions or the *on-a-cone* versions. The following notion is particularly important:

DEFINITION 2.2.14. A relation R is *structurally complete* if every structurally r.i.c.e. relation is structurally r.i. computable in R. R is *structurally r.i.c.e. complete* if it is also structurally r.i.c.e. itself.

We will see below that the linear dependence relation is structurally r.i.c.e. complete in $\mathbb{Q}$-vector spaces, and that the adjacency relation is structurally co-r.i.c.e. complete on linear orderings, among other examples. We will further analyze structurally complete relations in Section 10.1 once we have more tools at hand.

2.3. Examples of r.i.c.e. complete relations

In this section, we consider structures that have nice structurally complete relations. The first example, linear dependence on vector spaces, is rather simple. The proof for the second example, adjacency on linear orderings, is quite interesting.

LEMMA 2.3.1. *The relation LD of linear dependence on a $\mathbb{Q}$-vector space is structurally complete. Moreover $LD \oplus 0'$ is r.i.c.e. complete.*[19]

PROOF. The key point is that any $\mathbb{Q}$-vector space has a canonical computable copy, and using LD, one can find an isomorphism with that particular copy. One can then move c.e. relations through that isomorphism.

All the countable $\mathbb{Q}$-vector spaces are of the form $\mathbb{Q}^n$ for some $n \in \mathbb{N} \cup \{\infty\}$. Each $\mathbb{Q}^n$ has a standard, nicely behaved computable ω-presentation. Assume $n = \infty$ as the other cases are even simpler. Let $R \subseteq \mathbb{N} \times (\mathbb{Q}^\infty)^{<\mathbb{N}}$ be a r.i.c.e. relation. We want to show that R is r.i. computable in $LD \oplus 0'$.

Let $\mathcal{W}$ be a copy of $\mathbb{Q}^\infty$, and let $R^{\mathcal{W}}$ be the image of R. We need to show that $R^{\mathcal{W}}$ is computable from $LD^{\mathcal{W}} \oplus 0'$. We can use $LD^{\mathcal{W}}$ to find a basis for $\mathcal{W}$ and hence compute an isomorphism $g \colon \mathbb{Q}^\infty \to \mathcal{W}$, which maybe be different from the original isomorphism we had between $\mathcal{W}$ and $\mathbb{Q}^\infty$ that we used to define $R^{\mathcal{W}}$ from R. What we do have is that

$$(\mathbb{Q}^\infty, R) \cong (\mathcal{W}, \mathcal{R}^{\mathcal{W}}) \cong (\mathbb{Q}^\infty, g^{-1}(R^{\mathcal{W}})).$$

Since R is r.i.c.e., that $g^{-1}(R^{\mathcal{W}})$ is c.e. in $D(\mathbb{Q}^\infty)$, and hence c.e., and hence computable from $0'$. We then get that $R^{\mathcal{W}}$ is computable from $g \oplus 0'$, and hence from $LD^{\mathcal{W}} \oplus 0'$ as needed. □

---

[19] Recall from Example 2.1.1 that the field $\mathbb{Q}$ is not part of the structure and we use the vocabulary that includes a unary scalar multiplication symbol $q \cdot \_$ for each $q \in \mathbb{Q}$.

The same argument above can be used to show that the "algebraic dependence" relation is structurally complete on algebraically closed fields.

LEMMA 2.3.2. *Let $\mathcal{A} = (A; \leq)$ be a linear ordering. Then*

$$\mathsf{Adj} = \{\langle a, b \rangle \in A^2 : a < b \ \& \ \nexists c \, (a < c < b)\}.$$

is structurally complete. Furthermore, $\neg\mathsf{Adj} \oplus 0'$ is r.i.c.e. complete.

PROOF. The proof goes by showing that every Σ_1^c formula over the vocabulary $\{\leq\}$ is equivalent to a finitary universal formula over the vocabulary $\{\leq, \mathsf{Adj}\}$, and that $0'$ can find these equivalent $\forall$-formulas uniformly. We then get that every Σ_1^c-definable relation is co-r.i.c.e. in $\mathsf{Adj} \oplus 0'$ and hence r.i. computable in it. One could prove this in a purely syntactical way in the style of a quantifier-elimination argument. Instead, we give a more model-theoretic proof.

Let $\varphi(x_1, \ldots, x_k)$ be a Σ_1^c formula about linear orderings (i.e., over the vocabulary $\{\leq\}$). Let $\bar{c} = \langle c_1, \ldots, c_k \rangle$ be new constant symbols and $\tau' = \{\leq, c_1, \ldots, c_k\}$. We will use the term $\bar{c}$-*linear ordering* to refer to a linear ordering where the constants from $\bar{c}$ have been assigned. As a preview of the rest of the proof, let us mention that one of the key points is that the finite $\bar{c}$-linear orderings form a well-quasi-ordering under embeddability. The proof is divided into three claims:

CLAIM 2.3.3. *Two Σ_1^c τ'-sentences are equivalent on $\bar{c}$-linear orderings if and only if they hold on the same finite $\bar{c}$-linear orderings.*

The left-to-right direction is obvious; we prove the other direction. Let φ and ψ be two Σ_1^c sentences which hold on the same finite $\bar{c}$-linear orderings. Consider an infinite $\bar{c}$-linear ordering $\mathcal{L}$ where φ holds. Then one of the $\exists$-disjuncts of φ holds in $\mathcal{L}$, and hence holds on a finite τ'-substructure of $\mathcal{L}$. By the assumption, ψ holds on that same finite $\bar{c}$-linear ordering, and by upward-persistence of Σ_1^c formulas, ψ holds in $\mathcal{L}$ too.

CLAIM 2.3.4. *For every Σ_1^c τ'-sentence φ, there is a finite set of finite $\bar{c}$-linear orderings $\mathcal{L}_{d_1}, \ldots, \mathcal{L}_{d_\ell}$ such that, for any $\bar{c}$-linear ordering $\mathcal{A}$, $\mathcal{A} \models \varphi$ if and only if one of those finite $\bar{c}$-linear orderings $\mathcal{L}_{d_i}$ τ'-embeds[20] into $\mathcal{A}$. Furthermore, $0'$ can find those $\bar{c}$-linear orderings uniformly in φ.*

Given a permutation $\langle \pi_1, \ldots, \pi_k \rangle$ of $\langle 1, \ldots, k \rangle$ and $k + 1$ numbers $\bar{n} = \langle n_0, \ldots, n_k \rangle$, let $\mathcal{L}_{\pi, \bar{n}}$ be the finite $\bar{c}$-linear ordering with $c_{\pi_1} \leq c_{\pi_2} \leq \cdots \leq c_{\pi_k}$ that has exactly n_0 elements less than c_{π_1}, n_i elements between c_{π_i} and $c_{\pi_{i+1}}$, and n_k elements greater than c_{π_k}. Consider the ordering $\preceq$ on $S_k \times \mathbb{N}^{k+1}$ given by

$$\langle \pi, \bar{n} \rangle \preceq \langle \sigma, \bar{m} \rangle \iff \pi = \sigma \ \& \ (\forall i \leq k) \, n_i \leq m_i,$$

where S_k is the set of permutations of $\{1, \ldots, k\}$. We then have that

$$\langle \pi, \bar{n} \rangle \preceq \langle \sigma, \bar{m} \rangle \implies \mathcal{L}_{\pi, \bar{n}} \text{ embeds in } \mathcal{L}_{\sigma, \bar{m}}.$$

---

[20] By τ'-embed we meant that it embeds as a τ'-structure.

By upward-persistence of Σ_1^c formulas, it follows that the set D of $\langle \pi, \bar{n} \rangle \in S_k \times \mathbb{N}^{k+1}$ such that $\mathcal{L}_{\pi,\bar{n}} \models \varphi$ is $\preceq$-upwards closed. Now, by Dickson's Lemma, the ordering $\preceq$ is a *well-quasi-ordering*.[21] (See Definition 10.2.1.) That means that every subset of $S_k \times \mathbb{N}^{k+1}$ has a finite set of minimal elements, and hence that for every upward-closed subset $D \subseteq S_k \times \mathbb{N}^{k+1}$, there is a finite set of elements $d_1, \ldots, d_\ell \in D$ such that

$$f \in D \iff \bigvee_{j \le \ell} d_j \preceq f \quad \text{for all } f \in S_k \times \mathbb{N}^{k+1}.$$

The oracle $0'$ can find this finite set $\{d_1, \ldots, d_\ell\}$ because it can check that every $\mathcal{L}_{d_j}$ satisfies φ and that every $\mathcal{L}_f$ with $(\forall j \le \ell)\, d_j \not\preceq f$ does not satisfy φ. This proves our second claim.

Let $\psi_n(x, y)$ be the $\exists$-formula that says that there are at least n elements strictly in between x and y:

$$\psi_n(x, y) \equiv \exists z_1, \ldots, z_n\ (x < z_1 < z_2 < \cdots < z_n < y).$$

We write $\psi_n(-\infty, y)$ for the unary $\exists$-formula that says that there are at least n elements less than y, and analogously with $\psi_n(x, \infty)$. Given a permutation $\pi \in S_k$ and $\bar{n} = \langle n_0, \ldots, n_k \rangle \in \mathbb{N}^{k+1}$, we let

$$\psi_{\pi,\bar{n}}(x_1, \ldots, x_k) \equiv x_{\pi_1} \le \cdots \le x_{\pi_k} \ \wedge$$
$$\left(\psi_{n_0}(-\infty, x_{\pi_1}) \wedge \psi_{n_1}(x_{\pi_1}, x_{\pi_2}) \wedge \cdots \wedge \psi_{n_k}(x_{\pi_k}, \infty) \right).$$

A $\bar{c}$-linear ordering satisfies $\psi_{\pi,\bar{n}}(c_1, \ldots, c_k)$ if and only if $\mathcal{L}_{\pi,\bar{n}}$ embeds in it. Then we get from the claim that every Σ_1^c formula $\varphi(x_1, \ldots, x_k)$ is equivalent to a finite disjunction of formulas of the form $\psi_{\pi,\bar{n}}(x_{\pi_1}, \ldots, x_{\pi_k})$. Furthermore, $0'$ can find these formulas uniformly. The following claim is all that is left to prove the lemma.

CLAIM 2.3.5. *The formulas $\psi_n(x, y)$ are equivalent to $\forall\text{-}\{\le, \mathrm{Adj}\}$-formulas, and hence so are the formulas $\psi_{\pi,\bar{n}}(x_1, \ldots, x_k)$.*

Just observe that $\psi_n(x, y)$ is equivalent to the following universal formula over the adjacency predicate:

$$\psi_n(x, y) \iff$$

$$\bigwedge_{j \le n} \nexists z_0, \ldots, z_j \left(x = z_0 \le \cdots \le z_j = y \wedge \left(\bigwedge_{i=0}^{j-1} \left(\mathrm{Adj}(z_i, z_{i+1}) \right) \right) \right).$$

This proves the claim. We should still observe that the unary formulas $\psi_n(-\infty, x)$ are equivalent to $\forall\text{-}\{\le, \mathrm{Adj}\}$-formulas only once we know who the first element is, and if there is one. That is, using the first element as

---

[21] A *well-quasi-ordering* is a partial ordering which has no infinite descending sequences and no infinite antichains. Equivalently, it is a partial ordering on which every set has a finite subset of minimal elements. Dickson's Lemma states that $\mathbb{N}^n$ is well-quasi-ordered under the coordinate-wise ordering.

parameter, say f, we have that $\psi_n(-\infty, x) \equiv \psi_{n-1}(f, x)$, and if there is no first element, then $\psi_n(-\infty, x)$ is always true. Knowing what are the first and last elements, and if they exists, is non-uniform information that we need.

It follows that the formulas $\psi_{\pi,\bar{n}}(x_1, \ldots, x_k)$ are equivalent to $\forall$-formulas over the vocabulary $\{\leq, \mathsf{Adj}, \mathsf{fi}, \mathsf{la}\}$, where fi and la are unary relations identifying the first and last elements if they exist. Then so are all Σ_1^c formulas, though $0'$ is necessary to find the equivalent formula. Therefore, every Σ_1^c formula is uniformly r.i. computable in $\mathsf{Adj} \oplus \mathsf{fi} \oplus \mathsf{la} \oplus 0'$, and in particular r.i. computable in $\mathsf{Adj} \oplus 0'$, as fi and la are either empty or singletons, and thus (non-uniformly) r.i. computable. $\qquad \square$

The unary relations $\mathsf{fi}(y)$ and $\mathsf{la}(x)$ for first and last elements are, in a sense, extreme cases of the adjacency relation:

$$\mathsf{fi}(y) \leftrightarrow \mathsf{Adj}(-\infty, y) \quad \text{and} \quad \mathsf{la}(x) \leftrightarrow \mathsf{Adj}(x, +\infty).$$

Of course, $-\infty$ and $+\infty$ are not elements of the linear ordering, nor symbols of our vocabulary. When we use these symbols in a formula, an atomic sub-formula of the form $x < +\infty$ should always be read as true, and a sub-formula $+\infty \leq x$ should always be read as false.

DEFINITION 2.3.6. We define a new relation symbol $\bar{\mathsf{Adj}}$ that encapsulates these three uses of the adjacency relation into one:[22]

$$\bar{\mathsf{Adj}} \equiv \mathsf{Adj} \oplus \mathsf{fi} \oplus \mathsf{la}.$$

From the proof above we get that, if what we want is a relation that is structurally r.i.c.e. complete in a uniform way across all linear orderings, we need to consider $\bar{\mathsf{Adj}}$ instead of Adj.

This technique of proving the finite substructures with added constants are well-quasi-ordered by embeddability can be used on some other classes too. We will study this technique in more generality in Section 10.2.

An *equivalence structure* is a structure $\mathcal{E} = (D; E)$, where E is a equivalence relation on the domain D. Define the following relations on $\mathcal{E}$:

1. for $k \in \mathbb{N}$, $F_k = \{x \in D : \text{there are} \geq k \text{ elements equivalent to } x\}$, and

2. the *character* of E: $G = \{\langle n, k \rangle \in \mathbb{N}^2 : \text{there are} \geq n \text{ equivalence classes with} \geq k \text{ elements}\}$.

EXERCISE 2.3.7. (a) Show that the relation $\vec{F} = \bigoplus_{k \in \mathbb{N}} F_k \subseteq \mathbb{N} \times D$ is structurally complete.

(b) Show that $\vec{F} \oplus G \oplus 0'$ is r.i.c.e. complete. Hint in footnote.[23]

---

[22]We defined the join $\oplus$ of relations in Definition 2.1.27. In this case, on a linear ordering $\mathcal{A}$, $\bar{\mathsf{Adj}}$ can be seen as a subset of $(2 \times A) \sqcup A^2$, where $\langle 0, a \rangle \in \bar{\mathsf{Adj}}$ if $\mathsf{fi}(a)$, $\langle 1, a \rangle \in \bar{\mathsf{Adj}}$ if $\mathsf{la}(a)$, and $\langle a, b \rangle \in \bar{\mathsf{Adj}}$ if $\mathsf{Adj}(a, b)$, for $a, b \in A$.

[23]The proof of (b) follows a somewhat similar outline to that of Lemma 2.3.2. You need to use that the set of finite subsets of $\mathbb{N}^2$ ordered by $A \leq B \iff \forall \langle x, y \rangle \in A \exists \langle x', y' \rangle \in B$ $(x \leq x' \, \& \, y \leq y')$ is well-quasi-ordered.

EXERCISE 2.3.8. Show that the atom relation on a Boolean algebra is structurally complete. (An element in a Boolean algebra is an *atom* if it is non-zero and has no elements below it other than zero.)

EXERCISE 2.3.9. (Hard) (a) [Sho78, Theorem 2.2] Show that $LD_{n+1} \not\leq_{rT} LD_n$ in the ∞-dimensional $\mathbb{Q}$-vector space, where LD_n is the linear dependence relation on n tuples.

(b) [Mon12, Theorem 7.2] Show that no relation of fixed arity is structurally complete in the ∞-dimensional $\mathbb{Q}$-vector space.

2.4. Superstructures

The notion of r.i.c.e. relation is equivalent to other notions that were known many decades ago. In this section, we study one of them, the Σ-definable subsets of the hereditarily finite superstructure $\mathcal{HF}_A$. There are some advantages to working in this setting: One is that r.i.c.e. relations are now defined by finitary formulas instead of computably infinitary ones. Another one is that there is almost no coding required; while subsets of $(A^{<\mathbb{N}})^{<\mathbb{N}}$ can be coded by subsets of $A^{<\mathbb{N}}$ as in Section 2.1.5, subsets of $(\mathbb{HF}_A)^{<\mathbb{N}}$ are already subsets of $\mathbb{HF}_A$. Nevertheless, the advantage of working with $\mathbb{N} \times A^{<\mathbb{N}}$ is that it is easier to visualize. At the end of the day, all these advantages and disadvantages are purely aesthetic and not really significant.

2.4.1. The hereditarily finite superstructure. Another approach to the study of r.i.c.e. relations is using Σ-definability on admissible structures. We will not consider admissible structures in general, but just the hereditarily finite extension of an abstract structure $\mathcal{A}$. The elements of this extension are the finite sets of finite sets of ... of finite sets of elements of A.

DEFINITION 2.4.1. Let $\mathcal{P}_{\text{fin}}(X)$ denote the collection of finite subsets of X. Given a set A, we define:

1. $\mathrm{HF}_A(0) = \varnothing$,
2. $\mathrm{HF}_A(n + 1) = \mathcal{P}_{\text{fin}}(A \cup \mathrm{HF}_A(n))$, and
3. $\mathbb{HF}_A = \bigcup_{n \in \mathbb{N}} \mathrm{HF}_A(n)$.

Now, given a τ-structure $\mathcal{A}$, we define the $\tau \cup \{\in, D\}$-structure $\mathcal{HF}_A$ whose domain has two sorts, A and $\mathbb{HF}_A$, and where the symbols from τ are interpreted in the A-sort as in $\mathcal{A}$, '$\in$' is interpreted in the obvious way, and D is a unary relation coding the atomic diagram of $\mathcal{A}$ as defined below. The need for adding D is slightly technical, so we will explain it later.

A quantifier of the form $\forall x \in y$ or $\exists x \in y$ is called a *bounded quantifier*. A Σ-*formula* is a finitary $\tau \cup \{\in, D\}$-formula that is built out of atomic and negation-of-atomic formulas using disjunctions, conjunctions, bounded quantifiers, and existential unbounded quantifiers. A subset of $\mathcal{HF}_A$ is Δ-*definable* if it and its complement are both Σ-definable.

Clearly, on $\mathcal{HF}_\mathcal{A}$ we have the usual pairing function $\langle x, y \rangle = \{\{x\},$ $\{x, y\}\}$, and we can encode n-tuples, strings, etc. Notice also that $\mathcal{HF}_\mathcal{A}$ includes the finite von Neumann ordinals (denoted by $\underline{n}$, where $\underline{0} = \varnothing$ and $\underline{n+1} = \{\underline{0}, \dots, \underline{n}\}$). We use ω to denote the Δ-definable set of finite ordinals of $\mathcal{HF}_\mathcal{A}$. The operations of successor, addition, and multiplication on ω are also Δ-definable, and hence so is Kleene's T predicate. It follows that every c.e. subset of ω is Σ-definable, and every computable function is Δ-definable in $\mathcal{HF}_\mathcal{A}$ (for more details, see [Bar75, Theorem II.2.3]).

We define D to be the *satisfaction relation for atomic formulas*, that is

$$D = \{\langle i, \bar{a} \rangle : \mathcal{A} \models \varphi_i^{\text{at}}(\bar{a})\} \subseteq \mathbb{HF}_\mathcal{A},$$

where $\{\varphi_0^{\text{at}}, \varphi_1^{\text{at}}, \dots\}$ is an effective enumeration of all the atomic τ-formulas. Notice that if the vocabulary of $\mathcal{A}$ is finite and relational, this is a finite list of formulas, and hence D is Δ-definable in $\mathcal{HF}_\mathcal{A}$ without using D. In that case, there is no need to add D to the vocabulary $\mathcal{HF}_\mathcal{A}$. On the other hand, when τ is infinite, if we do not add D, Σ-formulas could only involve finitely many symbols from τ which would be too restrictive. An important consequence of having D in the vocabulary is that the $\exists$-diagram of $\mathcal{A}$ is Σ-definable in $\mathcal{HF}_\mathcal{A}$.[24]

Given any $R \subseteq \mathbb{N} \times A^{<\mathbb{N}}$, we can view it directly as a subset of $\mathbb{HF}(\mathcal{A})$. Conversely, there is also a natural way of going from relations in $\mathcal{HF}_\mathcal{A}$ to subsets of $\mathbb{N} \times A^{<\mathbb{N}}$. Let $X = \{x_0, x_1, \dots\}$ be a list of variable symbols. Every $t \in \mathbb{HF}_X$ is essentially a term over a finite set of variables, and we write $t(\bar{x})$ to show the variables that appear in t. Observe that $\mathbb{HF}_\mathcal{A} = \{t(\bar{a}) : t(\bar{x}) \in \mathbb{HF}_X, \bar{a} \in A^{|\bar{x}|}\}$. Let $\{t_i : i \in \mathbb{N}\}$ be an effective enumeration of $\mathbb{HF}_X \cup X$. Now, given $Q \subseteq \mathcal{HF}_\mathcal{A}$, we define

$$s(Q) = \{\langle i, \bar{a} \rangle : t_i(\bar{a}) \in Q\} \subseteq \mathbb{N} \times A^{<\mathbb{N}}.$$

Observation 2.4.2. The relation $\{\langle b, \underline{n}, \bar{a} \rangle : b \in \mathbb{HF}_\mathcal{A}, n \in \mathbb{N}, \bar{a} \in A^{<\mathbb{N}}$ & $b = t_n(\bar{a})\} \subseteq \mathbb{HF}_\mathcal{A} \times \omega \times A^{<\mathbb{N}}$ is Δ-definable in $\mathcal{HF}_\mathcal{A}$. This is not completely trivial, and is proved by recursion on terms. We leave the details to the reader.

THEOREM 2.4.3. *Given $R \in \mathbb{N} \times A^{<\mathbb{N}}$, the following are equivalent:*

1. *R is r.i.c.e. in $\mathcal{A}$.*
2. *R is Σ-definable in $\mathcal{HF}_\mathcal{A}$ with parameters.*

Given $Q \subseteq A \cup \mathbb{HF}_\mathcal{A}$, the following are equivalent:

1. *$s(Q)$ is r.i.c.e. in $\mathcal{A}$.*
2. *Q is Σ-definable in $\mathcal{HF}_\mathcal{A}$ with parameters.*

Historical Remark 2.4.4. This theorem is credited to Vaĭtsenavichyus [Vai89] in [Stu] and appears in some form in [BT79].

---

[24]The Σ-definition of the $\exists$-diagram of $\mathcal{A}$ says that, given an $\exists$-formula φ, there exists a variable assignment and a truth valuations of the sub-formulas of φ that makes φ true, using D on the atomic sub-formulas.

PROOF. We only prove the second part; the proof of the first part is very similar. Suppose first that $s(Q)$ is r.i.c.e. in $\mathcal{A}$. Using Theorem 2.1.16, we get a c.e. set W and a tuple $\bar{p} \in A^{<\mathbb{N}}$ such that

$$\langle i, \bar{a} \rangle \in s(Q) \iff \mathcal{A} \models \bigvee_{e:\langle i,e,|\bar{a}|\rangle \in W} \varphi^{\exists}_{e,|\bar{p}\bar{a}|}(\bar{p}, \bar{a})$$

$$\text{for all } i \in \mathbb{N} \text{ and } \bar{a} \in A^{<\mathbb{N}},$$

where $\{\varphi^{\exists}_{e,j} : e \in \mathbb{N}\}$ is an effective enumeration of the $\exists$-τ-formulas with j free variables. Then

$$b \in Q \iff \exists i, e \in \mathbb{N}\, \exists \bar{a} \in A^{<\mathbb{N}} (b = t_i(\bar{a}) \;\&\; \langle i, e, |\bar{a}| \rangle \in W \;\&\;$$
$$\mathcal{A} \models \varphi^{\exists}_{e,|\bar{p}\bar{a}|}(\bar{p}, \bar{a})).$$

Using that deciding whether $b = t_i(\bar{a})$ is Δ-definable and that both W and the existential diagram of $\mathcal{A}$ are Σ-definable, we get that Q is Σ-definable with parameters $\bar{p}$.

Conversely, suppose now that Q is Σ-definable in $\mathcal{HF}_{\mathcal{A}}$ with parameters; we want to prove that $s(Q)$ is r.i.c.e.. Let $\mathcal{B}$ be a copy of $\mathcal{A}$. Computably in $D(\mathcal{B})$, build $\mathbb{HF}_{\mathcal{B}}$ and a copy of $\mathcal{HF}_{\mathcal{B}}$, and then use the Σ-definition of Q to enumerate $Q^{\mathcal{HF}_{\mathcal{B}}}$. We end up with a $D(\mathcal{B})$-computable enumeration of $Q^{\mathcal{B}}$, which we can then use to produce a $D(\mathcal{B})$-computable enumeration of $s(Q)$. □

Historical Remark 2.4.5. In [Mos69], Moschovakis introduces what we now call the *Moschovakis enrichment* of a structure $\mathcal{A}$, denoted $\mathcal{A}^*$. For our purposes, there is no real difference between $\mathcal{A}^*$ and $\mathcal{HF}_{\mathcal{A}}$. The difference is that in the iterative definition of the domain of $\mathcal{A}^*$ we take pairs instead of finite subsets as we did for $\mathbb{HF}_{\mathcal{A}}$. Moschovakis [Mos69] then defines a class of partial multi-valued functions from $(\mathcal{A}^*)^n$ to $\mathcal{A}^*$ which he calls *search computable functions*. This class is defined as the least class closed under certain primitive operations, much in the style of Kleene's definition of primitive recursive and partial recursive functions, where instead of the Kleene's least-element operator μ, we have a multivalued search operator ν. A subset of A^* is *search computable* if its characteristic function is, and it is *semi-search computable* if it has a definition of the form $\exists y \, (f(x, y) = 1)$, where f is search computable.

The definition of search computable allows us to add a list of new primitive functions to our starting list (so long as they are given in an effective list, with computable arities), obtaining a sort of *relativized version* of search computability. If we have a structure $\mathcal{A}$, we would add to the list of primitive functions the characteristic functions of the relations in $\mathcal{A}$ to obtain a notion of partial, multi-valued, *search computable functions in $\mathcal{A}$*.

Much in the same way as we did for $\mathcal{HF}_{\mathcal{A}}$ above, we have a natural way of encoding relations $R \subseteq \mathbb{N} \times A^{<\mathbb{N}}$ by subsets of $\mathcal{A}^*$, and vice versa.

Maybe even more directly, one can go from subsets of A^* to subsets of $\mathcal{HF}_A$ and back. Gordon [Gor70] proved that the notions of search computable in $\mathcal{A}$ and semi-search computable in $\mathcal{A}$ for subsets of A^* coincide with the notions of Δ-definable and Σ-definable for subsets of $\mathcal{HF}_A$. Therefore, when we add parameters, they also coincide with the notions of r.i. computable and r.i.c.e. for relations in $\mathbb{N} \times A^{<\mathbb{N}}$.

Chapter 3

EXISTENTIALLY-ATOMIC MODELS

The key notion in this chapter is that of *existentially atomic structures*: these are atomic structures where all the types are generated by existential formulas. They are the best-behaved structures around. Given a structure, typical questions in computable structure theory include: How difficult is it to compute isomorphisms between different ω-presentations? How difficult is it to identify it syntactically? Can we characterize the set of oracles that can compute an ω-presentation of it? In these three senses, existentially atomic structures are the simplest ones. Not only are they simple, they are also general: every structure is $\exists$-atomic if one adds enough relations to the vocabulary, as for instance, if one adds enough jumps, as we will see in [MonP2]. This means that the results we present in this chapter apply to all structures relative to those relations.

In this chapter, we will also introduce a variety of tools that will be useful throughout the book, for instance, the Cantor back-and-forth argument and the notion of a structure having enumeration degree.

3.1. Definition

Let $\mathcal{A}$ be a τ-structure. The *automorphism orbit* of a tuple $\bar{a} \in A^{<\mathbb{N}}$ is the relation

$$\mathrm{orb}_{\mathcal{A}}(\bar{a}) = \{\bar{b} \in A^{|\bar{a}|} : \text{there is an automorphism of } \mathcal{A} \text{ mapping } \bar{a} \text{ to } \bar{b}\}.$$

DEFINITION 3.1.1. A structure $\mathcal{A}$ is $\exists$-*atomic* if, for every tuple $\bar{a} \in A^{<\mathbb{N}}$, there is an $\exists$-formula[25] $\varphi_{\bar{a}}(\bar{x})$ which defines the automorphism orbit of $\bar{a}$; that is,

$$\mathrm{orb}_{\mathcal{A}}(\bar{a}) = \{\bar{b} \in A^{|\bar{a}|} : \mathcal{A} \models \varphi_{\bar{a}}(\bar{b})\}.$$

We say that $\mathcal{A}$ is $\exists$-*atomic over parameters* if there is a finite tuple $\bar{a} \in A^{<\mathbb{N}}$ such that the structure $(\mathcal{A}, \bar{a})$ is $\exists$-atomic.

---

[25] Recall that an $\exists$-formula is one of the form $\exists \bar{y} \, \varphi(\bar{x}, \bar{y})$ where φ is quantifier free.

These structures were studied by Simmons [Sim76, Section 2], and he cites [Pou72] as their first occurrence in the literature.

The set $\{\varphi_{\bar{a}} : \bar{a} \in A^{<\mathbb{N}}\}$ of all these defining formulas makes what we call a Scott family:

DEFINITION 3.1.2. A *Scott family* for a structure $\mathcal{A}$ is a set S of formulas such that each $\bar{a} \in A^{<\mathbb{N}}$ satisfies some formula $\varphi(\bar{x}) \in S$, and if $\bar{a}$ and $\bar{b}$ satisfy the same formula $\varphi(\bar{x}) \in S$, they are automorphic.

Thus, a structure is $\exists$-atomic if and only if it has a Scott family of $\exists$-formulas. Having access to a Scott family for a structure $\mathcal{A}$ allows us to recognize the different tuples in $\mathcal{A}$ up to automorphism. This is exactly what one needs to build isomorphisms between different copies of $\mathcal{A}$. As we will see in Theorem 3.4.2, if we want to build a computable isomorphism, we need the Scott family to be computably enumerable.

DEFINITION 3.1.3. We say that a Scott family is *c.e.* if the set of indices for its formulas is c.e. A structure $\mathcal{A}$ is *effectively $\exists$-atomic* if it has a c.e. Scott family of $\exists$-formulas.

EXAMPLE 3.1.4. A linear ordering is $\exists$-atomic if and only if it is either finite or dense without end points:

If a linear ordering has n elements, the ith element can be characterized by the $\exists$-formula that says that there are $i - 1$ elements below it and $n - i - 1$ elements above it. If a linear ordering is dense without endpoints, then two tuples are automorphic if and only if they are ordered the same way.

Suppose now that we have a linear ordering that is neither dense nor finite. We claim that there must exist a tuple a, b, c such that: either $a < b < c$, a and b are adjacent, and there are infinitely elements to the right of c; or $c < b < a$, a and b are adjacent, and there are infinitely many elements to the left of c. To prove the claim, we consider three cases: If there is only one adjacency pair in the whole linear ordering, then the linear ordering must have a dense segment; let a and b be the elements of the adjacency pair and take c from the dense segment. If every element has finitely many elements to its right or to its left, then the linear ordering has either an initial segment isomorphic to ω or a final segment isomorphic to ω^*; either let $a < b < c$ be the first three elements, or let $c < b < a$ be the last three. If neither of the above is the case, take c so that it has infinitely many elements to both its left and its right, and let a, b be an adjacency pair disjoint from c. Now that we have proved that a, b, c always exist, we claim that no existential formula defines the orbit of the pair $\langle a, b \rangle$. For this, we notice that any $\exists$-formula true of $\langle a, b \rangle$ is true of $\langle a, c \rangle$: This follows from the analysis of $\exists$-formulas we did in Lemma 2.3.2, and the fact that the number of elements in each of the intervals $(-\infty, a)$, (a, b), and $(b, +\infty)$ is less than or equal to the number of elements in $(-\infty, a)$, (a, c), and $(c, +\infty)$ respectively. But $\langle a, b \rangle$ and $\langle a, c \rangle$ are not automorphic because a and b are adjacent, and a and c are not. This proves the claim that no $\exists$-formula defines the orbit of $\langle a, b \rangle$.

EXAMPLE 3.1.5. Augment the vocabulary $\{\leq\}$ of linear orderings by adding a symbol $\bar{\text{A}}\text{dj}$ for the adjacency relation.[26] Call these structures $(L; \leq, \bar{\text{A}}\text{dj})$, *adjacency linear orderings*. It follows from work of McCoy [McC03, Theorem 2.6] that the $\exists$-atomic adjacency linear orderings over parameters are exactly the ones of the form

$$\mathcal{A}_0 + \mathbf{1} + \mathcal{A}_1 + \mathbf{1} + \cdots + \mathbf{1} + \mathcal{A}_k,$$

where each $\mathcal{A}_i$ is isomorphic to one of the following: $\mathbf{0}, \omega, \omega^*, \omega + \omega^*$, or $\mathbf{m} \cdot \mathbb{Q}$, for $m \in \mathbb{N}$.

EXERCISE 3.1.6. (a) Prove that the adjacency linear ordering $\mathbf{m} \cdot \mathbb{Q}$ is $\exists$-atomic. (b) Prove that the adjacency linear orderings $\omega, \omega^*, \omega + \omega^*$ are $\exists$-atomic over parameters. (c) Prove that if $m_0 \neq m_1$, $\mathbf{m_0} \cdot \mathbb{Q} + \mathbf{m_1} \cdot \mathbb{Q}$ is not $\exists$-atomic even over parameters.

3.2. Existentially algebraic structures

We will see that fields of finite transcendence degree, graphs of finite valence with finitely many connected components, and torsion-free abelian groups of finite rank are all $\exists$-atomic over a finite set of parameters. The reason is that they are $\exists$-algebraic.

DEFINITION 3.2.1. An element $a \in A$ is $\exists$-*algebraic* in $\mathcal{A}$ if there is an $\exists$-formula $\varphi(x)$ true of a such that $\{b \in A : \mathcal{A} \models \varphi(b)\}$ is finite. A structure $\mathcal{A}$ is $\exists$-*algebraic* if all its elements are.

EXAMPLE 3.2.2. A *field that is algebraic over its prime sub-field* is $\exists$-algebraic because every element is one of finitely many that is a root of a polynomial over the prime field. We will develop this example further in Example 3.8.10.

A *connected graph of finite valence with a selected root vertex* is $\exists$-algebraic because every element is one of finitely many that are at a given distance from the root.

An *abelian torsion-free group with a selected basis* is $\exists$-algebraic because every element is the only one for which a certain non-trivial $\mathbb{Z}$-linear combination of it and the basis evaluates to 0.

We prove that $\exists$-algebraic structures are $\exists$-atomic in two lemmas. The core of the arguments is an application of König's lemma that appears in the first one.

---

[26] Recall from Definition 2.3.6, that $\bar{\text{A}}\text{dj}$ is the version of the adjacency relation that allows for the three uses of the adjacency relation: $\text{Adj}(x, y)$, $\text{Adj}(-\infty, y)$, and $\text{Adj}(x, +\infty)$, where $\text{Adj}(-\infty, y)$ only holds of the first element if there is any, and $\text{Adj}(x, +\infty)$ of the last.

DEFINITION 3.2.3. The $\exists$-theory of a structure $\mathcal{A}$, denoted $\exists\text{-}Th(\mathcal{A})$, is the set of (indices) of $\exists$-sentences true about $\mathcal{A}$.

Notice that, as opposed to the diagram, or even the $\exists$-diagram of a structure, the $\exists$-theory is independent of the presentation of the structure.

LEMMA 3.2.4. *Two structures that are $\exists$-algebraic and have the same $\exists$-theories are isomorphic.*

PROOF. Let $\mathcal{A}$ and $\mathcal{B}$ be $\exists$-algebraic structures with the same $\exists$-theories. To prove that $\mathcal{A}$ and $\mathcal{B}$ are isomorphic, we will define a tree of finite approximations to possible isomorphisms from $\mathcal{A}$ to $\mathcal{B}$, and then use König's lemma to show this tree has a path.

List the elements of A as $\{a_0, a_1, \ldots\}$. For each n, let $\varphi_n(x_0, \ldots, x_{n-1})$ be an $\exists$-formula which is true of tuple $\langle a_0, \ldots, a_{n-1}\rangle$ and has finitely many solutions. (A *solution* to a formula is a tuple that makes it true.) By taking conjunctions if necessary, we may assume that $\varphi_n(x_0, \ldots, x_{n-1})$ implies $\varphi_{n-1}(x_0, \ldots, x_{n-2})$. Let

$$T = \{\bar{b} \in B^{<\mathbb{N}} : D_{\mathcal{B}}(\bar{b}) = D_{\mathcal{A}}(a_0, \ldots, a_{|\bar{b}|-1}) \ \& \ \mathcal{B} \models \varphi_{|\bar{b}|}(\bar{b})\}.$$

We will prove that a path through T gives us an isomorphism from $\mathcal{A}$ to $\mathcal{B}$. But before that, let us prove T has a path.

T is clearly a tree in the sense that it is closed under taking initial segments of tuples. It is finitely branching because, for each n, φ_n has finitely many solutions in $\mathcal{A}$, say k many, and thus it cannot have more than k solutions in $\mathcal{B}$, as otherwise, the $\exists$-sentence saying that φ_n has at least $k + 1$ different solutions would be true in $\mathcal{B}$ but false in $\mathcal{A}$. To show that T is infinite, notice that, for each n,

$$\mathcal{A} \models \exists x_0, \ldots, x_{n-1}(D(\bar{x}) = \sigma \ \& \ \varphi_n(\bar{x})), \text{where } \sigma = D_{\mathcal{A}}(a_0, \ldots, a_{|\bar{b}|-1}),$$

as witnessed by $a_0, \ldots, a_{n-1}$. Since $\mathcal{A}$ and $\mathcal{B}$ have the same $\exists$-theories, $\mathcal{B}$ models this $\exists$-sentence too, and the witness is an n-tuple that belongs to T. König's lemma states that every infinite finitely branching tree must have an infinite path. Thus, T must have an infinite path $P \in B^{\mathbb{N}}$. This path determines a map $g: A \to B$ mapping a_n to $P(n)$. That map is an embedding as it preserves finite atomic diagrams: This is because since $\langle g(a_0), \ldots, (a_n)\rangle \in T$, $D_{\mathcal{B}}(g(a_0), \ldots, (a_n)) = D_{\mathcal{A}}(a_0, \ldots, a_n)$ for all n. This map must also be onto: If $b \in B$ is a solution of an $\exists$-formula φ with finitely many solutions, then φ must have the same number of solutions in $\mathcal{A}$ (because $\exists\text{-}Th(\mathcal{A}) = \exists\text{-}Th(\mathcal{B})$), and since $\exists$-formulas are preserved under embeddings, one of those solutions has to be mapped to b. □

LEMMA 3.2.5. *Every $\exists$-algebraic structure is $\exists$-atomic.*

PROOF. Let $\mathcal{A}$ be $\exists$-algebraic and take $\bar{a} \in A^{<\mathbb{N}}$. Let $\varphi(\bar{x})$ be an $\exists$-formula true of $\bar{a}$ with the least possible number of solutions, say k solutions. We claim that every solution to φ is automorphic to $\bar{a}$. Suppose, toward a contradiction, that $\bar{b}$ satisfies φ but is not automorphic to $\bar{a}$.

Then there must be an $\exists$-formula $\psi(\bar{x})$ that is true of either $\bar{a}$ or $\bar{b}$, but not of both. This is because if $(\mathcal{A}, \bar{a})$ and $(\mathcal{A}, \bar{b})$ satisfied the same $\exists$-formulas, the previous lemma would imply they are isomorphic. If $\psi(\bar{x})$ is true of $\bar{a}$, then $\varphi(\bar{x}) \wedge \psi(\bar{x})$ would be true of $\bar{a}$ and have fewer solutions than φ, contradicting our choice of φ. If $\psi(\bar{x})$ is not true of $\bar{a}$ and it is true of i out of the k solutions of φ, then the formula of $\bar{x}$ saying

"$\varphi(\bar{x})$ and there are i solutions to $\varphi \wedge \psi$ all different from $\bar{x}$"

is an $\exists$-formula true of $\bar{a}$ with $k - i$ solutions, getting the desired contradiction.   □

Historical Remark 3.2.6. The statements of the lemmas in this section are new, but the ideas behind them are not. Proofs like that of Lemma 3.2.4 using König's lemma have appeared in many other places before, for instance in [HLZ99]. The ideas for the proof of Lemma 3.2.5 are similar to those one would use in a proof that algebraic structures are atomic (without the $\exists$-), except that here one has to be slightly more careful.

3.3. Cantor's back-and-forth argument

Before we move on with more on $\exists$-atomic structures, we take an interlude to introduce a tool we will use throughout the book.

DEFINITION 3.3.1. Given structures $\mathcal{A}$ and $\mathcal{B}$, we say that a set $I \subseteq A^{<\mathbb{N}} \times B^{<\mathbb{N}}$ has the *back-and-forth property* if, for every $\langle \bar{a}, \bar{b} \rangle \in I$,

- $D_\mathcal{A}(\bar{a}) = D_\mathcal{B}(\bar{b})$ (i.e., $|\bar{a}| = |\bar{b}|$ and $\bar{a}$ and $\bar{b}$ satisfy the same $\tau_{|\bar{a}|}$-atomic formulas);
- for every $c \in A$, there exists $d \in B$ such that $\langle \bar{a}c, \bar{b}d \rangle \in I$; and[27]
- for every $d \in B$, there exists $c \in A$ such that $\langle \bar{a}c, \bar{b}d \rangle \in I$.

The canonical example is the following. If $\mathcal{A}$ and $\mathcal{B}$ are isomorphic, then the set

$$\{ \langle \bar{a}, \bar{b} \rangle \in A^{<\mathbb{N}} \times B^{<\mathbb{N}} : (\mathcal{A}, \bar{a}) \cong (\mathcal{B}, \bar{b}) \},$$

has the back-and-forth property. We let the reader verify this fact.

Observation 3.3.2. It follows immediately from the example above, that if $\mathcal{A}$ and $\mathcal{B}$ are isomorphic and S is a Scott family for $\mathcal{A}$, then the set

$$I_{\mathcal{A},\mathcal{B}} = \{ \langle \bar{a}, \bar{b} \rangle \in A^{<\mathbb{N}} \times B^{<\mathbb{N}} : \text{(for some } \varphi \in S) \; \mathcal{A} \models \varphi(\bar{a}) \; \& $$
$$\mathcal{B} \models \varphi(\bar{b}) \}$$

has the back-and-forth property.

---

[27] Recall that we are using the notation $\bar{a}c$ for the concatenation $\bar{a}^\frown c$.

LEMMA 3.3.3. *If $I \subseteq A^{<\mathbb{N}} \times B^{<\mathbb{N}}$ has the back-and-forth property, then for every $\langle \bar{a}, \bar{b} \rangle \in I$, there is an isomorphism $g\colon \mathcal{A} \to \mathcal{B}$ mapping $\bar{a}$ to $\bar{b}$. Moreover, such an isomorphism can be computed from an enumeration of I.*

PROOF. The map $g\colon \mathcal{A} \to \mathcal{B}$ is defined by stages. Let $\bar{a}_0 = \bar{a}$ and $\bar{b}_0 = \bar{b}$. At each stage $s + 1$, we define tuples $\bar{a}_{s+1} \in A^{<\mathbb{N}}$ and $\bar{b}_{s+1} \in B^{<\mathbb{N}}$ with $\bar{a}_s \subseteq \bar{a}_{s+1}$, $\bar{b}_s \subseteq \bar{b}_{s+1}$, and $\langle \bar{a}_{s+1}, \bar{b}_{s+1} \rangle \in I$. The back-and-forth property will allow us to build such sequences in a way that, for every $c \in A$, there is some s such that c is one of the entries of $\bar{a}_s$, and, for every $d \in B$, there is some s such that d is one of the entries of $\bar{b}_s$: All we have to do is take turns choosing elements from $\mathcal{A}$ and $\mathcal{B}$ in such a way that we eventually choose them all. At the end of stages, we define $g\colon \mathcal{A} \to \mathcal{B}$ so that $g(\bar{a}_s) = \bar{b}_s$. Since $\bar{a}_s$ and $\bar{b}_s$ satisfy the same $\tau_{|\bar{a}_s|}$-atomic formulas, we get that g preserves all the relations, functions, and constants and hence that it is an isomorphism. (Notice that $D_{\mathcal{A}}(\bar{a}_s) = D_{\mathcal{B}}(\bar{b}_s)$ also implies that, if two entries in $\bar{a}_s$ are equal, so are the corresponding ones in $\bar{b}_s$, and hence there is no issue defining g so that it maps $\bar{a}_s$ to $\bar{b}_s$.)

It is clear that g can be computed from an enumeration of I. □

EXERCISE 3.3.4. Let I be a subset of $A^{<\mathbb{N}} \times A^{<\mathbb{N}}$ which (a) has the back-and-forth property, (b) is an equivalence relation, and (c) satisfies the following condition: for every $n, m \in \mathbb{N}$ and $\pi\colon n \to m$, if $\langle \langle a_0, \ldots, a_{m-1} \rangle, \langle b_0, \ldots, b_{m-1} \rangle \rangle \in I$, then $\langle \langle a_{\pi(0)}, \ldots, a_{\pi(n-1)} \rangle, \langle b_{\pi(0)}, \ldots, b_{\pi(n-1)} \rangle \rangle \in I$. Prove that there is a relation $R \subseteq \mathbb{N} \times A^{<\mathbb{N}}$ such that

$$I = \{ \langle \bar{a}, \bar{b} \rangle : (\mathcal{A}, R, \bar{a}) \cong (\mathcal{A}, R, \bar{b}) \}.$$

3.4. Uniform computable categoricity

An issue we have to be constantly aware of when working with computable structures is that different copies of the same structure may behave differently computationally. Computably categorical structures are the ones where this issue does not show up. They are the ones whose computable copies all have the same computability theoretic properties. We will study them in Chapter 8. For now, we consider the stronger notion of uniform computable categoricity.

DEFINITION 3.4.1. A computable structure $\mathcal{A}$ is *uniformly computably categorical* if there is a computable operator that, when given the atomic diagram $D(\mathcal{B})$ of a computable copy $\mathcal{B}$ of $\mathcal{A}$ as an oracle, outputs an isomorphism from $\mathcal{B}$ to $\mathcal{A}$. A computable structure $\mathcal{A}$ is *uniformly relatively computably categorical* if there is a computable operator that, when given $D(\mathcal{B})$ for a (not necessarily computable) copy $\mathcal{B}$ of $\mathcal{A}$, outputs an isomorphism from $\mathcal{B}$ to $\mathcal{A}$.

Notice that if a structure $\mathcal{A}$ has a c.e. Scott family of $\exists$-formulas, and $\mathcal{B}$ is a copy of $\mathcal{A}$, then the set $I_{\mathcal{A},\mathcal{B}}$ from Observation 3.3.2 is c.e. in $D(\mathcal{B})$ and has the back-and-forth property. Then, by Lemma 3.3.3, we get that $\mathcal{A}$ and $\mathcal{B}$ are $D(\mathcal{B})$-computably isomorphic. Furthermore, the definition of $I_{\mathcal{A},\mathcal{B}}$, and the construction of the isomorphism in Lemma 3.3.3 are completely uniform, and produce a computable operator as needed in the definition of uniform relative computable categoricity.

THEOREM 3.4.2 (Ventsov [Ven92]). *Let $\mathcal{A}$ be a computable structure. The following are equivalent:*

1. $\mathcal{A}$ *is effectively $\exists$-atomic (Definition 3.1.3).*
2. $\mathcal{A}$ *is uniformly relatively computably categorical.*
3. $\mathcal{A}$ *is uniformly computably categorical.*

PROOF. That (1) implies (2) was observed in the previous paragraph. It is just a back-and-forth construction using the set $I_{\mathcal{A},\mathcal{B}}$ from Observation 3.3.2. It is obvious that (2) implies (3). The proof that (3) implies (1) is quite a bit more elaborate.

Suppose Γ is a computable operator such that $\Gamma^{D(\mathcal{B})}$ is an isomorphism from $\mathcal{B}$ to $\mathcal{A}$ for every computable copy $\mathcal{B}$ of $\mathcal{A}$. We need to find $\exists$-formulas defining each tuple in $\mathcal{A}$. Here is the key observation: suppose that for $\bar{q} \in A^{<\mathbb{N}}$ we have that $\Gamma^{D_{\mathcal{A}}(\bar{q})}$ converges on $0, \dots, k-1$ for some $k \in \mathbb{N}$, then, the automorphism orbit of $\bar{q} \upharpoonright k$ is determined by $D_{\mathcal{A}}(\bar{q}) \in 2^{<\mathbb{N}}$ in the sense that if $D_{\mathcal{A}}(\bar{p}) = D_{\mathcal{A}}(\bar{q})$ then $\bar{p} \upharpoonright k$ is automorphic to $\bar{q} \upharpoonright k$. To prove this, we first claim that for every $k \in \mathbb{N}$ and every tuple $\bar{q} \in A^{<\mathbb{N}}$ such that $\Gamma^{D_{\mathcal{A}}(\bar{q})}$ converges on $0, \dots, k-1$, we have that[28]

$$\bar{q} \upharpoonright k \text{ is automorphic to } \Gamma^{D_{\mathcal{A}}(\bar{q})} \upharpoonright k.$$

To see this, extend $\bar{q}$ to a computable onto map $g \colon \mathbb{N} \to \mathcal{A}$. For $\mathcal{B} = g^{-1}(\mathcal{A})$, $\Gamma^{D(\mathcal{B})}$ is an isomorphism from $\mathcal{B}$ to $\mathcal{A}$. Since g is also an isomorphism from $\mathcal{B}$ to $\mathcal{A}$, the two images of $\langle 0, \dots, k-1 \rangle$ through those isomorphisms must be automorphic; namely $g \upharpoonright k$ and $\Gamma^{D(\mathcal{B})} \upharpoonright k$ (see figure below). Since $g \supset \bar{q}$, $g \upharpoonright k = \bar{q} \upharpoonright k$, and since $D(\mathcal{B}) \supset D_{\mathcal{A}}(\bar{q})$, $\Gamma^{D(\mathcal{B})} \upharpoonright k = \Gamma^{D_{\mathcal{A}}(\bar{q})} \upharpoonright k$.

$$\mathcal{A} \xleftarrow[\cong]{g} \mathcal{B} \xrightarrow[\cong]{\Gamma^{D(\mathcal{B})}} \mathcal{A}$$

$$\bar{q} \upharpoonright k \longleftarrow\!\!\mid \langle 0, \dots, k-1 \rangle \longmapsto \Gamma^{D_{\mathcal{A}}(\bar{q})} \upharpoonright k$$

Now, given a tuple $\bar{a} \in A^{<\mathbb{N}}$, we need to produce an $\exists$-formula defining its orbit, and we need to find this formula computably. Let $k = |\bar{a}|$. Search

---

[28]As the reader may expect: $\Gamma^{D_{\mathcal{A}}(\bar{q})} \upharpoonright k = \langle \Gamma^{D_{\mathcal{A}}(\bar{q})}(0), \Gamma^{D_{\mathcal{A}}(\bar{q})}(1), \dots, \Gamma^{D_{\mathcal{A}}(\bar{q})}(k-1) \rangle \in A^k$.

for $\bar{q} \in A^{<\mathbb{N}}$ extending $\bar{a}$ such that $\Gamma^{D_A(\bar{q})}$ converges on $0, \ldots, k - 1$. We now claim that the following $\exists$-formula defines the orbit of $\bar{a}$:[29]

$$\varphi_{\bar{a}}(\bar{x}) \equiv (\exists \bar{y} \supseteq \bar{x}) \text{``} D(\bar{y}) = \sigma\text{''}, \quad \text{where } \sigma = D_A(\bar{q}) \in 2^{<\mathbb{N}}.$$

Clearly, $\bar{a}$ satisfies $\varphi_{\bar{a}}$ using $\bar{y} = \bar{q}$. Suppose now that $\mathcal{A} \models \varphi_{\bar{a}}(\bar{c})$; we need to show that $\bar{a}$ and $\bar{c}$ are automorphic. From $\varphi_{\bar{a}}(\bar{c})$ we get a tuple $\bar{p} \supseteq \bar{c}$ such that $D_A(\bar{p}) = \sigma = D_A(\bar{q})$. So

$$\Gamma^{D_A(\bar{p})} \upharpoonright k = \Gamma^{D_A(\bar{q})} \upharpoonright k.$$

By our first claim above, the left-hand-side is automorphic to $\bar{p} \upharpoonright k = \bar{c}$, and the right-hand-side is automorphic to $\bar{q} \upharpoonright k = \bar{a}$.

We conclude that $\{\varphi_{\bar{a}}(\bar{x}) : \bar{a} \in A^{<\mathbb{N}}\}$ is a c.e. Scott family of $\exists$-formulas for $\mathcal{A}$.   □

EXERCISE 3.4.3. Build the uniformly computably categoricity operator explicitly for the case of a computable connected graph of finite valance with a root node (identified with a constant symbol).

Historical Remark 3.4.4. The theorem above is due to Ventsov [Ven92]. Other notions of uniform categoricity were studied by Kudinov [Kud96a, Kud96b, Kud97] and by Downey, Hirschfeldt, and Khoussainov [DHK03].

3.5. Existential atomicity in terms of types

The usual definition of atomic models in model theory is in terms of types (as in (A2) below). We show in this section that, for $\exists$-atomic models, it is enough to look at $\forall$-types instead of full first-order types.

We need to review some basic definitions. A $\forall$-*type* on the variables $x_1, \ldots, x_n$ is a set $p(\bar{x})$ of $\forall$-formulas with free variables among $x_1, \ldots, x_n$ that is *consistent*, i.e., that is satisfied by some tuple $a_1, \ldots, a_n$ in some structure. We say that a $\forall$-type is *realized in* a structure $\mathcal{A}$ if it is satisfied by some tuple in $\mathcal{A}$. Given $\bar{a} \in A^{<\mathbb{N}}$, the $\forall$-*type of $\bar{a}$ in $\mathcal{A}$* is the set of $\forall$-formulas true of $\bar{a}$:[30]

$$\forall\text{-}tp_A(\bar{a}) = \{\varphi(\bar{x}) : \varphi \text{ is a } \forall\text{-formula and } \mathcal{A} \models \varphi(\bar{a})\}.$$

The reason we allow types to be partial is that $\forall$-types are never complete, as we could not add the negation of $\forall$-formulas. For the same reason, instead of principal types, we have to deal with supported types.

DEFINITION 3.5.1. A type $p(\bar{x})$ is $\exists$-*supported within a class* $\mathbb{K}$ *of structures* if there exists an $\exists$-formula $\varphi(\bar{x})$ which is realized in some structure in $\mathbb{K}$ and which implies all of $p(\bar{x})$ within $\mathbb{K}$; that is, $\mathcal{A} \models \forall \bar{x}(\varphi(\bar{x}) \rightarrow \psi(\bar{x}))$

---

[29]Recall from Observation 1.1.10 that, for each $\sigma \in 2^{\ell|\bar{z}|}$, there is a quantifier-free formula $\varphi_\sigma^{at}(\bar{z})$ which holds if and only if $D(\bar{z}) = \sigma$.

[30]The obvious assumption here is that $|\bar{x}| = |\bar{a}|$.

for every $\psi(\bar{x}) \in p(\bar{x})$ and every $\mathcal{A} \in \mathbb{K}$. We say that $p(\bar{x})$ is $\exists$-*supported in a structure* $\mathcal{A}$ if it is $\exists$-supported in $\mathbb{K} = \{\mathcal{A}\}$.

THEOREM 3.5.2. *For every structure* $\mathcal{A}$, *the following are equivalent*:

(A1) $\mathcal{A}$ *is* $\exists$-*atomic*.
(A2) *Every elementary first-order type realized in* $\mathcal{A}$ *is* $\exists$-*supported in* $\mathcal{A}$.
(A3) *Every* $\forall$-*type realized in* $\mathcal{A}$ *is* $\exists$-*supported in* $\mathcal{A}$.

PROOF. It is not hard to see that (A1) implies (A2) as the $\exists$-formula defining the orbit of $\bar{a}$ supports its type. Clearly (A2) implies (A3). Let us prove that (A3) implies (A1).

For each $\bar{a} \in A^{<\mathbb{N}}$, let $\varphi_{\bar{a}}(\bar{x})$ be an $\exists$-formula supporting the $\forall$-type of $\bar{a}$. We show that $S = \{\varphi_{\bar{a}} : \bar{a} \in A^{<\mathbb{N}}\}$ is a Scott family for $\mathcal{A}$. We start by noticing that $\mathcal{A} \models \varphi_{\bar{a}}(\bar{a})$. This is because, otherwise, $\neg\varphi_{\bar{a}}$ would be part of the $\forall$-type of $\bar{a}$, and hence implied by $\varphi_{\bar{a}}$, which cannot be the case because $\varphi_{\bar{a}}$ is realizable in $\mathcal{A}$. Consider the set

$$I_{\mathcal{A}} = \{\langle \bar{a}, \bar{b} \rangle \in A^{<\mathbb{N}} \times A^{<\mathbb{N}} : \mathcal{A} \models \varphi_{\bar{a}}(\bar{b})\}.$$

First, let us prove $I_{\mathcal{A}}$ is symmetric; that is, that if $\mathcal{A} \models \varphi_{\bar{a}}(\bar{b})$, then $\mathcal{A} \models \varphi_{\bar{b}}(\bar{a})$. If not, then $\neg\varphi_{\bar{b}}(\bar{x})$ would be part of the $\forall$-type of $\bar{a}$, and hence implied by $\varphi_{\bar{a}}$. But we know this is not the case because $\bar{b}$ models both $\varphi_{\bar{a}}$ and $\varphi_{\bar{b}}$.

Second, we now claim that $I_{\mathcal{A}}$ has the back-and-forth property (Definition 3.3.1). Suppose $\langle \bar{a}, \bar{b} \rangle \in I_{\mathcal{A}}$. Observe $\bar{a}$ and $\bar{b}$ must satisfy the same $\forall$-types as they both satisfy $\varphi_{\bar{a}}$ and $\varphi_{\bar{b}}$ which support their respective $\forall$-types. In particular, they satisfy the same $\tau_{|\bar{a}|}$-atomic formulas and hence have the same atomic diagrams. To show the second condition in Definition 3.3.1, take $c \in A$. If there was no $d \in A$ with $\langle \bar{a}c, \bar{b}d \rangle \in I_{\mathcal{A}}$, we would have that $\mathcal{A} \models \neg\exists y \varphi_{\bar{a}c}(\bar{b}, y)$. This formula would be part of the $\forall$-type of $\bar{b}$, and hence implied by $\varphi_{\bar{b}}$. But then, since $\mathcal{A} \models \varphi_{\bar{b}}(\bar{a})$, we would have $\mathcal{A} \models \neg\exists y \varphi_{\bar{a}c}(\bar{a}, y)$, which is not true as witnessed by c. The third condition of the back-and-forth property follows from the symmetry of $I_{\mathcal{A}}$.

Finally, to see that S is a Scott family for $\mathcal{A}$, notice that if $\varphi_{\bar{a}}(\bar{b})$ and $\varphi_{\bar{a}}(\bar{c})$ both hold, then, by Lemma 3.3.3, both $\bar{b}$ and $\bar{c}$ are automorphic to $\bar{a}$, and hence automorphic to each other. $\square$

EXERCISE 3.5.3. [DKLT13, Theorem 1.6] (a) Prove that the index set of all computable structures that are effectively $\exists$-atomic after adding some parameters, is Σ_3^0.

(b) Prove that it is Σ_3^0-complete. Hint in footnote.[31]

---

[31] Use $\mathbb{Q}$-vector spaces.

3.6. Building structures and omitting types

Before we continue studying the properties of $\exists$-atomic structures, we need to make another stop to prove some general lemmas that will be useful in future sections. First, we prove a lemma that will allow us to find computable structures in a given class of structures. Second, using similar techniques, we prove the type omitting lemma for $\forall$-types, and its effective version.

We need one more level of the hierarchy of infinitary formulas:

DEFINITION 3.6.1. An *infinitary Π_2 formula* (denoted Π_2^{in}) is a countable infinite (or finite) conjunction of formulas of the form $\forall \bar{y}\psi(\bar{y}, \bar{x})$, where each formula ψ is Σ_1^{in}, and $\bar{x}$ is a fixed tuple of free variables. That is, a Π_2^{in} formula is one of the form

$$\bigwedge_{i \in \mathbb{N}} \forall \bar{y}_i \bigvee_{j \in \mathbb{N}} \exists \bar{z}_j \; \varphi_{i,j}(\bar{x}, \bar{y}_i, \bar{z}_j),$$

where the formulas $\varphi_{i,j}$ are finitary and quantifier free. Such a formula is *computable infinitary Π_2* (denoted Π_2^c) if the formulas ψ are Σ_1^c and the list of indices of the formulas ψ is computably enumerable, or equivalently, if the matrix $\{\ulcorner \varphi_{i,j} \urcorner : i, j \in \mathbb{N}\}$ is computable. A class of structures is Π_2^c if it is the class of all the ω-presentations that satisfy a certain Π_2^c sentence. Given an oracle X, we use Π_2^{cX} to denote the X-computable infinitary Π_2 formulas.

As the reader may expect, an *infinitary Σ_2 formula* (denoted Σ_2^{in}) is a countable disjunction of formulas of the form $\exists \bar{y}\psi(\bar{y}, \bar{x})$, where each formula ψ is Π_1^{in}, and $\bar{x}$ is a fixed tuple of free variables.

Observe that every Π_2^c formula on a structure $\mathcal{A}$ is equivalent to a Π_1^c formula on $(\mathcal{A}, \vec{K}^{\mathcal{A}})$, where $\vec{K}^{\mathcal{A}}$ is Kleene's relation defined in 2.2.3.

DEFINITION 3.6.2. Assume, without loss of generality, we are working with a relational vocabulary τ. Given a class of structures $\mathbb{K}$, we let $\mathbb{K}^{fin}$ be—essentially—the set of all the finite substructures of the structures in $\mathbb{K}$:[32]

$$\mathbb{K}^{fin} = \{D(\mathcal{A}) : \mathcal{A} \text{ a finite } \tau_{|.|}\text{-substructure of some } \mathcal{B} \in \mathbb{K}\}$$
$$= \{D_{\mathcal{B}}(\bar{a}) : \mathcal{B} \in \mathbb{K}, \bar{a} \in \mathcal{B}^{<\mathbb{N}}\} \subseteq 2^{<\mathbb{N}}.$$

Observation 3.6.3. An $\exists$-sentence ψ holds of a structure $\mathcal{A}$ on a relational vocabulary if and only if it holds on some finite substructure of $\mathcal{A}$.

EXERCISE 3.6.4. Show that

$$\mathbb{K}^{fin} \equiv_{pos} \bigcup \{\exists\text{-}Th(\mathcal{A}) : \mathcal{A} \in \mathbb{K}\},$$

---

[32] Recall from Definition 1.1.6 that a $\tau_{|.|}$-structure is a τ_s-structure for $s = |\mathcal{A}|$ where τ_s consists, usually, of the first s symbols of τ when τ is infinite and $\tau_s = \tau$ when τ is finite.

where $\equiv_{pos}$ is *positive equivalence* defined in page xviii. In particular, they are both Turing and enumeration equivalent.

EXERCISE 3.6.5. Prove that $\mathbb{K}^{fin}$ is the set of diagrams of all the finite $\tau_{|.|}$-structures satisfying $\forall\text{-}Th(\mathbb{K})$.[33] Recall that $\forall\text{-}Th(\mathbb{K})$ is the set of $\forall$-sentences true in all structures in $\mathbb{K}$.

LEMMA 3.6.6. *Let $\mathbb{K}$ be the class of models of a Π_2^c sentence and suppose that $\mathbb{K}^{fin}$ is c.e. Then there is at least one computable structure in $\mathbb{K}$.*

PROOF. We build a structure in $\mathbb{K}$ by building a finite approximation to it as in Definition 1.1.6. That is, we build an increasing chain of finite structures $\mathcal{A}_s$, $s \in \mathbb{N}$, over increasing vocabularies. Each $\mathcal{A}_s$ is a $\tau_{|.|}$-structure whose domain is an initial segment of $\mathbb{N}$. Furthermore, we require that each $\mathcal{A}_s$ be in $\mathbb{K}^{fin}$ (i.e., the diagram of $\mathcal{A}_s$ be in $\mathbb{K}^{fin}$), and that $\mathcal{A}_s \subseteq \mathcal{A}_{s+1}$ (as $\tau_{|\mathcal{A}_s|}$-structures). At the end of stages, we define the τ-structure $\mathcal{A} = \bigcup_{s \in \mathbb{N}} \mathcal{A}_s$.

Let $\bigwedge_{i \in I} \forall \bar{y}_i \psi_i(\bar{y}_i)$ be the Π_2^c sentence that axiomatizes $\mathbb{K}$, where each ψ_i is Σ_1^c. To get $\mathcal{A} \in \mathbb{K}$, we need to guarantee that, for each i and each $\bar{a} \in A^{|\bar{y}_i|}$, we have $\mathcal{A} \models \psi_i(\bar{a})$. For this, when we build $\mathcal{A}_{s+1}$, we will make sure that,

$$\text{for every } i < s \text{ and every } \bar{a} \in A_s^{|\bar{y}_i|}, \mathcal{A}_{s+1} \models \psi_i(\bar{a}). \qquad (\star)$$

Notice that since ψ_i is Σ_1^c, $\mathcal{A}_{s+1} \models \psi_i(\bar{a})$ implies $\mathcal{A} \models \psi_i(\bar{a})$. Thus, we would end up with $\mathcal{A} \models \bigwedge_{i \in I} \forall \bar{y}_i \psi_i(\bar{y}_i)$.

Now that we know what we need to do, let us build the sequence of $\mathcal{A}_s$'s. Suppose we have already built $\mathcal{A}_0, \ldots, \mathcal{A}_s$ and we want to define $\mathcal{A}_{s+1} \supseteq \mathcal{A}_s$. All we need to do is search for a finite structure in $\mathbb{K}^{fin}$ satisfying $(\star)$, which we can check computably. We need to show that at least one such structure exists. Since $\mathcal{A}_s \in \mathbb{K}^{fin}$, there is some $\mathcal{B} \in \mathbb{K}$ which has a substructure $\mathcal{B}_s$ $\tau_{|.|}$-isomorphic to $\mathcal{A}_s$. Since $\mathcal{B} \models \bigwedge_{i \in I} \forall \bar{y}_i \psi_i(\bar{y}_i)$, for every $i < s$ and every $\bar{b} \in \mathcal{B}_s^{|\bar{y}_i|}$, there exists a tuple in $\mathcal{B}$ witnessing that $\mathcal{B} \models \psi_i(\bar{b})$. Let $\mathcal{B}_{s+1}$ be a finite $\tau_{|.|}$-substructure of $\mathcal{B}$ containing $\mathcal{B}_s$ and all those witnessing tuples, and large enough so that all the symbols in the $\exists$-disjunct of the ψ_i witnessing $\mathcal{B} \models \psi_i(\bar{b})$ for $i < s$ appear in $\tau_{|\mathcal{B}_{s+1}|}$. Then $\mathcal{B}_{s+1}$ satisfies $(\star)$ with respect to $\mathcal{B}_s$ as needed.   □

COROLLARY 3.6.7. *Let $\mathbb{K}$ be a Π_2^c class of structures, and S be the $\exists$-theory of some structure in $\mathbb{K}$. If S is c.e. in a set X, then there is an X-computable ω-presentation of a structure in $\mathbb{K}$ with $\exists$-theory S.*

---

[33] Of course, we refer only to the sentences that use the vocabulary of the finite structure. Thus, if $\mathcal{A}$ is a $\tau_{|\mathcal{A}|}$-structure, the claim is that $D(\mathcal{A}) \in \mathbb{K}^{fin}$ if and only if all $\forall\text{-}\tau_{|\mathcal{A}|}$-sentences in $\forall\text{-}Th(\mathbb{K})$ are true in $\mathcal{A}$.

Proof. Add to the Π_2^c axiom for $\mathbb{K}$ the Π_2^{cX} sentence saying that the structure must have $\exists$-theory S:

$$\left(\bigwedge_{\varphi \in S} \varphi \right) \wedge \left(\bigwedge_{\varphi \ \exists\text{-formula: } \varphi \notin S} \neg\varphi \right)$$

where the left-hand-side says of a structure $\mathcal{A}$ that $S \subseteq \exists\text{-}Th(\mathcal{A})$, and the right-hand-side that $\exists\text{-}Th(A) \subseteq S$. Let $\mathbb{K}_S$ be the new Π_2^{cX} class of structures. It is nonempty because we are assuming that some structure in $\mathbb{K}$ has $\exists$-theory S. All the structures in $\mathbb{K}_S$ have $\exists$-theory S, and hence $\mathbb{K}_S^{fin}$ is enumeration reducible to S, and hence it is c.e. in X too. Applying Lemma 3.6.6 relative to X, we get an X-computable structure in $\mathbb{K}_S$ as wanted. □

Not only can we build a computable structure in such a class $\mathbb{K}$, we can build one omitting certain types. The first one to prove an effective version of the omitting types theorem was Terry Millar [Mil83]. Our version below is different than hers, as she used first-order types over complete decidable first-order theories.

Lemma 3.6.8 (The $\forall$-type omitting theorem). *Let $\mathbb{K}$ be an Π_2^{in} class of structures. Let $\{p_i(\bar{x}_i) : i \in \mathbb{N}\}$ be a sequence of $\forall$-types which are not $\exists$-supported in $\mathbb{K}$. Then there is a structure $\mathcal{A} \in \mathbb{K}$ which omits all the types $p_i(\bar{x}_i)$ for $i \in \mathbb{N}$.*

Furthermore, if $\mathbb{K}$ is Π_2^c, $\mathbb{K}^{fin}$ is c.e. and the list $\{p_i(\bar{x}_i) : i \in \mathbb{N}\}$ is c.e., we can make $\mathcal{A}$ computable.

Proof. We construct $\mathcal{A}$ by stages as in the proof of Lemma 3.6.6, the difference being that now we need to omit the types p_i. So, on the even stages s, we do exactly the same thing we did in Lemma 3.6.6, and we use the odd stages to omit the types. That is, we build a sequence of finite $\tau_{|\cdot|}$-structures $\mathcal{A}_0 \subseteq \mathcal{A}_1 \subseteq \cdots$ and at even stages we define $\mathcal{A}_{s+1}$ so that it satisfies $(\star)$ from Lemma 3.6.6 guaranteeing that $\mathcal{A}$ belongs to $\mathbb{K}$. At odd stages $s + 1 = 2\langle i, j \rangle + 1$, we ensure that the jth tuple $\bar{a}$ does not satisfy p_i as follows. We are given $\mathcal{A}_s$ and we need to define $\mathcal{A}_{s+1}$ so that $\bar{a}$ satisfies some $\exists$-formula whose negation is in p_i. Let $\bar{c}$ be the tuple of elements of $\mathcal{A}_s$ which are not in $\bar{a}$, and let $\sigma = D_{\mathcal{A}_s}(\bar{a}, \bar{c})$. Then, $\bar{a}$ satisfies $\exists\bar{y}\,(D(\bar{a}, \bar{y}) = \sigma)$. Since p_i is not $\exists$-supported in $\mathbb{K}$, there exists a $\forall$-formula $\psi(\bar{x}) \in p_i$ which is not implied by $\exists\bar{y}D(\bar{x}, \bar{y}) = \sigma$ within $\mathbb{K}$. That means that, for some $\mathcal{B} \in \mathbb{K}$ and $\bar{b} \in B^{<\mathbb{N}}$,

$$\mathcal{B} \models \exists\bar{y}D(\bar{b}, \bar{y}) = \sigma \ \& \ \neg\psi(\bar{b}).$$

Then, there is a finite substructure $\widetilde{\mathcal{B}} \in \mathbb{K}^{fin}$ of $\mathcal{B}$ containing such $\bar{y}$ the witnesses for $\neg\psi(\bar{b})$ which also satisfies $\exists\bar{y}D(\bar{b}, \bar{y}) = \sigma$ and $\neg\psi(\bar{b})$. Since $\widetilde{\mathcal{B}} \models \exists\bar{y}D(\bar{b}, \bar{y}) = \sigma$, we can assume that $\mathcal{A}_s \subseteq_{\tau_{|\mathcal{A}_s|}} \widetilde{\mathcal{B}}$ and that $\bar{b} = \bar{a}$.

Since such $\widetilde{\mathcal{B}}$ and ψ exist, we can wait until we find them and then define $\mathcal{A}_{s+1}$ accordingly.   □

We will see how the classical first-order type omitting theorem is a corollary of the $\forall$-type omitting theorem in [MonP2] once we see Morley-izations.

3.7. Scott sentences of existentially atomic structures

Existentially atomic structures are also among the simplest ones in terms of the complexity of their Scott sentences.

DEFINITION 3.7.1. A sentence ψ is a *Scott sentence* for a structure $\mathcal{A}$ if $\mathcal{A}$ is the only countable structure satisfying ψ. That is, ψ is true on a structure $\mathcal{B}$ if and only if $\mathcal{B}$ is isomorphic to $\mathcal{A}$.

We will see in [MonP2] that every countable structure has a Scott sentence in the infinitary language $\mathcal{L}_{\omega_1,\omega}$. For now, we prove it only for $\exists$-atomic structures.

LEMMA 3.7.2. *Every $\exists$-atomic structure has an Π_2^{in} Scott sentence. Furthermore, every effectively $\exists$-atomic computable structure has a Π_2^c Scott sentence.*

PROOF. Let S be a Scott family of $\exists$-formulas for $\mathcal{A}$. For each $\bar{a} \in A^{<\mathbb{N}}$, let $\varphi_{\bar{a}}(\bar{x})$ be the $\exists$-formula defining the orbit of $\bar{a}$. For the empty tuple, let $\varphi_{\langle\rangle}()$ be a sentence that is always true. Given any other structure $\mathcal{B}$, consider the set

$$I_{\mathcal{B}} = \{\langle \bar{a}, \bar{b}\rangle \in A^{<\mathbb{N}} \times B^{<\mathbb{N}} : \mathcal{B} \models \varphi_{\bar{a}}(\bar{b})\}.$$

If $I_{\mathcal{B}}$ had the back-and-forth property, then, by Lemma 3.3.3, we would know that $\mathcal{B}$ is isomorphic to $\mathcal{A}$ because $\langle\langle\rangle, \langle\rangle\rangle \in I_{\mathcal{B}}$. Recall from the proof of Theorem 3.5.2 that $I_{\mathcal{A}}$ has the back-and-forth property. Thus, if $\mathcal{B}$ is isomorphic to $\mathcal{A}$, then $I_{\mathcal{B}}$ also has the back-and-forth property. Therefore, we get that $I_{\mathcal{B}}$ has the back-and-forth property if and only if $\mathcal{B}$ is isomorphic to $\mathcal{A}$. The Scott sentence for $\mathcal{A}$ says of a structure $\mathcal{B}$ that $I_{\mathcal{B}}$ has the back-and-forth property:

$$\bigwedge_{\bar{a}\in A^{<\mathbb{N}}} \forall x_1,\ldots,x_{|\bar{a}|} \left(\varphi_{\bar{a}}(\bar{x}) \implies \left(\text{``}D(\bar{x}) = D_{\mathcal{A}}(\bar{a})\text{''} \right) \wedge \right.$$
$$\left. \left(\forall y \bigvee_{b\in A} \varphi_{\bar{a}b}(\bar{x}y) \right) \wedge \left(\bigwedge_{b\in A} \exists y \varphi_{\bar{a}b}(\bar{x}y) \right) \right),$$

where "$D(\bar{x}) = D_{\mathcal{A}}(\bar{a})$" stands for $\varphi^{at}_{D_{\mathcal{A}}(\bar{a})}(\bar{x})$ as in Observation 1.1.10. As for the effectivity claim, if $\mathcal{A}$ is a computable ω-presentation and S

is c.e., then the map $\bar{a} \mapsto \varphi_{\bar{a}}$ is computable, and the conjunctions and disjunctions in the Scott sentence above are all computable. $\qquad\square$

To prove the other direction, we need to go through the type omitting theorem for $\forall$-types.

THEOREM 3.7.3. *Let A be a structure. The following are equivalent:*

1. A *is* $\exists$*-atomic.*
2. A *has an* Π_2^{in} *Scott sentence.*

PROOF. We already know that (1) implies (2). For the other direction, suppose ψ is an Π_2^{in} Scott sentence for A, but that A is not $\exists$-atomic. By Theorem 3.5.2, there is a $\forall$-type realized in A which is not $\exists$-supported. But then, by Lemma 3.6.8, there exists a model of ψ which omits that type. This structure could not be isomorphic to A, contradicting that ψ was a Scott sentence for A. $\qquad\square$

LEMMA 3.7.4. *Let A be a structure. The following are equivalent:*

1. A *is* $\exists$*-atomic over a finite tuple of parameters.*
2. A *has an* Σ_3^{in} *Scott sentence.*

As the reader might be able to guess by now, an Σ_3^{in} *formula* is a countable disjunction of formulas of the form $\exists \bar{y} \psi(\bar{y}, \bar{x})$, where ψ is Π_2^{in} and $\bar{x}$ is a fixed tuple of variables.

PROOF. If A is $\exists$-atomic over a finite tuple of parameters $\bar{a}$, then $(A, \bar{a})$ has an Π_2^{in} Scott sentence $\varphi(\bar{c})$. Then $\exists \bar{y} \varphi(\bar{y})$ is a Scott sentence for A.

Suppose now that A has a Scott sentence $\bigvee_{i \in \mathbb{N}} \exists \bar{y}_i \psi_i(\bar{y}_i)$. A must satisfy one of the disjuncts, and that disjunct must then also be a Scott sentence for A. So, suppose the Scott sentence for A is $\exists \bar{y}\ \psi(\bar{y})$, where ψ is Π_2^{in}. Let $\bar{c}$ be a new tuple of constants of the same size as $\bar{y}$. If $\psi(\bar{c})$ were a Scott sentence for $(A, \bar{a})$, we would know A is $\exists$-atomic over $\bar{a}$; but this might not be the case. Suppose $(B, \bar{b}) \models \psi(\bar{c})$. Then B must be isomorphic to A as it satisfies $\exists \bar{y}\ \psi(\bar{y})$, but we could have $(B, \bar{b}) \not\cong (A, \bar{a})$. However, it is enough for us to show that one of the models $(B, \bar{b})$ of $\psi(\bar{c})$ is $\exists$-atomic over $\bar{b}$. There are only countably many models of $\psi(\bar{c})$ because there are only countably many tuples in $A^{<\mathbb{N}}$ to which we can assign $\bar{c}$. Therefore, there are countably many $\forall$-types among the models of $\psi(\bar{c})$. Thus, we can omit the non-$\exists$-supported ones while satisfying $\psi(\bar{c})$. The resulting structure would be $\exists$-atomic over $\bar{c}$ by Theorem 3.5.2 and isomorphic to A because it satisfies $\exists \bar{y}\ \psi(\bar{y})$. $\qquad\square$

3.8. Turing degree and enumeration degree

To measure the computational complexity of a structure, the most common tool is its degree spectrum, which we will study in Chapter 5. A much more natural attempt to measure the computational complexity of a structure is given in the following definition; unfortunately, it does not always apply.

DEFINITION 3.8.1 (Jockusch and Richter [Ric81]). A structure $\mathcal{A}$ has *Turing degree* $X \in 2^{\mathbb{N}}$ if X computes a copy of $\mathcal{A}$, and every copy of $\mathcal{A}$ computes X.

It turns out that if we look at a similar definition, but on the enumeration degrees, we obtain a better behaved notion.

DEFINITION 3.8.2. A structure $\mathcal{A}$ has *enumeration degree* $X \subseteq \mathbb{N}$ if every enumeration of X computes a copy of $\mathcal{A}$, and every copy of $\mathcal{A}$ computes an enumeration of X. Recall that an enumeration of X is an onto function $f : \mathbb{N} \to X$.

Equivalently, $\mathcal{A}$ has enumeration degree X if and only if, for every Y,

$$Y \text{ computes a copy of } \mathcal{A} \iff X \text{ is c.e. in } Y.$$

Notice that the enumeration degree of a structure is unique up to enumeration equivalence. (See page xviii.)

EXAMPLE 3.8.3. Given $X \subseteq \mathbb{N}$, the standard example of a structure with enumeration degree X is the *graph* $\mathcal{G}_X$, which is made out of disjoint cycles of different lengths and which contains a cycle of length $n + 3$ if and only if $n \in X$. It is not hard to see that every presentation of this graph can enumerate X: Whenever we find a cycle of length $n + 3$, we enumerate n into X. For the other direction, if we can enumerate X, we can build a copy of $\mathcal{G}_X$ by enumerating a cycle of length $n + 3$ every time we see a number n enter X.

EXAMPLE 3.8.4. Given $X \subseteq \mathbb{N}$, consider the group $\mathcal{G}_X = \bigoplus_{i \in X} \mathbb{Z}_{p_i}$ as in Example 2.1.21, where p_i is the ith prime number. Then $\mathcal{G}_X$ has enumeration degree X: We can easily build $\mathcal{G}_X$ out of an enumeration of X, and for the other direction, we have that $n \in X$ if and only if there exists $g \in \mathcal{G}_X$ of order p_n.

EXERCISE 3.8.5. Show that both the graph and the group from the previous examples are $\exists$-atomic. Hint in the footnote.[34]

Note that $\mathcal{A}$ has Turing degree X if and only if has enumeration degree $X \oplus X^c$.[35] This is because $X \leq_T Y \iff X \oplus X^c$ is c.e. in Y. So, in either of the examples above, we can get a graph or a group of Turing degree X by considering $\mathcal{G}_{X \oplus X^c}$. A set X is said to have *total enumeration degree* if it is enumeration equivalent to a set of the form $Z \oplus Z^c$. There are sets which do not have total enumeration degree (cf. [Med55]). Those are exactly the sets X for which the set $\{Y \in 2^{\mathbb{N}} : X \text{ is c.e. in } Y\}$ has no

---

[34]Show they are $\exists$-algebraic.

[35]X^c denotes the complement of X.

least Turing degree.[36] It follows that if a structure has enumeration degree X and X does not have total enumeration degree, then the structure does not have Turing degree.

The enumeration degree of a structure is indeed a good way to measure its computational complexity. Unfortunately, in general, structures need not have enumeration degree. Furthermore, there are whole classes of structures, like linear orderings for instance, where no structure has enumeration degrees unless it is already computable (Section 5.1). Before getting into that, the rest of the section is dedicated to classes whose structures all have enumeration degree.

THEOREM 3.8.6. *Let $\mathbb{K}$ be a Π_2^c class, all whose structures are $\exists$-atomic. Then every structure in $\mathbb{K}$ has enumeration degree given by its $\exists$-theory.*

The proof of Theorem 3.8.6 needs a couple of lemmas that are interesting on their own right.

LEMMA 3.8.7. *Let S be the $\exists$-theory of a structure $\mathcal{A}$. If $\mathcal{A}$ belongs to some Π_2^c class $\mathbb{K}$ where $\mathcal{A}$ is the only structure with $\exists$-theory S, then $\mathcal{A}$ has enumeration degree S.*

PROOF. By Corollary 3.6.7, if X can compute an enumeration of S, then it can compute an ω-presentation of a structure $\mathcal{B} \in \mathbb{K}$ with $\exists$-theory S. By the assumption on $\mathbb{K}$, $\mathcal{A}$ and $\mathcal{B}$ must be isomorphic. So, X can compute a copy of $\mathcal{A}$. Of course, every copy of $\mathcal{A}$ can enumerate S, and hence $\mathcal{A}$ has enumeration degree S.   □

LEMMA 3.8.8. *If $\mathcal{A}$ and $\mathcal{B}$ are $\exists$-atomic and have the same $\exists$-theory, then they are isomorphic.*

Recall that we already proved this lemma for $\exists$-algebraic structures; cf. Lemma 3.2.4.

PROOF. We prove that $\mathcal{A}$ and $\mathcal{B}$ are isomorphic using a back-and-forth construction. Let

$$I = \{\langle \bar{a}, \bar{b} \rangle : \forall\text{-}tp_\mathcal{A}(a_0, \ldots, a_s) = \forall\text{-}tp_\mathcal{B}(b_0, \ldots, b_s)\}.$$

We need to show that I has the back-and-forth property (Definition 3.3.1). Clearly, $\forall\text{-}tp_\mathcal{A}(a_0, \ldots, a_s) = \forall\text{-}tp_\mathcal{B}(b_0, \ldots, b_s)$ implies $D_\mathcal{A}(a_0, \ldots, a_s) = D_\mathcal{B}(b_0, \ldots, b_s)$. By assumption, $\langle \langle \rangle, \langle \rangle \rangle \in I$. For the second condition in Definition 3.3.1, suppose $\langle \bar{a}, \bar{b} \rangle \in I$, and let $c \in A$. Let ψ be the principal $\exists$-formula satisfied by $\bar{a}c$. Since $\forall\text{-}tp_\mathcal{A}(\bar{a}) = \forall\text{-}tp_\mathcal{B}(\bar{b})$, there is a d in $\mathcal{B}$ satisfying the same formula. We need to show that $\forall\text{-}tp_\mathcal{A}(\bar{a}c) = \forall\text{-}tp_\mathcal{B}(\bar{b}d)$. Let us remark that since we do not know $\mathcal{A}$ and $\mathcal{B}$ are isomorphic yet, we do not know that ψ generates a $\forall$-type in $\mathcal{B}$.

---

[36] If Z is the least Turing degree of the set $\{Y \in 2^\mathbb{N} : X$ is c.e. in $Y\}$, then X would be enumeration equivalent to $Z \oplus Z^c$. This is because we have that, for all Y, X is c.e. in Y if and only if $Z \leq_T Y$, if and only if $Z \oplus Z^c$ is c.e. in Y, and thus the same Y's can enumerate both X and $Z \oplus Z^c$.

First, to show $\forall\text{-}tp_A(\bar{a}c) \subseteq \forall\text{-}tp_B(\bar{b}d)$, take $\theta(\bar{x}y) \in \forall\text{-}tp_A(\bar{a}c)$. Then

$$\text{``}\forall y(\psi(\bar{x}y) \to \theta(\bar{x}y))\text{''} \in \forall\text{-}tp_A(\bar{a}) = \forall\text{-}tp_B(\bar{b}),$$

and hence $\theta \in \forall\text{-}tp_B(\bar{b}d)$. Let us now prove the other inclusion. Let $\widetilde{\psi}(\bar{x}y)$ be the $\exists$-formula generating $\forall\text{-}tp_B(\bar{b}d)$. Then since $\neg\widetilde{\psi} \notin \forall\text{-}tp_B(\bar{b}d)$, by our previous argument, $\neg\widetilde{\psi} \notin \forall\text{-}tp_A(\bar{a}c)$ either, and hence $A \models \widetilde{\psi}(\bar{a}c)$. The rest of the proof that $\forall\text{-}tp_B(\bar{b}d) \subseteq \forall\text{-}tp_A(\bar{a}c)$ is now symmetrical to the one of the other inclusion: For $\theta(\bar{x}y) \in \forall\text{-}tp_B(\bar{b}d)$, we have that $\text{``}\forall y(\widetilde{\psi}(\bar{x}y) \to \widetilde{\theta}(\bar{x}y))\text{''} \in \forall\text{-}tp_B(\bar{a}c) = \forall\text{-}tp_A(\bar{a}c)$, and hence $\theta \in \forall\text{-}tp_A(\bar{a}c)$.   □

Proof of Theorem 3.8.6. The proof is immediate from Lemmas 3.8.7 and 3.8.8.   □

The following exercise gives a structural property that is sufficient for a structure to have enumeration degree. The property is far from necessary though.

Exercise 3.8.9. Suppose that a structure A has a Σ_3^c Scott sentence. Prove that A has enumeration degree. Hint in footnote.[37]

Example 3.8.10 (Frolov, Kalimullin, and Miller [FKM09]). Consider the class $\mathbb{K}$ of fields of finite transcendence degree over $\mathbb{Q}$. We claim that every such field has enumeration degree. This class is not Π_2^c, but if we consider $\mathbb{K}_n$ to be the class of fields of transcendence degree n, and add n constant symbols to name a transcendence basis, $v_1, \ldots, v_n$, then we do get a Π_2^c class. Since all these fields are algebraic over $\mathbb{Q}(v_1, \ldots, v_n)$, they are $\exists$-algebraic, and hence $\exists$-atomic. It then follows from Theorem 3.8.6 that every such field has enumeration degree, namely the enumeration degree of the $\exists$-type of a transcendence basis.

Conversely, we claim that for every set X, there is an algebraic field whose $\exists$-theory is enumeration-equivalent to X: Take the field that contains the p_nth roots of unity if and only if $n \in X$, where p_n is the nth prime number. From an enumeration of X, one can build such a field, and hence enumerate its $\exists$-theory, and conversely, the $\exists$-theory of that field can enumerate X.

Example 3.8.11 (Calvert, Harizanov, and Shlapentokh [CHS07]). Torsion-free abelian groups of finite rank always have enumeration degree. If we add a basis of the group as parameters, then the class of torsion-free abelian groups generated by such a basis is Π_2^c. All these groups are clearly $\exists$-algebraic and $\exists$-atomic, as every element is generated as a $\mathbb{Q}$-linear combination of the base. Thus, they have enumeration degree.

Furthermore, for every set X there is a torsion-free abelian group of rank one with enumeration degree X: Consider the subgroup of $\mathbb{Q}$ generated by $1/p_n$ for $n \in X$.

---

[37]Use Corollary 3.6.7.

EXAMPLE 3.8.12 (Steiner [Ste13]). Graphs of finite valence with finitely many connected components always have enumeration degree and can have all possible enumeration degrees: Let $\mathcal{G}$ be such a graph. Add a constant element for each connected component. Recall from Example 3.2.2 that, with the added constants, $\mathcal{G}$ becomes $\exists$-algebraic and hence $\exists$-atomic. Saying that every element is connected to one of these constants is Π_2^c. However, saying that $\mathcal{G}$ has finite valence is not. But the $\forall$-theory of $\mathcal{G}$ says that it has finite valence: for each constant element, and for each $k \in \mathbb{N}$, there is a $\forall$-formula that says that exists no more than a certain finite number of nodes at distance k from that constant. Since different $\exists$-atomic structures must have different $\exists$-theories, the isomorphism type of $\mathcal{G}$ is determined by the Π_2^c sentence saying every element is connected to one of the constants and its $\exists$-theory. It then follows from Lemma 3.8.7 that $\mathcal{G}$ has enumeration degree.

One can show that for every X there is a connected graph of finite valence and enumeration degree X. The graphs $\mathcal{G}_X$ from 3.8.3 are not connected, but a small modification would work: make all of these cycles sharing exactly one common node.

EXERCISE 3.8.13. Show that if a structure $\mathcal{A}$ has enumeration degree, that degree is the enumeration degree of some $\exists$-type of some tuple in $\mathcal{A}$.

EXERCISE 3.8.14. Show that if $\mathcal{A}$ is $\exists$-atomic and has enumeration degree, then its enumeration degree is given by its $\exists$-theory. Hint in footnote.[38]

EXERCISE 3.8.15. Give an example of a structure which has enumeration degree, but whose enumeration degree is not that of its $\exists$-theory.

---

[38]Show that every $\exists$-type is e-reducible to the $\exists$-theory of $\mathcal{A}$.

Chapter 4

GENERIC PRESENTATIONS

Forcing and generics are useful tools all over computability theory. The first forcing-style argument in computability theory can be traced back to the Kleene–Post construction of two incomparable degrees (cf. [KP54]), published a decade before the invention of forcing. In this chapter, we give an introduction to forcing in computable structure theory. We will develop a more general framework for forcing in [MonP2], once we gain more familiarity with infinitary languages. For now, instead of looking at fully generic objects, we consider 1-generics, which have relatively low computational complexity.

The notion of forcing was introduced by Cohen to prove that the continuum hypothesis does not follow from the ZFC axioms of set theory. Soon after, forcing became one of the main tools in set theory to prove independence results of all kinds. Generic objects are "generic" or "typical" in the sense that they do not have any property that is satisfied by a meager class of objects, where meagerness is viewed as a notion of smallness. This implies that if a generic satisfies a particular property, it must belong to a class where most objects have that property, and hence there is a clear reason why it has that the property. Our forcing arguments will essentially have that form: If a generic ω-presentation has a certain computational property, then there must be a structural reason for it.

Generic objects come in different shapes and sizes, but here, we will only consider Cohen generics. A *Cohen generic real* is a real in $\mathbb{N}^\mathbb{N}$ that does not belong to any meager set, where a subset of $\mathbb{N}^\mathbb{N}$ is *meager* if it is contained in a countable union of nowhere-dense closed sets, and a set is *nowhere dense* if it is not dense when restricted to any open set. Meager sets are considered to be *small sets*; for instance, Baire's category theorem states that no countable union of meager sets can cover all of $\mathbb{N}^\mathbb{N}$. If a real belongs to a particular meager set, belonging to this set would be a property of this real that most reals do not have. As we will see, the feature characterizing generics is the following: If $G \in \mathbb{N}^\mathbb{N}$ is generic, $P \subseteq \mathbb{N}^\mathbb{N}$ is a definable set viewed as a property, and $G \in P$, then there is a finite initial segment $\sigma \subseteq G$ which *forces* G to belong to P in the sense that every generic extending σ belongs to P. Behind this is Baire's theorem that says that if P is Borel, then there is an open set such that, restricted to that open set, P is either meager or co-meager. One problem that arises is that every real belongs to a meager set, namely the singleton that contains itself. That is why in set theory one has to work with generic reals that live outside the universe of sets. For the purposes of computability theory, we do not need to consider all meager sets, but only countably many of them. Since countable unions of meager sets are still meager, we can find object that are generic enough for our purposes.

We start this chapter by introducing 1-*generic reals*; these are the ones that avoid all nowhere-dense closed sets given as the boundaries of effectively open sets (Definition 4.1.1). The notion of 1-generic was isolated by Jockusch [Joc80], though Kleene–Post's construction in [KP54] already gives 1-generic reals 26 years earlier. See Exercise 4.1.8 below for a proof of Kleene–Post's result that every countable partial ordering embeds into the Turing degrees using 1-generics. They were then used in all kinds of embeddability results into the Turing degrees and other kinds of degrees. They are also often used in effective randomness and in reverse mathematics.

The objective of this chapter, though, is to introduce 1-*generic enumerations* and 1-*generic presentations* of structures. We will develop a more general notion of forcing and generics later in [MonP2], which is similar to the notion originally considered independently by Knight [Kni86], and by Manasse and Slaman (later published in [AKMS89]). For now, 1-generic presentations are enough for the results in this first part of the book. We will use them in the next chapter to prove Richter's theorem 5.1.9, Knight et al.'s theorem 5.3.1, Andrews and Miller's theorem 5.3.6, and other results later on.

4.1. Cohen generic reals

We review the standard notion of 1-genericity for reals and prove some of their basic properties. (For more background, see [Ler83, Section IV.2] or [Soa16, Section 6.3].) We will extend these proofs to generic enumerations of structures in the next sections.

For $R \subseteq \mathbb{N}^{<\mathbb{N}}$, define the *open subset of* $\mathbb{N}^{\mathbb{N}}$ *generated by* R to be

$$[R]^{\mathsf{C}} = \{X \in \mathbb{N}^{\mathbb{N}} : \exists \sigma \in R \ (\sigma \subset X)\}.$$

In other words, if $[\sigma]^{\mathsf{C}}$ denotes the clopen set of extensions of σ, namely $\{X \in \mathbb{N}^{\mathbb{N}} : \sigma \subset X\}$, then $[R]^{\mathsf{C}} = \bigcup_{\sigma \in R}[\sigma]^{\mathsf{C}}$. A subset of $\mathbb{N}^{\mathbb{N}}$ is *effectively open* if it is of the form $[R]^{\mathsf{C}}$ for some c.e. $R \subseteq \mathbb{N}^{<\mathbb{N}}$. A real $G \in \mathbb{N}^{\mathbb{N}}$ is 1-*generic* if and only if it avoids the boundaries of all effectively open sets. Thus, for every effectively open set, either G is well inside it or well outside it. Here is the equivalent definition we will actually use:

DEFINITION 4.1.1 (Jockusch [Joc80]). Let $R \subseteq \mathbb{N}^{<\mathbb{N}}$ be closed upwards, that is, if $\sigma \subseteq \tau$ and $\sigma \in R$, then $\tau \in R$ too. We say that a string $\gamma \in \mathbb{N}^{<\mathbb{N}}$ *decides* R if either $\gamma \in R$ or $\sigma \notin R$ for all $\sigma \supseteq \gamma$. If R is not closed upwards, we say that γ *decides* R if it decides its upward closure. A real $G \in \mathbb{N}^{\mathbb{N}}$ is 1-*generic* if for every upward-closed c.e. subset R of $\mathbb{N}^{<\mathbb{N}}$, there is an initial string of G, $G \upharpoonright k$ for some k, which decides R.

The reason we use the words "decide" and "force" is the following: Let G be 1-generic and $[R]^{\complement}$ be an effectively open set for R upward closed. For $\gamma \subset G$, if $\gamma \in R$, we say that γ *forces* G *to be in* $[R]^{\complement}$, while if $(\forall \sigma \supseteq \gamma)\ \sigma \notin R$, then we say that γ *forces* G *to be outside* $[R]^{\complement}$. In either case, γ *decides* whether G belongs to $[R]^{\complement}$ or not.

One can require more genericity by requiring G to decide more sets, e.g., α-generics decide all Σ_α^0 sets R, as we will see in [MonP2]. *Cohen generics* decide all sets R in the universe. We will not deal with these in this book.

Observation 4.1.2. 1-generic reals are not computable: For each computable $C \in \mathbb{N}^{\mathbb{N}}$, consider $R_C = \{\sigma \in \mathbb{N}^{<\mathbb{N}} : \sigma \not\subset C\}$. Since there is not enough room to force out of R_C in the sense that there is no $\gamma \in \mathbb{N}^{<\mathbb{N}}$ all whose extensions are outside R_C, any 1-generic must be forced to be in $[R_C]^{\complement}$ and be hence different from C.

LEMMA 4.1.3. *There is a 1-generic real computable from* $0'$.

PROOF. This is essentially an effective version of the Baire category theorem.

We build a 1-generic G as the union of an increasing sequence of finite strings $\bar{p}_0 \subseteq \bar{p}_1 \subseteq \cdots \in \mathbb{N}^{<\mathbb{N}}$. Let $\bar{p}_0$ be the empty string. At stage $s + 1 = e$, we define $\bar{p}_{s+1}$ so that it decides the eth c.e. set $W_e \subseteq \mathbb{N}^{<\mathbb{N}}$: If there is a $\bar{q} \supseteq \bar{p}_s$ with $\bar{q} \in W_e$, we let $\bar{p}_{s+1} = \bar{q}$. Otherwise, we let $\bar{p}_{s+1} = \bar{p}_s$. At the end of stages, we define $G = \bigcup_s \bar{p}_s$. It is not hard to check that G is 1-generic. (To see that the lengths of the $\bar{p}_s$'s go to infinity, notice that, for each n, the set $\{\sigma \in \mathbb{N}^{<\mathbb{N}} : |\sigma| \geq n\}$ is c.e. and hence is eventually considered during the construction as one of the W_e's.)

The only step in the construction that was not computable was checking whether there existed $\bar{q} \supseteq \bar{p}_s$ with $\bar{q} \in W_e$. This is a question $0'$ can answer, and hence the whole construction is computable in $0'$.   □

For the next lemma, we need to consider the relativized version of 1-genericity. Given $X \in \mathbb{N}^{\mathbb{N}}$, we say that $G \in \mathbb{N}^{\mathbb{N}}$ is X-1-*generic* if every X-c.e. subset of $\mathbb{N}^{<\mathbb{N}}$ is decided by an initial segment of G. The next lemma implies that the only sets that are c.e. in all generic sets are the ones that are already c.e.

LEMMA 4.1.4. *Let* $G, X \in \mathbb{N}^{\mathbb{N}}$. *Suppose that* G *is* X-1-*generic. Then* X *is not c.e. in* G, *unless* X *is c.e. already.*

PROOF. Suppose that $X = W_e^G$ for some $e \in \mathbb{N}$; we will show that X is already c.e. Consider the set of strings which "force '$W_e^G \not\subseteq X$'," by which we mean:

$$Q = \{\bar{q} \in \mathbb{N}^{<\mathbb{N}} : \exists n \, (n \in W_e^{\bar{q}} \wedge n \notin X)\}.$$

Notice that Q is c.e. in X, and hence it is decided by some initial segment of G, say by $G \upharpoonright k$. If we had $G \upharpoonright k \in Q$, we would get $n \in W_e^G$ and $n \notin X$, contradicting our assumption that $X = W_e^G$. Thus, $G \upharpoonright k$ must force out of $[Q]^{\complement}$ and no extension of $G \upharpoonright k$ is in Q.

We now claim that

$$X = \{n \in \mathbb{N} : (\exists \bar{q} \supseteq G \upharpoonright k) \, n \in W_e^{\bar{q}}\}.$$

Notice that this would show that X is c.e. as needed. As for the claim: For the left-to-right inclusion, if $n \in X$, since $X = W_e^G$, there is some initial segment $\bar{q}$ of G satisfying $n \in W_e^{\bar{q}}$. For the other inclusion, if there exists $\bar{q} \supseteq G \upharpoonright k$ with $n \in W_e^{\bar{q}}$, then n must belong to X as otherwise $\bar{q}$ would be an extension of $G \upharpoonright k$ in Q. □

In particular, we get that if G is X-1-generic, then G computes X if and only if X is computable.[39] Thus, if G is X-1-generic, G and X form a *minimal pair*, i.e., there is no non-computable set computable from both. This is because if $Y \leq_T X$, then G is Y-1-generic too, so if also $Y \leq_T G$, Y must be computable.

The following lemma shows that 1-generics do not code much information on their jumps: For $Z \in \mathbb{N}^{\mathbb{N}}$, basic properties of the Turing jump imply that $Z' \geq_T Z \oplus 0'$. We say that Z is *generalized low* if $Z' \equiv_T Z \oplus 0'$.

LEMMA 4.1.5. *Every 1-generic real G is* generalized low.

PROOF. That $G' \geq_T G \oplus 0'$ is true for all reals G. Let us prove that $G' \leq_T G \oplus 0'$. Take $e \in \mathbb{N}$; we want to decide if $e \in G'$, that is, if $\Phi_e^G(e)\downarrow$ using $G \oplus 0'$ as oracle, uniformly in e. Consider the set

$$R_e = \{\bar{q} \in \mathbb{N}^{<\mathbb{N}} : \Phi_e^{\bar{q}}(e)\downarrow\}.$$

Since R_e is c.e., it is decided by G. Notice that $0'$ can tell if a string γ forces in (which is a Σ_1^0 question), forces out (which is a Π_1^0 question), or does not decide $[R_e]^{\complement}$. Then, using $G \oplus 0'$, we can find $k \in \mathbb{N}$ such that $G \upharpoonright k$ decides R_e. If $G \upharpoonright k \in R_e$, we know that $\Phi_e^G(e)\downarrow$ and hence $e \in G'$. If no extension of $G \upharpoonright k$ is in R_e, then $\Phi_e^G(e)\uparrow$ and hence $e \notin G'$. □

An important application of 1-generics is Friedberg's jump inversion theorem which implies that every Turing degree above $0'$ is the jump of some degree.

THEOREM 4.1.6 (Friedberg's jump inversion theorem [Fri57a]). *For every $A \in \mathbb{N}^{\mathbb{N}}$ with $A \geq_T 0'$, there is a 1-generic G such that*

$$A \equiv_T G' \equiv_T G \oplus 0'.$$

---

[39] Here we are using that computable is equivalent to c.e. and co-c.e.

PROOF. We follow the construction of a 1-generic computable from $0'$ (Lemma 4.1.3) but with extra steps to encode A into G.

We build G as the union of an increasing sequence of finite strings $\bar{p}_0 \subseteq \bar{p}_1 \subseteq \cdots \in \mathbb{N}^{<\mathbb{N}}$. Let $\bar{p}_0$ be the empty string. At an odd stage $s + 1 = 2e + 1$ we define $\bar{p}_{s+1}$ so that it decides the eth c.e. set $W_e \subseteq \mathbb{N}^{<\mathbb{N}}$: Ask $0'$ if there is a $\bar{q} \supseteq \bar{p}_s$ with $\bar{q} \in W_e$. If yes, let $\bar{p}_{s+1}$ be the least such $\bar{q}$ according to some enumeration of $\mathbb{N}^{<\mathbb{N}}$, insisting that $\bar{q}$ was least was not necessary in Lemma 4.1.3. If no, let $\bar{p}_{s+1} = \bar{p}_s$. At an even stage $s + 1 = 2e$, we let $\bar{p}_{s+1} = \bar{p}_s {}^\frown A(e)$.

Since A computes $0'$, A can run the construction and thus $G \leq_T A$. Conversely, $G \oplus 0'$ can recover all the steps of the construction and recover the sequence $\bar{p}_0 \subseteq \bar{p}_1 \subseteq \cdots$: This is because $0'$ can figure out how $\bar{p}_{s+1}$ was defined at odd stages—using that $\bar{q}$ was chosen to be least—and, at even stages, $\bar{p}_{s+1}$ is just one bit longer than $\bar{p}_s$, i.e., $\bar{p}_{s+1} = G \upharpoonright |\bar{p}_s| + 1$. We can then recover A, as $A(e)$ is the last entry of $\bar{p}_{2e}$. This shows that $A \equiv_T G \oplus 0'$. Since we made G 1-generic, $G' \equiv_T G \oplus 0'$.   □

The following lemma shows that if we split a 1-generic in two pieces, we get two 1-generics. Furthermore, the pieces are 1-generic relative to each other.

LEMMA 4.1.7. *Let* $G, H \in \mathbb{N}^{\mathbb{N}}$. *Then* $G \oplus H$ *is* 1-*generic if and only if* G *is* 1-*generic and* H *is* G-1-*generic.*

This is essentially an effective version of the Kuratowski-Ulam Theorem, that says that a set $P \subseteq \mathbb{N}^{\mathbb{N}} \times \mathbb{N}^{\mathbb{N}}$ is comeager if and only if the set $\{x \in \mathbb{N}^{\mathbb{N}} : P \cap \{x\} \times \mathbb{N}^{\mathbb{N}}$ is comeager in $\{x\} \times \mathbb{N}^{\mathbb{N}}\}$ is comeager.

PROOF. Suppose first that $G \oplus H$ is 1-generic. Consider a c.e. operator W which outputs subsets of $\mathbb{N}^{<\mathbb{N}}$. To prove that H is G-1-generic, we need to show that H decides W^G using the genericity of $G \oplus H$. Consider the c.e. set of pairs of strings that "force $H \in [W^G]^{\complement}$," by which we mean:

$$R = \{\gamma \oplus \delta \in \mathbb{N}^{<\mathbb{N}} : \delta \in W^\gamma\}.$$

$G \oplus H$ must decide R. If we have $\gamma \oplus \delta \subset G \oplus H$ with $\gamma \oplus \delta \in R$, then $\delta \in W^G$ and H is forced into $[W^G]^{\complement}$. If we have that $(\forall \tau \supseteq \gamma \oplus \delta)\, \tau \notin R$, then $(\forall \sigma \supseteq \delta)\, \sigma \notin W^G$ and H is forced out of $[W^G]^{\complement}$.

In exactly the same way we can show that G is H-1-generic, and in particular 1-generic.

For the other direction, suppose G is 1-generic and H is G-1-generic. Let R be an upwards-closed c.e. subset of $\mathbb{N}^{<\mathbb{N}}$; we must prove that $G \oplus H$ decides it. Define

$$S_1 = \{\delta \in \mathbb{N}^{<\mathbb{N}} : (G \upharpoonright |\delta|) \oplus \delta \in R\}.$$

S_1 is c.e. in G and thus H must decide it. If there is a $\delta_1 \subset H$ with $\delta \in S_1$, then $G \upharpoonright |\delta_1| \oplus \delta_1$ forces $G \oplus H$ to be in $[R]^{\complement}$. Otherwise, suppose there is

$\delta_1 \subset H$ no extension of which is in S_1. Define

$$S_0 = \{\gamma \in \mathbb{N}^{<\mathbb{N}} : \exists \delta \in \mathbb{N}^{|\gamma|} \ (\delta \supseteq \delta_1 \ \& \ \gamma \oplus \delta \in R)\}.$$

G must decide S_0. There cannot be a $\gamma \subseteq G$ with $\gamma \in S_0$, because the witness δ would be an extension of δ_1 in S_1. So, there is a $\gamma_1 \subseteq G$ no extension of which is in S_0. Thus, for $\gamma \supseteq \gamma_1$ and $\delta \supseteq \delta_1$, $\gamma \oplus \delta \notin R$. We get that $\gamma_1 \oplus (H \upharpoonright |\gamma_1|)$ forces $G \oplus H$ out of $[R]^{\complement}$. $\square$

Such H and G are said to be *mutually generic*. Similarly, we can get an infinite sequence of mutually generic reals by taking the columns $\{G^{[n]} : n \in \mathbb{N}\}$ of a 1-generic G.

EXERCISE 4.1.8. Prove Kleene–Post's theorem that every countable partial ordering embeds into the Turing degrees. To prove it, given a partial ordering $(P; \preccurlyeq)$, consider a bijection $f : P \times \mathbb{N} \to \mathbb{N}$, and consider the pull-back $H = f^{-1}(G)$ of a 1-generic real $G \subseteq \mathbb{N}$. Show that the map $p \mapsto \bigoplus_{q \preccurlyeq p} H^{[q]}$ from P to $\mathbb{N}^{\mathbb{N}}$ induces the desired embedding.

EXERCISE 4.1.9. Prove that the countable atomless Boolean algebra embeds into the Turing degrees preserving joins and meets. Hint in footnote.[40]

EXERCISE 4.1.10. Prove that given a real $A \geq_T 0'$, there exist reals G and H such that $G' \equiv_T A \equiv_T H'$, and G and H form a minimal pair, meaning that there is no non-computable set computable from both G and H.

4.2. Generic enumerations of sets

Before diving into generic enumerations of structures, let us take a quick look at generic enumerations of sets and give a proof of Selman's theorem about enumeration reducibility. (See page xviii.) Recall that an *enumeration* of a set Z is nothing more than a function $g \colon \mathbb{N} \to Z$ that is onto Z. We say that $g \in Z^{\mathbb{N}}$ is a 1-generic enumeration of Z if for every subset R of $Z^{<\mathbb{N}}$ that is enumeration reducible to Z, there is an initial segment of g, $g \upharpoonright k$ for some k, that decides it in the sense that either $g \upharpoonright k \in R$ or no extension of $g \upharpoonright k$ is in R. Notice that a 1-generic enumeration of Z must be onto Z, because for each $z \in Z$, the set $R = \{\sigma \in Z^{<\mathbb{N}} : \exists i < |\sigma| \ \sigma(i) = z\}$ is dense and enumeration reducible to Z and hence must be forced in. We relativize this notion in the obvious way: $g \in Z^{\mathbb{N}}$ is an X-1-generic enumeration of Z if for every

---

[40]Consider a 1-generic subset H of $\mathbb{Q}$ and then map an element a of the interval algebra of $\mathbb{Q}$ to $a \cap H$.

subset R of $Z^{<\mathbb{N}}$ that is X-enumeration reducible to Z,[41] there is an initial segment of g that decides it.

The next lemma implies that the only sets that are c.e. in all generic enumerations of Z are the ones that are already enumeration reducible to Z. It is the analog of Lemma 4.1.4, and the proof is almost the same verbatim.

LEMMA 4.2.1. *Consider sets $Z \subseteq \mathbb{N}$ and $X \subseteq \mathbb{N}$. Suppose that g is an X-1-generic enumeration of Z. Then X is not c.e. in g, unless X is enumeration reducible to Z.*

PROOF. Suppose that $X = W_e^g$ for some $e \in \mathbb{N}$; we will show that $X \le_e Z$. Consider the set of strings which "force '$W_e^g \not\subseteq X$'," by which we mean:

$$Q = \{\bar{q} \in Z^{<\mathbb{N}} : \exists n \, (n \in W_e^{\bar{q}} \wedge n \notin X)\}.$$

Notice that Q is X-enumeration reducible to Z, and hence it is decided by some initial segment of g, say by $g \upharpoonright k$. If we had $g \upharpoonright k \in Q$, we would get $n \in W_e^g$ and $n \notin X$, contradicting our assumption that $X = W_e^g$. Thus, $g \upharpoonright k$ must force out of $[Q]^C$ and we have that no extension of $g \upharpoonright k$ is in Q.

We now claim that

$$X = \{n \in \mathbb{N} : (\exists \bar{q} \in Z^{<\mathbb{N}}) \, \bar{q} \supseteq g \upharpoonright k \, \& \, n \in W_e^{\bar{q}}\}.$$

If $n \in X$, then, since $X = W_e^g$, there is some initial segment $\bar{q}$ of g satisfying $n \in W_e^{\bar{q}}$. For the other inclusion, if there exists $\bar{q} \in Z^{<\mathbb{N}}$ with $\bar{q} \supseteq g \upharpoonright k$ and $n \in W_e^{\bar{q}}$, then n must belong to X as otherwise $\bar{q}$ would be an extension of $g \upharpoonright k$ in Q.

Now, let us observe that the claim implies that $X \le_e Z$ as needed. Define the following c.e. enumeration operator:

$$\Theta = \{\langle \ulcorner D \urcorner, n \rangle : D \subset_{\text{fin}} \mathbb{N}, n \in \mathbb{N}, (\exists \bar{q} \in D^{<\mathbb{N}}) \, \bar{q} \supseteq g \upharpoonright k \, \& \, n \in W_e^{\bar{q}}\}.$$

From the claim above we get that $X = \Theta^Z$. □

Selman's theorem proves the equivalence between the different definitions of enumeration reducibility:

THEOREM 4.2.2 (Selman [Sel71]). *Let Y, Z be subsets of $\mathbb{N}$. The following are equivalent:*

1. *Every enumeration of Z computes an enumeration of Y.*
2. *There is a single computable operator that maps every enumeration of Z into an enumeration of Y.*
3. *There is a c.e. enumeration operator Θ such that $Y = \Theta^Z$.*

PROOF. The implications (3) $\Longrightarrow$ (2) and (2) $\Longrightarrow$ (1) are quite straightforward. The interesting direction is (1) $\Longrightarrow$ (3). For this, let g by a Y-1-generic enumeration of Z. By (1), Y is c.e. in g, and hence from the

---

[41] We say that R is *X-enumeration reducible* to Z if there is an X-c.e. enumeration operator Θ such that $R = \Theta^Z$, where $\Theta^Z = \{n : (\exists D \subseteq_{\text{fin}} Z) \, \langle \ulcorner D \urcorner, n \rangle \in \Theta\}$.

proof of the previous lemma we get a c.e. enumeration operator Θ such that $Y = \Theta^Z$. $\Box$

4.3. Generic enumerations of structures

We now turn to consider 1-generic enumerations of structures. The main difference with 1-generic reals is that, instead of deciding the c.e. subsets of $\mathbb{N}^{<\mathbb{N}}$, we now decide the r.i.c.e. subsets of $A^{<\mathbb{N}}$.

We assume throughout the rest of the chapter that $\mathcal{A}$ is an ω-presentation of a τ-structure, and, of course, that τ is a computable vocabulary. Given a set A, let $A^\star$ be the set of all finite strings from A whose entries are all different:

$$A^\star = \{\sigma \in A^{<\mathbb{N}} : (\forall i \neq j < |\sigma|)\, \sigma(i) \neq \sigma(j)\}.$$

DEFINITION 4.3.1. We say that $\gamma \in A^\star$ *decides* an upward-closed subset $R \subseteq A^\star$ if either $\gamma \in R$ or $\sigma \notin R$ for all $\sigma \supseteq \gamma$. We say that a one-to-one function $g \in A^{\mathbb{N}}$ is a 1-*generic enumeration of* $\mathcal{A}$ if, for every r.i.c.e. set $R \subseteq A^\star$, there is an initial segment of g that decides R.[42]

The existence of 1-generic enumerations follows from the Baire category theorem. As in Lemma 4.1.3, we can build a 1-generic enumeration of $\mathcal{A}$ computably in $D(\mathcal{A})'$ by finite approximations deciding all $D(\mathcal{A})$-c.e. sets, and hence all r.i.c.e. subsets of $A^{<\mathbb{N}} = \mathbb{N}^{<\mathbb{N}}$. Since we only need to decide the r.i.c.e. sets, we can do this with less than $D(\mathcal{A})'$: The lemma below says that $\vec{K}^{\mathcal{A}}$ is enough. See 2.2.3 for the definition of the complete r.i.c.e. set $\vec{K}^{\mathcal{A}}$, and recall that we always have $\vec{K}^{\mathcal{A}} \leq_T D(\mathcal{A})'$, and that sometimes $\vec{K}^{\mathcal{A}} <_T D(\mathcal{A})'$ (Exercise 2.2.6).

LEMMA 4.3.2. *Every ω-presentation $\mathcal{A}$ has a 1-generic enumeration computable in $\vec{K}^{\mathcal{A}}$.*

PROOF. We build g as the union of a strictly increasing sequence $\{\bar{p}_s : s \in \mathbb{N}\}$ with $\bar{p}_s \in A^\star$. Using Remark 2.2.5 we get a $D(\mathcal{A})$-computable enumeration $\{R_0, R_1, \dots\}$ of the r.i.c.e. subsets of $A^\star$. At stage $s + 1 = e$, we define $\bar{p}_{s+1}$ to decide the eth r.i.c.e. set $R_e \subseteq A^\star$ as follows: If there is a $\bar{q} \supseteq \bar{p}_s$ with $\bar{q} \in R_e$, we let $\bar{p}_{s+1} = \bar{q}$. Otherwise, we let $\bar{p}_{s+1} = \bar{p}_s$. Finally, we let $g = \bigcup_s \bar{p}_s \in A^{\mathbb{N}}$. It is not hard to check that g is one-to-one and 1-generic.

To carry on this construction, we need to check at each stage $s + 1$ whether there exists $\bar{q} \supseteq \bar{p}_s$ with $\bar{q} \in R_e$ or not. The set of $\bar{p}$'s such that $\exists \bar{q} \supseteq \bar{p}\ (\bar{q} \in R_e)$, namely the downward closure of R_e, is Σ_1^c-definable and its index can be obtained uniformly from e. Hence, $\vec{K}^{\mathcal{A}}$ can decide

---

[42]Let us remind the reader that we identify tuples $\gamma \in A^{<\mathbb{N}}$ with functions $\gamma : \{0, \dots, |\gamma| - 1\} \to A$. For instance, when we say that a tuple is an initial segment of a function $g : \mathbb{N} \to A$, we are viewing the tuple as a function.

whether $\bar{p}_s$ belongs to the downward closure of R_e or not, and thus, the whole construction is computable in $\vec{K}^{\mathcal{A}}$. □

It is not hard to see that a 1-generic enumeration must be onto: Given $a \in A$, the set $\{\bar{p} \in A^\star : \bigvee_{i<|\bar{p}|} \bar{p}(i) = a\}$ is r.i.c.e. and *dense*, in the sense that every $\bar{q} \in A^\star$ has a extension in it. Thus a 1-generic enumeration must force into it. 1-generic enumerations are then bijections between $\mathbb{N}$ and A. Using the pull-back (see Section 1.1.7), each 1-generic enumeration induces what we call a 1-generic presentation:

DEFINITION 4.3.3. A 1-*generic presentation* of $\mathcal{A}$ is the pull-back $g^{-1}(\mathcal{A})$ of some 1-generic enumeration g of $\mathcal{A}$.

The reason we defined 1-generic enumerations of $\mathcal{A}$ using r.i.c.e. sets, instead of $D(\mathcal{A})$-c.e. sets, is that we get a notion that is independent of the given ω-presentation of $\mathcal{A}$:

LEMMA 4.3.4. *Let* $\mathcal{A}$ *and* $\mathcal{B}$ *be isomorphic. Any* 1-*generic presentation of* $\mathcal{A}$ *is also a* 1-*generic presentation of* $\mathcal{B}$.

PROOF. Let $h: \mathcal{A} \to \mathcal{B}$ be an isomorphism. The key point is that h preserves Σ_1^c-definable sets.

Suppose that $g: \mathbb{N} \to A$ is a 1-generic enumeration of $\mathcal{A}$, and let $\mathcal{C} = g^{-1}(\mathcal{A})$. We want to show that $\mathcal{C}$ is a 1-generic presentation of $\mathcal{B}$ too. Since $\mathcal{C} = (h \circ g)^{-1}(\mathcal{B})$, it is enough to show that $h \circ g$ is a 1-generic enumeration of $\mathcal{B}$. Let $R \subseteq B^\star$ be Σ_1^c-definable in $\mathcal{B}$ with parameters; we need to show that $h \circ g$ decides it. Since h is an isomorphism, $h^{-1}(R) \subseteq A^\star$ is Σ_1^c-definable in $\mathcal{A}$ with parameters, and hence decided by g. Let $k \in \mathbb{N}$ be such that either $g \upharpoonright k \in h^{-1}(R)$ or $\sigma \notin h^{-1}(R)$ for all $\sigma \in A^\star$ with $\sigma \supseteq g \upharpoonright k$. Applying h, we get that $(h \circ g) \upharpoonright k$ decides R, as wanted. □

Remark 4.3.5. In particular, a 1-generic presentation of a structure $\mathcal{A}$ is also a 1-generic presentation of itself. An ω-presentation $\mathcal{C}$ is a 1-generic presentation if and only if every r.i.c.e. set $R \subseteq C^\star = \mathbb{N}^\star$ is decided by some tuple of the form $\langle 0, 1, \ldots, k-1 \rangle$.

EXERCISE 4.3.6. Suppose the vocabulary τ consists of only one unary relation symbol R and $\mathcal{A}$ is a τ-structure where $\mathsf{R}^{\mathcal{A}}$ is infinite and co-infinite. Prove an ω-presentation $\mathcal{C}$ of $\mathcal{A}$ is a 1-generic presentation if and only if the characteristic function of $\mathsf{R}^{\mathcal{C}}$, viewed as a real in $2^{\mathbb{N}}$, is a 1-generic real.

EXERCISE 4.3.7. Let X compute a copy of a structure $\mathcal{A}$ and let G be X-1-generic. Prove that $X \oplus G$ can compute a 1-generic presentation of $\mathcal{A}$.

4.4. Relations on generic presentations

Generic presentations are useful because whatever happens to them happens for a reason. For instance, we will see that if a relation is c.e. on

a generic presentation, it is because it was r.i.c.e. already (assuming the
ω-presentation is generic relative to the relation too). In Theorem 2.1.16,
we showed that a relation $R \subseteq A^{<\mathbb{N}}$ is Σ_1^c-definable with parameters if and
only if $R^{\mathcal{B}}$ is c.e. in $D(\mathcal{B})$ for every $(\mathcal{B}, R^{\mathcal{B}}) \cong (\mathcal{A}, R)$ (i.e., it is r.i.c.e.).
The following theorem, which is the analogue of Lemma 4.1.4, shows that
we do not need to consider all the copies of $(\mathcal{A}, R)$, but just one that is
1-generic. The construction in the proof of Ash, Knight, Manasse, Slaman,
and Chisholm's Theorem 2.1.16 (cf. [AKMS89, Chi90]) essentially already
builds a 1-generic copy of $(\mathcal{A}, R)$.

THEOREM 4.4.1. *Let $\mathcal{A}$ be a structure and $R \subseteq A^{<\mathbb{N}}$. Suppose $(\mathcal{A}, R)$ is a
1-generic presentation. Then R is c.e. in $D(\mathcal{A})$ if and only if R is r.i.c.e.*[43]

PROOF. Clearly, if R is r.i.c.e. it is c.e. in $D(\mathcal{A})$. Let us prove the other
direction.

Suppose that $R = W_e^{D(\mathcal{A})}$ for some $e \in \mathbb{N}$. Consider the same set used
in the proof of Theorem 2.1.16, in which we were trying to build a copy $\mathcal{C}$
of $\mathcal{A}$ satisfying $W_e^{D(\mathcal{C})} \not\subseteq R^{\mathcal{C}}$:

$$Q = \{\bar{q} \in A^\star : \exists \ell, j_1, \dots, j_\ell < |\bar{q}|(\langle j_1, \dots, j_\ell \rangle \in W_e^{D_{\mathcal{A}}(\bar{q})} \text{ and}$$
$$\langle q_{j_1}, \dots, q_{j_\ell} \rangle \notin R)\}.$$

It is not hard to see that Q is r.i.c.e. in $(\mathcal{A}, R)$. So Q is decided by some
tuple of the form $\langle 0, \dots, k-1 \rangle \in A^\star$ as in Remark 4.3.5 above. We cannot
have $\langle 0, \dots, k-1 \rangle \in Q$, as otherwise there would be a tuple $\langle j_1, \dots, j_\ell \rangle \in$
$W_e^{D(\mathcal{A})}$ with $\langle j_1, \dots, j_\ell \rangle \notin R$, contradicting that $R = W_e^{D(\mathcal{A})}$. Thus, no
extension of $\langle 0, \dots, k-1 \rangle$ is in Q. Let $\bar{p} = \langle 0, \dots, k-1 \rangle$. It now follows
from Claim 1 from the proof of Theorem 2.1.16 that R is Σ_1^c-definable in
$\mathcal{A}$ with parameters $\bar{p}$ as needed. To be more explicit, recall that the proof
of Claim 1 went through proving that

$$R = \{\langle q_{j_1}, \dots, q_{j_\ell} \rangle : \text{ for } \bar{q} \in A^\star \text{ and } \ell, j_1, \dots, j_\ell < |\bar{q}|,$$
$$\text{with } \bar{q} \supseteq \bar{p} \text{ and } \langle j_1, \dots, j_\ell \rangle \in W_e^{D_{\mathcal{A}}(\bar{q})}\}. \qquad \square$$

Recall that a set $X \subseteq \mathbb{N}$ is c.e.-coded by $\mathcal{A}$ if and only if it is c.e. in every
presentation of $\mathcal{A}$ (see Subsection 2.1.4). This is equivalent to saying that
X is r.i.c.e. in $\mathcal{A}$. Recall that we can view X as a subset of $\mathbb{N} \times A^{<\mathbb{N}}$ as in
Section 2.1.4. Let us also remark that saying that $(\mathcal{A}, X)$ is a 1-generic
presentation is equivalent to saying that $\mathcal{A}$ is X-1-generic, as the r.i.c.e.
relations on $(\mathcal{A}, X)$ are exactly the X-r.i.c.e. relations on $\mathcal{A}$, which are the
ones that need to be decided by the X-1-generic enumeration of $\mathcal{A}$.[44]

---

[43]The same is true of relations $R \subseteq \mathbb{N} \times A^{<\mathbb{N}}$. By Remark 2.1.28 it is enough to consider
$R \subseteq A^{<\mathbb{N}}$.

[44]A relation $R \subseteq \mathbb{N} \times A^{<\mathbb{N}}$ is X-r.i.c.e. if it is c.e. in $D(\mathcal{B}) \oplus X$ for all copies $\mathcal{B}$ of $\mathcal{A}$. A
one-to-one enumeration g of $\mathcal{A}$ is X-1-*generic* if, for every X-r.i.c.e. set $R \subseteq A^\star$, there is an
initial segment of g that decides it.

COROLLARY 4.4.2. *Let $X \subseteq \mathbb{N}$ and suppose $\mathcal{A}$ is a X-1-generic presenta-tion. Then X is c.e. in $D(\mathcal{A})$ if and only if it is c.e.-coded by $\mathcal{A}$.*

PROOF. Immediate from the previous theorem.   □

COROLLARY 4.4.3. *For every ω-presentation $\mathcal{B}$, there is another ω-presen-tation $\mathcal{A} \cong \mathcal{B}$ such that, a set $X \subseteq \mathbb{N}$ is c.e.-coded by $\mathcal{B}$ if and only if it is c.e. in both $D(\mathcal{A})$ and $D(\mathcal{B})$.*

PROOF. If X is c.e.-coded by $\mathcal{B}$, by definition it is c.e. in $D(\mathcal{B})$ and in $D(\mathcal{A})$ for every copy $\mathcal{A}$ of $\mathcal{B}$.

For the other direction, let $Y = D(\mathcal{B})'$ and let $\mathcal{A}$ be Y-1-generic presentation of $\mathcal{B}$. If X is c.e. in $D(\mathcal{B})$, then $X \leq_T Y$ and hence $\mathcal{A}$ is X-1-generic too. If X is also c.e. in $D(\mathcal{A})$, by the previous corollary, X is c.e.-coded by $\mathcal{A}$ and hence also by $\mathcal{B}$.   □

The next lemma is the analogue of Lemma 4.1.5 that says that 1-generics are generalized low. Recall that always $\vec{K}^{\mathcal{B}} \leq_T D(\mathcal{B})$ and there are ω-presentations $\mathcal{B}$ with $\vec{K}^{\mathcal{B}} <_T D(\mathcal{B})'$ (Exercise 2.2.6).

LEMMA 4.4.4 (Vatev [Vat11]). *If $\mathcal{B}$ is 1-generic, then $D(\mathcal{B})' \equiv_T \vec{K}^{\mathcal{B}}$.*

PROOF. We already know that $\vec{K}^{\mathcal{B}} \leq_T D(\mathcal{B})'$ for every presentation $\mathcal{B}$. Let us prove that $D(\mathcal{B})' \leq_T \vec{K}^{\mathcal{B}}$. Take $e \in \mathbb{N}$; we want to decide if $\Phi_e^{D(\mathcal{B})}(e)\!\downarrow$ using $\vec{K}^{\mathcal{B}}$ as an oracle uniformly in e. Consider the set

$$R_e = \{\bar{q} \in B^{\star} : \Phi_e^{D_{\mathcal{B}}(\bar{q})}(e)\!\downarrow\}.$$

It is not hard to see that R_e is r.i.c.e. The set of tuples which force into $[R_e]^{\subseteq}$, namely

$$\{\bar{p} \in B^{\star} : \exists \bar{q} \subseteq \bar{p} \; (\bar{q} \in R_e)\},$$

is r.i.c.e.. The set of tuples which force out of $[R_e]^{\subseteq}$, namely

$$\{\bar{p} \in B^{\star} : \forall \bar{q} \supseteq \bar{p} \; (\bar{q} \notin R_e)\},$$

is co-r.i.c.e.. Since $\mathcal{B}$ is 1-generic, R_e is decided by some tuple of the form $\langle 0, \dots, k-1 \rangle$ as in Remark 4.3.5. Using $\vec{K}^{\mathcal{B}}$, we can then find such a k and decide whether $\langle 0, \dots, k-1 \rangle$ forces into $[R_e]^{\subseteq}$ and hence that $e \in D(\mathcal{B})'$, or $\langle 0, \dots, k-1 \rangle$ forces out of $[R_e]^{\subseteq}$ and hence that $e \notin D(\mathcal{B})'$.   □

Vatev [Vat11] used this lemma to give the first proof that $\vec{K}^{\mathcal{B}}$ is never r.i. computable in $\mathcal{B}$.

Chapter 5

DEGREE SPECTRA

Among the main objectives of computable structure theory is measuring the computational complexity of structures. There are various ways of doing this. The most common one is through degree spectra.

We already know how to assign a Turing degree to an ω-presentation (namely $D(\mathcal{A})$, as in Subsection 1.1.1), but a structure may have many ω-presentations with different Turing degrees. We want a measure of complexity that is independent of the particular ω-presentation.

DEFINITION 5.0.5. The *degree spectrum* of a structure $\mathcal{M}$ is the set

$$DgSp(\mathcal{M}) = \{X \in 2^{\mathbb{N}} : X \text{ computes a copy of } \mathcal{M}\}.$$

Degree spectra are closed upward under Turing reduction, and in particular under Turing equivalence. Thus, we can think of them as sets of Turing degrees rather than sets of reals. As it follows from Knight's Theorem 1.2.1, $DgSp(\mathcal{A})$ is the set of Turing degrees of all the copies of $\mathcal{A}$, provided $\mathcal{A}$ is non-trivial.

Understanding which subsets of the Turing degrees can be realized as degree spectra is an important open problem in the area.

5.1. The c.e. embeddability condition

Our first approach to measuring the complexity of a structure was by assigning it an enumeration degree (Definition 3.8.2), if possible. Linda Richter showed in her Ph.D. thesis [Ric77] that there are many structures that do not have enumeration degree. She gave a general sufficient condition for this to happen:

DEFINITION 5.1.1 (Richter [Ric81, Section 3]). A structure $\mathcal{A}$ has the *computable embeddability condition* if each $\exists$-type realized in $\mathcal{A}$ is computable. A structure $\mathcal{A}$ has the *c.e. embeddability condition* if each $\exists$-type realized in $\mathcal{A}$ is c.e.[45]

---

[45]Recall that the $\exists$-type of a tuple is the set indices of all $\exists$-formulas true about it.

The reason Richter introduced this notion was to prove Theorem 5.1.6 and Corollary 5.1.8 below.

EXAMPLE 5.1.2. Richter then showed that linear orderings have the computable embeddability condition. This is because the $\exists$-type of a tuple $a_1, \ldots, a_k$, or equivalently the set of finite extensions of $\bar{a}$, namely the finite $\bar{a}$-linear orderings as in Example 2.3.2, is determined by the ordering among the elements of the tuple $\bar{a}$, how many elements are in between each pair from the tuple, how many elements are to the left of the whole tuple, and how many are to the right. By "how many," we mean either a finite number or infinity. Thus, the $\exists$-type of a k-tuple is determined by a permutation σ of $\{1, 2, \ldots, k\}$, and a $k + 1$ tuple from $\mathbb{N} \cup \{\infty\}$, in the sense that given that information, one can computably decide if an $\exists$-formula belongs to the type or not.

EXAMPLE 5.1.3. Richter also showed that trees when viewed as partial orderings (i.e., as $\{\leq\}$-structures) also have the computable embeddability condition. We defer this proof to Chapter 10, where we will prove that the class of trees is Σ-small.

Other examples include Boolean algebras, $\mathbb{Q}$-vector spaces, algebraically closed fields, differentially closed fields, abelian p-groups, and equivalence structures.

Historical Remark 5.1.4. Richter's original definition was not in terms of types, but in terms of finite structures embeddable in $\mathcal{A}$ and extending a fixed tuple, as in the following exercise. She defined the computable embeddability condition and not the c.e. one. However, Theorem 5.1.6 below has a more rounded statement when we consider the latter notion. In Russia, structures with the c.e. embeddability condition are called *locally constructivizable*.

EXERCISE 5.1.5. For each tuple $\bar{a} \in A^{<\mathbb{N}}$, prove that the set

$$\{D_A(\bar{a}\bar{b}) : \bar{b} \in A^{<\mathbb{N}}\} \subseteq 2^{<\mathbb{N}}$$

is positive equivalent to $\exists\text{-}tp_A(\bar{a})$. In particular, they are both Turing and enumeration equivalent. (For the definition of positive reducibility, see page xviii.)

THEOREM 5.1.6 (Richter). *Let $\mathcal{A}$ be any structure. The following are equivalent*:

1. $\mathcal{A}$ *has the c.e. embeddability condition.*
2. *Every set $X \subseteq \mathbb{N}$ that c.e.-coded by $\mathcal{A}$ is already c.e.*

Recall from Section 2.1.4 that X is c.e.-coded by $\mathcal{A}$ if X is c.e. in every presentation of $\mathcal{A}$.

PROOF. To show that (1) implies (2), recall Knight's Theorem 2.1.24 that if X is c.e.-coded by $\mathcal{A}$, it must be enumeration reducible to some $\exists$-type realized in $\mathcal{A}$. Since these are all c.e., X must be c.e. too.

For the other direction, notice that every $\exists$-type realized in $\mathcal{A}$ is c.e.-coded by $\mathcal{A}$, and hence (2) implies they are all c.e. $\square$

Recall that a structure $\mathcal{A}$ has enumeration degree Y if and only if $DgSp(\mathcal{A}) = \{X \in 2^{\mathbb{N}} : Y$ is c.e. in X$\}$, and $\mathcal{A}$ has Turing degree Y if $DgSp(\mathcal{A}) = \{X \in 2^{\mathbb{N}} : Y \geq_T X\}$. Richter's theorem allows us to conclude many degree spectra do not have that shape.

COROLLARY 5.1.7. *If $\mathcal{A}$ has the c.e. embeddability condition, then it does not have enumeration degree, unless it has a computable copy and enumeration degree $\emptyset$.*

PROOF. If $\mathcal{A}$ has enumeration degree Y, then Y must be c.e. coded in $\mathcal{A}$, and hence c.e. $\square$

COROLLARY 5.1.8. *If $\mathcal{A}$ has the c.e. embeddability condition, then it does not have Turing degree, unless it has a computable copy and Turing degree $\emptyset$.*

PROOF. Apply the previous corollary to $X \oplus X^c$. $\square$

Richter's original result is actually stronger than Theorem 5.1.6. We say that X and Y from a *c.e.-minimal pair* if no set is c.e. in both X and Y, unless it is already c.e. Notice that a c.e.-minimal pair is also a *minimal pair* in the sense that whenever a set is computable in both X and Y, it is already computable.

THEOREM 5.1.9 (Richter). *Let $\mathcal{A}$ have the c.e. embeddability condition. Then, for every non-computable set X, there is a copy $\mathcal{B}$ of $\mathcal{A}$ that forms a c.e.-minimal pair with X.*

PROOF. Let $\mathcal{B}$ be an X'-1-generic presentation of $\mathcal{A}$. Let Y be c.e. in both X and $D(\mathcal{B})$. Since Y is c.e. in X, $\mathcal{B}$ is Y-1-generic. Then, since Y is c.e. in $D(\mathcal{B})$, Y must be c.e.-coded by $\mathcal{A}$ (Corollary 4.4.2) and thus be c.e. $\square$

5.2. Co-spectra

The degree spectrum of a structure measures how difficult it is to present the structure. If instead we want to measure how much information is encoded in a structure, the first approach is to use co-spectra. This is not the only approach because, as we will see later, information can be coded within a structure in many different ways, as for instance, it can be coded in the jump of the structure without getting reflected in the co-spectra.

DEFINITION 5.2.1. The *co-spectrum* of a structure $\mathcal{A}$ is the set

$$coDgSp(\mathcal{A}) = \{X \subseteq \mathbb{N} : X \text{ is c.e.-coded by } \mathcal{A}\}.$$

Recall that X is c.e.-coded by $\mathcal{A}$ if and only if X is r.i.c.e. in $\mathcal{A}$, if and only if $X \leq_e \exists\text{-}tp_{\mathcal{A}}(\bar{p})$ for some $\bar{p} \in A^{<\mathbb{N}}$, and if and only if X is c.e. in every $Y \in DgSp(\mathcal{A})$ (see Section 2.1.4). Note that a structure has a trivial co-spectrum (i.e., the class of just the c.e. sets) if and only if it has the c.e. embeddability condition.

DEFINITION 5.2.2. A set $\mathcal{S} \subseteq \mathcal{P}(\mathbb{N})$ is an *ideal in the enumeration degrees* if it is closed downward under enumeration reducibility and closed under joins.

Co-spectra are always ideals in the enumeration degrees. The reverse is also true.

LEMMA 5.2.3 (Soskov [Sos04]). *Every countable ideal in the enumeration degrees $\mathcal{S} \subseteq \mathcal{P}(\mathbb{N})$ is the co-spectrum of some structure.*

PROOF. Given a set X, let $\mathcal{G}_X$ be the graph from Example 3.8.3 with one modification: $\mathcal{G}_X$ is made out of cycles of length $n + 3$ for $n \in X$ and all of these cycles share exactly one common node. We call it a *flower graph* because the cycles look like petals coming out of a center node. Recall that in Example 3.8.3 we showed that $\mathcal{G}_X$ has enumeration degree X. For a set $\mathcal{S} \subseteq \mathcal{P}(\mathbb{N})$, let $\mathcal{G}_{\mathcal{S}}^{\infty}$ be the graph formed by the disjoint and disconnected union of the graphs $\mathcal{G}_X$ for $X \in \mathcal{S}$, each one repeated infinitely often. We call it a bouquet graph. Clearly $\mathcal{S} \subseteq \mathrm{co}DgSp(\mathcal{G}_{\mathcal{S}}^{\infty})$, as for every $X \in \mathcal{S}$, X

FIGURE 1. Bouquet Graph

is c.e. in every copy of $\mathcal{G}_X$. Conversely, we claim that the $\exists$-type of any tuple $\bar{p} \in \mathcal{G}_{\mathcal{S}}^{\infty <\mathbb{N}}$ is e-reducible to a finite join of X's in $\mathcal{S}$, which would imply that every set c.e.-coded in $\mathcal{G}_{\mathcal{S}}^{\infty}$ is in $\mathcal{S}$. To see this, let $X_1, \ldots, X_n \in \mathcal{S}$ be such that the elements of $\bar{p}$ are in $\bigcup_{i=1}^n \mathcal{G}_{X_i}$. Let $\widetilde{\mathcal{G}}$ consist of $\bigcup_{i=1}^n \mathcal{G}_{X_i}$ and infinitely many copies of $\mathcal{G}_{\mathbb{N}}$ (i.e., G_Y for $Y = \mathbb{N}$), and let $\bar{q}$ be the tuple in $\widetilde{\mathcal{G}}$ corresponding to $\bar{p}$ (i.e., under the isomorphism between the pieces of the form $\bigcup_{i=1}^n \mathcal{G}_{X_i}$). We claim that $\exists\text{-}tp_{\mathcal{G}_{\mathcal{S}}^{\infty}}(\bar{p}) = \exists\text{-}tp_{\widetilde{\mathcal{G}}}(\bar{q})$: For the left-to-right inclusion, observe that there is an embedding $\mathcal{G}_{\mathcal{S}}^{\infty} \to \widetilde{\mathcal{G}}$ matching $\bar{p}$ and $\bar{q}$, mapping each $\mathcal{G}_X$ for $X \neq X_i$ into a copy of $\mathcal{G}_{\mathbb{N}}$, and recall that $\exists$-formulas are preserved forward under embeddings. For the other inclusion observe that $\widetilde{\mathcal{G}}$ is a sub-graph of $\mathcal{G}_{\mathcal{S}}^{\infty}$, just because $\mathbb{N} \in \mathcal{S}$ (since $\mathcal{S}$ contains all c.e. sets), and recall that $\exists$-formulas are preserved upwards. One can easily build a copy of $\widetilde{\mathcal{G}}$ from an enumeration of $X_1 \oplus \cdots \oplus X_n$, and hence enumerate the $\exists$-type of $\bar{q}$. Thus $\exists\text{-}tp_{\widetilde{\mathcal{G}}}(\bar{q}) \leq_e X_1 \oplus \cdots \oplus X_n \in \mathcal{S}$. We conclude that $\exists\text{-}tp_{\mathcal{G}_{\mathcal{S}}^{\infty}}(\bar{p}) \in \mathcal{S}$. $\square$

Richter's Theorem 5.1.9 can be generalized to arbitrary structures without the c.e. embeddability condition because, in a sense, every structure has the c.e. embeddability condition relative to its co-spectra:

LEMMA 5.2.4. *Suppose that every set in $coDgSp(\mathcal{A})$ is c.e. in Y. Then there is a copy $\mathcal{B}$ of $\mathcal{A}$ such that $D(\mathcal{B})$ and Y are a c.e.-exact pair for $coDgSp(\mathcal{A})$; that is, such that, for $Z \subseteq \mathbb{N}$, $Z \in coDgSp(\mathcal{A})$ if and only if Z is c.e. in both $D(\mathcal{B})$ and Y.*

PROOF. Let $\mathcal{B}$ be a Y'-1-generic copy of $\mathcal{A}$. Suppose now that X is c.e. in both Y and $D(\mathcal{B})$. Since X is a column in Y', $\mathcal{B}$ is also X-1-generic. Then, by Corollary 4.4.2, X must be c.e.-coded by $\mathcal{B}$, and hence belong to $coDgSp(\mathcal{A})$. $\square$

Recall from Corollary 4.4.3, that we can actually get two copies $\mathcal{B}$ and $\mathcal{C}$ of a structure $\mathcal{A}$ such that a set Z is c.e. in both if and only if it is in $coDgSp(\mathcal{A})$. That is, $D(\mathcal{B})$ and $D(\mathcal{C})$ form a c.e.-exact pair for $coDgSp(\mathcal{A})$.

5.3. Degree spectra that are not possible

In this section, we look at degree-theoretic properties degree spectra must have.

The first observation along these lines is that degree spectra are always Borel. This will follow from results in [MonP2], where we prove that every structure has an infinitary Scott sentence. But among upward closed Borel sets of Turing degrees, we know very little about which ones can be degree spectra and which ones cannot.

5.3.1. No two cones. One of the best-known results in this vein is due to Knight and her group in the 90's and says that no degree spectrum can be a non-trivial union of two upper cones of Turing degrees—not even the union of countably many upper cones. Her result also applies to cones of the following kind: the *enumeration upper cone with base* X, namely the set $\{Z \in 2^{\mathbb{N}} : X \text{ is c.e. in } Z\}$.

THEOREM 5.3.1 (Knight et al.). *No degree spectrum is the union of countably many enumeration upper cones, unless it is equal to just one enumeration upper cone.*

PROOF. Suppose that we have $X_1, X_2, \cdots \subseteq \mathbb{N}$ and a structure $\mathcal{A}$ with

$$DgSp(\mathcal{A}) = \bigcup_{n \in \mathbb{N}} \{Z \in 2^{\mathbb{N}} : X_n \text{ is c.e. in } Z\}.$$

Let $X = \bigoplus_n X_n$. Let $\mathcal{C}$ be a copy of $\mathcal{A}$ such that $\mathcal{C}$ is X-1-generic. Since $D(\mathcal{C}) \in DgSp(\mathcal{A})$, there exists an n such that X_n is c.e. in $D(\mathcal{C})$. From Lemma 4.4.2, we get that X_n is c.e.-coded by $\mathcal{C}$. But then $DgSp(\mathcal{A}) \subseteq \{Z \in 2^{\mathbb{N}} : X_n \text{ is c.e. in } Z\}$, and hence $DgSp(\mathcal{A}) = \{Z \in 2^{\mathbb{N}} : X_n \text{ is c.e. in } Z\}$ is a single enumeration upper cone. $\square$

Observation 5.3.2. No degree spectrum is the union of countably many Turing upper cones, unless it is equal to just one Turing upper cone: To see this, replace X_n by $X_n \oplus X_n^c$ in the proof of the theorem above.

5.3.2. Upward closure of F_σ. We can generalize Observation 5.3.2 quite a bit by extending some ideas of Andrews and Miller [AM15]. Recall that we give $\mathbb{N}^{\mathbb{N}}$ and $2^{\mathbb{N}}$ the product topology of the discrete topology on $\mathbb{N}$ and 2 respectively. Thus, the topology on $\mathbb{N}^{\mathbb{N}}$ is generated by the *basic open* sets $[\sigma]^{\subset} = \{X \in \mathbb{N}^{\mathbb{N}} : \sigma \subset X\}$ for $\sigma \in \mathbb{N}^{<\mathbb{N}}$, and similarly on $2^{\mathbb{N}}$. *Open* set are then of the form $[R]^{\subset} = \bigcup_{\sigma \in R}[\sigma]^{\subset}$ for some $R \subseteq \mathbb{N}^{<\mathbb{N}}$. The complement of $[R]^{\subset}$ can then be viewed as the set of paths through the tree $T = \{\tau \in \mathbb{N}^{<\mathbb{N}} : (\forall \sigma \subseteq \tau)\, \sigma \notin R\}$. We thus have that a set $P \subseteq \mathbb{N}^{\mathbb{N}}$ is *closed* if and only if it is the set of paths $[T]$ through some tree $T \subseteq \mathbb{N}^{<\mathbb{N}}$. A closed set can be defined by a boldface $\mathbf{\Pi}_1^0$ formula $\varphi(X)$ of arithmetic with a parameter for the tree T. Conversely, one can show that every boldface $\mathbf{\Pi}_1^0$ formula $\varphi(X)$ of arithmetic defines a closed set. Furthermore, a set $P \subseteq \mathbb{N}^{\mathbb{N}}$ can be defined by a lightface Π_1^0 formula $\varphi(X)$, i.e., without real parameters, if and only if it is the set of paths through a computable tree T, and the tree T can be computed uniformly from φ. We call such sets Π_1^0 *classes*:

DEFINITION 5.3.3. A set $P \subseteq \mathbb{N}^{\mathbb{N}}$ is a Π_1^0 *class* if there exists a computable tree T such that $P = [T]$.

A subset of $\mathbb{N}^{\mathbb{N}}$ is said to be F_σ if it is a countable union of closed sets, or equivalently, if it can be defined by a boldface $\mathbf{\Sigma}_2^0$ formula $\varphi(X)$ of arithmetic. For $F \subseteq \mathbb{N}^{\mathbb{N}}$, we define the *Turing-upward closure of* F to be $\{X \in \mathbb{N}^{\mathbb{N}} : \exists Y \leq_T X \, (Y \in F)\}$.

THEOREM 5.3.4. *A degree spectrum is never the Turing-upward closure of an F_σ set of reals in $\mathbb{N}^\mathbb{N}$, unless it is an enumeration-cone.*

We will prove this theorem on page 71. Let us first notice that we get Observation 5.3.2 as a corollary:

COROLLARY 5.3.5 (Knight et al.). *A degree spectrum is never the countable union of countably many Turing cones, unless it is a single Turing cone.*

PROOF OF COROLLARY. Every countable set is F_σ, as singleton sets are closed. So, by the theorem, if a degree spectra is the Turing-upper closure of a countable set, it must be an e-cone. But no e-cone is the Turing-upper closure of a countable set unless it is a Turing cone: To see this, consider the e-cone above a set $Z \subseteq \mathbb{N}$, and suppose that $X_0, X_1, \ldots$ are such that, for all $Y \in \mathbb{N}^\mathbb{N}$, Z is c.e. in Y if and only if $Y \geq_T X_n$ for some n. This is equivalent to: $Z \leq_e X_n \oplus X_n^c$ for all n, and, for all $Y \geq_e Z$, $Y \geq_e X_n \oplus X_n^c$ for some n. Let $g \in Z^\mathbb{N}$ be a $\bigoplus_n X_n$-1-generic enumeration of Z as in Section 4.2. One of the X_n's must be computable in g. In other words, $X_n \oplus X_n^c$ is c.e. in g for some $n \in \mathbb{N}$, say n_0. Since g is X_{n_0}-1-generic, using Lemma 4.2.1, we get that $X_{n_0} \oplus X_{n_0}^c \leq_e Z$. Therefore, $Z \equiv_e X_{n_0} \oplus X_{n_0}^c$, and the e-cone above Z is the Turing-cone above X_{n_0}. $\square$

Another corollary we will see below is that the following familiar classes of degrees are not degree spectra: DNC degrees, ML-random degrees, and PA degrees; they are all F_σ classes of reals. We will get this and a bit more below in Corollary 5.3.8, after we prove the following theorem, which contains some of the main ideas needed for Theorem 5.3.4.

THEOREM 5.3.6 (Andrews and Miller [AM15, Proposition 3.9]). *Let $\mathcal{A}$ satisfy the c.e. embeddability condition. Then $\mathcal{A}$ has a copy $\mathcal{B}$ such that, for every Π_1^0 class $P \subseteq \mathbb{N}^\mathbb{N}$ without computable members, $D(\mathcal{B})$ computes no real $X \in P$.*

To prove this theorem we need to introduce 2-generic enumerations; they are not really a new concept:

DEFINITION 5.3.7. An enumeration g of $\mathcal{A}$ is said to be 2-*generic* if it is a 1-generic enumeration of $(\mathcal{A}, \vec{K}^\mathcal{A})$. (Recall that $\vec{K}^\mathcal{A}$ is a complete r.i.c.e. relation on $\mathcal{A}$. The r.i.c.e. subsets of $(\mathcal{A}, \vec{K}^\mathcal{A})$ are exactly the ones that are Σ_2^c-definable with parameters over $\mathcal{A}$.)

PROOF OF THEOREM 5.3.6. Let g be a 2-generic enumeration of $\mathcal{A}$ and let $\mathcal{B}$ be the 2-generic presentation obtained as the pull-back of $\mathcal{A}$ through g. Consider a Π_1^0 class P and let $T \subseteq \mathbb{N}^{<\mathbb{N}}$ be a computable tree with $P = [T]$. Suppose $D(\mathcal{B})$ computes a path through T; we need to prove that T has a computable path.

Let Φ be a computable operator such that $\Phi^{D(\mathcal{B})}$ is a path through T, i.e., $\Phi^{D(\mathcal{B})}(n) \in \mathbb{N}^n \cap T$ and $\Phi^{D(\mathcal{B})}(n) \subseteq \Phi^{D(\mathcal{B})}(n+1)$ for every n. Let us start by finding an initial segment of g that forces $\Phi^{D(\mathcal{B})}$ to output the right kind of values. For this, consider the set of strings that do not:

$$Q_0 = \{\bar{p} \in A^* : \exists n \in \mathbb{N} \, (\Phi^{D_\mathcal{A}(\bar{p})}(n)\!\downarrow \, \& \, \Phi^{D_\mathcal{A}(\bar{p})}(n) \notin \mathbb{N}^n \cap T)\}.$$

(Remember that A^* is the set of strings from $\mathcal{A}^{<\mathbb{N}}$ without repetition.) The set Q_0 is r.i.c.e. in $\mathcal{A}$, and hence decided by some initial segment of the enumeration g. No initial segment of g is in Q_0 because $\Phi^{D(\mathcal{B})}(n)\!\downarrow \, \in \mathbb{N}^n \cap T$, and recall that $D(\mathcal{B}) = \bigcup_{k \in \mathbb{N}} D_\mathcal{A}(g \!\upharpoonright\! k)$. So there must be an initial segment $\bar{b}_0 \in A^*$ of g such that no extension of $\bar{b}_0$ is in Q_0. This means that whenever $\bar{p} \in A^*$ extends $\bar{b}_0$, if $\Phi^{D_\mathcal{A}(\bar{p})}(n)\!\downarrow$, then $\Phi^{D_\mathcal{A}(\bar{p})}(n) \in \mathbb{N}^n \cap T$.

Second, we force the values of $\Phi^{D(\mathcal{B})}$ to be compatible. For this, consider the set of strings that force them to be incompatible:

$$Q_1 = \{\bar{p} \in A^* : \exists n \in \mathbb{N} \, (\Phi^{D_\mathcal{A}(\bar{p})}(n)\!\downarrow \, \& \, \Phi^{D_\mathcal{A}(\bar{p})}(n+1)\!\downarrow \, \&$$
$$\Phi^{D_\mathcal{A}(\bar{p})}(n) \not\subseteq \Phi^{D_\mathcal{A}(\bar{p})}(n+1))\}.$$

The set Q_1 is r.i. computable in $\mathcal{A}$, and hence decided by some initial segment of the enumeration g. Again, since $\Phi^{D(\mathcal{B})} \in [T]$, no initial segment of g is in Q_1, and hence there must be an initial segment, $\bar{b}_1 \in A^*$, of g none of whose extensions is in Q_1. We may assume $\bar{b}_1 \supseteq \bar{b}_0$.

Third, we force that $\Phi^{D(\mathcal{B})}$ is total: For this, consider the set of strings which force $\Phi^{D(\mathcal{B})}$ to be undefined at some $n \in \mathbb{N}$:

$$Q_2 = \{\bar{p} \in A^* : \exists n \in \mathbb{N} \, \forall \bar{q} \in A^* \, (\bar{q} \supseteq \bar{p} \to \Phi^{D_\mathcal{A}(\bar{q})}(n)\!\uparrow)\}.$$

The set Q_2 is Σ_2^c in $\mathcal{A}$, and hence r.i.c.e. in $(\mathcal{A}, \vec{K}^\mathcal{A})$ and decided by an initial segment of g.[46] We cannot have an initial segment of g in Q_2 because we would have that $\Phi^{D(\mathcal{B})}(n)\!\uparrow$ for some n. So, for some initial segment $\bar{b}_2$ of g, we have that for every $\bar{p} \in A^*$ extending $\bar{b}_2$ and every n, there is a $\bar{q} \in A^*$ extending $\bar{p}$ for which $\Phi^{D_\mathcal{B}(\bar{q})}(n)\!\downarrow$. We may assume $\bar{b}_2 \supseteq \bar{b}_1$.

Now, using $\exists$-$tp_\mathcal{B}(\bar{b}_2)$, which we know is c.e., we define a computable path through P. We define a path $\{\sigma_n : n \in \mathbb{N}\} \subseteq T$ step by step as follows. Let σ_0 be the empty string. Given σ_n, chose $\sigma_{n+1} \in \mathbb{N}^{n+1} \cap T$ with $\sigma_{n+1} \supseteq \sigma_n$ so that

$$\mathcal{A} \models \exists \bar{x}(\Phi^{D_\mathcal{A}(\bar{b}_2 \frown \bar{x})}(n+1) = \sigma_{n+1}),$$

or, in other words, wait to find $\tau \in 2^{<\mathbb{N}}$ with $\Phi^\tau(n+1) = \sigma_{n+1}$ for which the formula $\exists \bar{x}(D(\bar{b}_2 \frown \bar{x}) = \tau)$ is in $\exists$-$tp_\mathcal{B}(\bar{b}_2)$, and then let $\sigma_{n+1} = \Phi^\tau(n+1)$. We know σ_{n+1} exists because, if $\bar{a}_n$ was the witness to define σ_n (i.e., $\Phi^{D_\mathcal{A}(\bar{b}_2 \frown \bar{a}_n)}(n) = \sigma_n$), then we know there is an extension $\bar{a}_{n+1}$ of $\bar{a}_n$ such that $\Phi^{D_\mathcal{A}(\bar{b}_2 \frown \bar{a}_{n+1})}(n+1)\!\downarrow$. We also know that $\Phi^{D_\mathcal{A}(\bar{b}_2 \frown \bar{a}_{n+1})}(n+1)$ must be in $\mathbb{N}^{n+1} \cap T$ and must extend $\Phi^{D_\mathcal{A}(\bar{b}_2 \frown \bar{a}_n)}(n)$. That is our σ_{n+1}. $\bigcup_n \sigma_n$ is then a computable path through T. $\square$

---

[46] $\bar{p} \in Q_2 \iff \mathcal{A} \models \bigvee_{n \in \mathbb{N}} (\forall \bar{q} \supseteq \bar{p}) \bigwedge_{\sigma \in 2^{<\mathbb{N}} : \Phi^\sigma(n)\downarrow} D_\mathcal{A}(\bar{q}) \neq \sigma.$

The following corollary is for the reader familiar with the following notions. A real $X \in 2^{\mathbb{N}}$ is said to be *diagonally non-computable* (DNC) if $\forall n(X(n) \neq \Phi_n(n))$; a real is *ML*-random if it does not belong to any effectively-null G_δ set; and a real is PA if it computes a complete, consistent theory extending the axioms of Peano arithmetic. See [Nie09] for more background on these classes.

COROLLARY 5.3.8 (Andrews and Miller [AM15]). *The class of DNC degrees, the class of ML-random degrees, and the class of PA degrees are not degree spectra. Furthermore, if a structure has the c.e. embeddability property, its degree spectrum is not contained in any of these classes.*

PROOF. All these classes are easily seen to be F_σ, and hence they cannot be degree spectra by Theorem 5.3.4. Furthermore, The classes of DNC and PA reals are both Π_1^0 classes without computable members, and the class of *ML*-random reals is an effective countable union of Π_1^0 classes without computable members. We will refer to [AM15, Proposition 3.6] for a proof that if a set is c.e. in all members of a given non-empty Π_1^0 class, it is c.e. already. It follows that neither the DNC, the PA, nor the ML-random degrees are contained in any proper e-cone.

The second part of the corollary follows from Theorem 5.3.6. □

Let us now give the proof of 5.3.4; we recommend reading the proof of Theorem 5.3.6 first.

PROOF OF THEOREM 5.3.4. Suppose $\mathcal{A}$ is a structure whose degree spectrum is the Turing-upper closure of an F_σ set $F \subseteq 2^{\mathbb{N}}$. Assume $F = \bigcup_{i \in \mathbb{N}} P_i$ where each $P_i = [T_i]$ for trees $T_i \subseteq 2^{<\mathbb{N}}$. Let g be a $(\bigoplus_{i \in \mathbb{N}} T_i)$-2-generic enumeration of $\mathcal{A}$ and let $\mathcal{B}$ be the pull-back structure. There is a computable functional Φ and an i such that $\Phi^{D(\mathcal{B})}$ is a path through T_i, i.e., $\Phi^{D(\mathcal{B})}(n) \in \mathbb{N}^n \cap T_i$ and $\Phi^{D(\mathcal{B})}(n) \subseteq \Phi^{D(\mathcal{B})}(n+1)$ for every n. As in the proof of Theorem 5.3.6, there is an initial segment $\bar{b} \in A^\star$ of the enumeration g which has no extensions in Q_0, Q_1, and Q_2. That is, the tuple $\bar{b}$ satisfies:

1. $(\forall \bar{q} \supseteq \bar{b}, \bar{q} \in A^\star)$, if $\Phi^{D_{\mathcal{A}}(\bar{q})}(n)\downarrow$, then $\Phi^{D_{\mathcal{A}}(\bar{q})}(n) \in T_i \cap 2^n$.
2. $(\forall \bar{q} \supseteq \bar{b}, \bar{q} \in A^\star)$, if $\Phi^{D_{\mathcal{A}}(\bar{q})}(n)\downarrow$ & $\Phi^{D_{\mathcal{A}}(\bar{q})}(n+1)\downarrow$, then $\Phi^{D_{\mathcal{A}}(\bar{q})}(n) \subset \Phi^{D_{\mathcal{A}}(\bar{q})}(n+1)$.
3. $(\forall n \in \mathbb{N}\, \forall \bar{q} \supseteq \bar{b}, \bar{q} \in A^\star)(\exists \bar{p} \supseteq \bar{q}, \bar{p} \in A^\star)\, \Phi^{D_{\mathcal{A}}(\bar{p})}(n)\downarrow$.

Consider now the tree of possible values of Φ:

$$S = \{\sigma \in 2^{<\mathbb{N}} : (\exists \bar{q} \supseteq \bar{b}, \bar{q} \in A^\star)\, \sigma \subseteq \Phi^{D_{\mathcal{A}}(\bar{q})}\}.$$

We claim that $\mathcal{A}$ has e-degree S. From its definition we get that S is r.i.c.e. in $\mathcal{A}$. On the other hand, by the assumptions on $\bar{b}$, we get that S is a subtree of T_i without dead ends. Thus, every enumeration of S can compute a path through S, and hence a path through T_i, which must then compute a copy of $\mathcal{A}$. □

Let us remark that Theorem 5.3.4 cannot be improved by replacing Turing-cone with enumeration-cone in its statement. That is, there are enumeration cones, and hence degree spectra, that are the Turing upward closure of closed sets but are not Turing cones. J. Miller and M. Soskova proved this is the case for all *continuous enumeration degrees* which are not total. (The continuous degrees are a sub-class of enumerations degrees larger than the total degrees introduced by Miller [Mil04].) Furthermore, McCarthy [McC] later characterized the enumeration degrees whose upper cone is the Turing upper closure of a closed set as exactly the *co-total* degrees, which have been recently shown to be a robust class within the enumeration degrees (cf. [AGK⁺]).

EXERCISE 5.3.9. (a) Prove that no degree spectrum can be the upward closure of a lightface Π_2^0 subset of $\mathbb{N}^{\mathbb{N}}$, unless it is an enumeration cone. Hint in footnote.[47]

(b) Furthermore, show that it cannot even be the upward closure of a countable union of lightface Π_2^0 subsets of $\mathbb{N}^{\mathbb{N}}$, unless it is an enumeration upper cone.

Notice that the degree spectrum of any ∃-atomic structure is always the Turing upward closure of a boldface Π_2^0 set, namely the set of presentations satisfying the Π_2^{in} Scott sentence of the structure. See Exercise 5.4.5 for the existence of an ∃-atomic structure whose degree spectrum is not an enumeration cone.

5.4. Some particular degree spectra

We already saw that all upper cones and enumeration cones can be realized as degree spectra (Examples 3.8.3 and 3.8.4). In this section, we look at another easy-to-describe though more surprising degree spectrum.

5.4.1. The Slaman–Wehner Family. The Slaman–Wehner structure is one that has no computable copy, but is computable in any non-computable set. The easiest way to describe it is using families of sets.

DEFINITION 5.4.1. We say that X can *enumerate a family* $\mathcal{S} \subseteq \mathcal{P}(\mathbb{N})$ if there is an X-c.e. set W such that $\mathcal{S} = \{ W^{[n]} : n \in \mathbb{N} \}$.[48]

Observation 5.4.2. For every countable family $\mathcal{S} \subseteq \mathcal{P}(\mathbb{N})$, there is a graph $\mathcal{G}_{\mathcal{S}}^{\infty}$ such that, for every oracle X, X can compute a copy of $\mathcal{G}_{\mathcal{S}}^{\infty}$ if and only if X can enumerate $\mathcal{S}$: As in the proof of Lemma 5.2.3, consider the *bouquet graph* $\mathcal{G}_{\mathcal{S}}^{\infty} = \bigcup_{Y \in \mathcal{S}, i \in \mathbb{N}} \mathcal{G}_Y$, where $\mathcal{G}_Y$ is the flower graph coding Y, that is $\mathcal{G}_Y$ contains a cycle of length $n + 3$ for each $n \in Y$, and all the

---

[47] Try to make the Π_2^0 set into a Π_1^0 one.
[48] Note that columns may be repeated.

cycles intersect at one node, a node common to all cycles in $\mathcal{G}_Y$. Notice that each $\mathcal{G}_Y$ appears infinitely often in $\mathcal{G}_S$.

THEOREM 5.4.3 (Slaman [Sla98], Wehner [Weh98]). *There is a structure* $\mathcal{W}$ *whose degree spectrum is* $\{X \in 2^{\mathbb{N}} : X \text{ not computable}\}$.

PROOF. Consider the family

$$\mathcal{F} = \{F \oplus \{n\} : F \subseteq \mathbb{N} \text{ finite } \& \ F \neq W_n\},$$

and let $\mathcal{W}$ be the bouquet graph $\mathcal{G}_{\mathcal{F}}^{\infty}$ as in the observation above. Then,

$$DgSp(\mathcal{W}) = \{X \in \mathbb{N}^{\mathbb{N}} : X \text{ can enumerate } \mathcal{F}\}.$$

We claim that X can enumerate $\mathcal{F}$ if and only if X is not computable.

Suppose $\mathcal{F}$ had a computable enumeration. We could then build a function g that, on input n, outputs the c.e. index of a finite set $W_{g(n)}$ with $W_{g(n)} \neq W_n$: given n, just look through the enumeration of $\mathcal{F}$ until you find a column of the form $F \oplus \{n\}$ for some F and let $g(n)$ be the c.e. index of that F.[49] This contradicts the recursion theorem (see page xv).

For the other direction, suppose X is not computable. We need to define an X-computable enumeration of $\mathcal{F}$. Let $Y = X \oplus X^c$, which we know is not c.e. since X is not computable. Here is the general intuition: At the beginning of stage t, enumerate into $\mathcal{F}$ all the sets of the form $F \oplus \{n\}$ for all $F \subseteq t$, and all $n < t$. If, among the columns that have been enumerated so far, one is of the form $F \oplus \{n\}$ with $F = W_n[t]$ (the stage-t approximation to W_n), we take it as a threat, and we add to F the least element of Y that is not in F already. The idea is that no column can be threatened infinitely often because that would imply that $W_n = F \cup Y$, which we know is not c.e.

More formally: Fix $n \in \mathbb{N}$; we want to enumerate the family $\mathcal{F}_n = \{F : F \subseteq \mathbb{N} \text{ finite } \& \ F \neq W_n\}$ uniformly in n. For each finite set F and every $s \in \mathbb{N}$, we will enumerate a set $R_{F,s}$ with the objective of having

$$\{R_{F,s} : F \subseteq \mathbb{N} \text{ finite}, s \in \mathbb{N}\} = \mathcal{F}_n.$$

We define $R_{F,s}$ by stages as $R_{F,s} = \bigcup_{t \in \mathbb{N}} R_{F,s}[t]$, where each $R_{F,s}[t]$ is finite. For $t \leq s$, let $R_{F,s}[t] = F$. At stage $t + 1 > s$, if $R_{F,s}[t] = W_n[t]$, we take it as a threat and let $R_{F,s}[t+1] = R_{F,s}[t] \cup \{y\}$, where y is the least element of $Y \smallsetminus R_{F,s}[t]$. The threats to $R_{F,s}$ must eventually stop, as otherwise we would have $W_n = \bigcup_{t \in \mathbb{N}} R_{F,s}[t] = F \oplus Y$, which is not c.e. Thus, $R_{F,s}$ will end up being finite and not equal to W_n, and hence $R_{F,s}$ belongs to $\mathcal{F}_n$. On the other hand, for every finite set $F \neq W_n$, we have $R_{F,s} = F$ for large enough s: Take s so that $(\forall t > s) \ W_n[t] \neq F$.   □

Kalimullin [Kal08] showed that the non-Δ_2^0 degrees are a degree spectrum (see Exercise 7.3.8). On the other hand, Andrews, Cai, Kalimullin, Lempp,

---

[49]That is, if $F \subseteq \mathbb{N}^2$ is the enumeration of $\mathcal{F}$, look for a column $c \in \mathbb{N}$ such that $2n + 1 \in F^{[c]}$ and let $g(n)$ be such that $W_{g(n)} = \{m \in \mathbb{N} : 2m \in F^{[c]}\}$.

Miller, and Montalbán [ACK$^+$16] showed that the class of non-Δ_n^0 degrees cannot be a degree spectrum, for $n \geq 3$. It remains open whether the non-arithmetic degrees from a degree spectrum.

EXERCISE 5.4.4. Instead of $\mathcal{G}_\mathcal{S}^\infty$, there are many other structure that code a countable family of sets $\mathcal{S} \subseteq \mathcal{P}(\mathbb{N})$. We define a structure $\mathcal{A}_\mathcal{S}$ in the vocabulary with three unary relations S, N, and C, a binary relation R, and a constant symbols c_n for each $n \in \mathbb{N}$. The relations S, N, and C partition the domain of $\mathcal{A}_\mathcal{S}$ in three sets. Every constant c_n belongs to N and every member of N is one of the constants. We define an ω-presentation of $\mathcal{A}_\mathcal{S}$ as follows: Let $N = \mathbb{N}$, $S = \mathcal{S} \times \omega$, $C = \{\langle\langle X, m\rangle, n\rangle : \langle X, m\rangle \in S, n \in X\} \subseteq S \times \mathbb{N}$, and $R = \{\langle\langle x, n\rangle, n\rangle : \langle x, n\rangle \in C\} \cup \{\langle\langle x, n\rangle, x\rangle : \langle x, n\rangle \in C\}$. Prove that an oracle can compute a copy of $\mathcal{A}_\mathcal{S}$ if and only if it can enumerate $\mathcal{S}$.

EXERCISE 5.4.5 (Hirschfeldt [Hir06]). A tree $T \subseteq 2^{<\mathbb{N}}$ is said to be a *PAC tree* if it has no dead ends and all its paths are computable. (PAC is for "paths all computable.") Goncharov and Nurtazin, and Millar proved the existence of a computable PAC tree for which there is no computable listing of its isolated paths.

(a) Prove that every non-computable real can compute a listing of all the isolated paths through a PAC tree. Hint in footnote.[50]

(b) Use this to build a structure whose degree spectrum is the non-computable degrees.

(c) Build such a structure so that it is also $\exists$-atomic.

---

[50]From any given point in the tree, try to climb up the tree following the direction of the non-computable real.

Chapter 6

COMPARING STRUCTURES AND CLASSES OF STRUCTURES

A common way to measure the complexity of an object is comparing it to other objects. If our objects are sets of natural numbers, there are various ways to compare their complexity: Turing reducibility, enumeration reducibility, many-one reducibility, etc. For structures, the situation is a bit more complicated due to the fact that structures have many different presentations. In this chapter we will look into Muchnik reducibility, Medvedev reducibility, effective interpretability (also known as Σ-definability), and effective bi-interpretability.

6.1. Muchnik and Medvedev reducibilities

Let us start by defining these reducibilities on classes of reals:

DEFINITION 6.1.1. A class $\mathcal{R} \subseteq 2^{\mathbb{N}}$ is *Muchnik reducible* to a class $\mathcal{S} \subseteq 2^{\mathbb{N}}$ if every real in $\mathcal{S}$ computes a real in $\mathcal{R}$ (cf. [Muc63]). If so, we write $\mathcal{R} \leq_w \mathcal{S}$, where the '$w$' stands for "weak," in contrast to the following stronger reducibility. A class $\mathcal{R} \subseteq 2^{\mathbb{N}}$ is *Medvedev reducible* to a class $\mathcal{S} \subseteq 2^{\mathbb{N}}$ if there is a computable operator Φ such that $\Phi^X \in \mathcal{R}$ for every $X \in \mathcal{S}$ (cf. [Med55]). If so, we write $\mathcal{R} \leq_s \mathcal{S}$, where the 's' is for strong.

Here is the idea behind these notions. Suppose we have two problems, R and S, which consist of finding reals with certain properties. Let $\mathcal{R}$ and $\mathcal{S}$ be the sets of reals which are solutions to R and S respectively. For either of the two reductions above, $\mathcal{R}$ reduces to $\mathcal{S}$ if and only if we can produce a solution for R using a solution for S. In the case of Medvedev reducibility, we need to produce a solution to R uniformly from one to S, while for Muchnik we can use different procedures for different solutions to S.

Both notions generalize both Turing reducibility and enumeration reducibility: For $X, Y \subseteq \mathbb{N}$, we have that $X \leq_T Y$ if and only if $\{X\} \leq_w \{Y\}$, and also if and only if $\{X\} \leq_s \{Y\}$. We have that $X \leq_e Y$ if and only if the set of enumerations of X (i.e., the set of onto functions $f : \mathbb{N} \to X$) is Muchnik reducible to the set of enumerations of Y, and also, but less trivially, if and only if the set of enumerations of X is Medvedev reducible to the set of enumerations of Y (cf. Selman [Sel71]; see Theorem 4.2.2).

EXAMPLE 6.1.2. Observe that a structure $\mathcal{A}$ c.e.-codes a set $X \subseteq \mathbb{N}$ if and only if the set of enumerations of X is Muchnik reducible to the set of ω-presentations of $\mathcal{A}$. This is just a re-writing of the definition of c.e. coding a set (see page 19).

When we are considering countable structures, we apply these reducibilities to the set of their ω-presentations.

DEFINITION 6.1.3. A structure $\mathcal{A}$ is *Muchnik reducible* to a structure $\mathcal{B}$ if every ω-presentation of $\mathcal{B}$ computes an ω-presentation of $\mathcal{A}$ or, more precisely, the atomic diagram of every ω-presentation of $\mathcal{B}$ computes the atomic diagram of an ω-presentation of $\mathcal{A}$. If so, we write $\mathcal{A} \leq_w \mathcal{B}$. A structure $\mathcal{A}$ is *Medvedev reducible* to a structure $\mathcal{B}$ if there is a computable operator Φ such that, for every ω-presentation $\widehat{\mathcal{B}}$ of $\mathcal{B}$, $\Phi^{D(\widehat{\mathcal{B}})} = D(\widehat{\mathcal{A}})$ for some ω-presentation $\widehat{\mathcal{A}}$ of $\mathcal{A}$. If so, we write $\mathcal{A} \leq_s \mathcal{B}$. We denote the respective notions of equivalence by $\equiv_w$ and $\equiv_s$ (i.e., $\mathcal{A} \equiv_w \mathcal{B} \iff \mathcal{A} \leq_w \mathcal{B}$ & $\mathcal{B} \leq_w \mathcal{A}$ and $\mathcal{A} \equiv_s \mathcal{B} \iff \mathcal{A} \leq_s \mathcal{B}$ & $\mathcal{B} \leq_s \mathcal{A}$).

Observation 6.1.4. Muchnik reducibility is nothing more than comparability of the degree spectra:

$$\mathcal{A} \leq_w \mathcal{B} \iff DgSp(\mathcal{A}) \supseteq DgSp(\mathcal{B}).$$

EXAMPLE 6.1.5. Given a linear ordering $\mathcal{A}$, every segment $[a, b]_{\mathcal{A}}$ of $\mathcal{A}$, is Muchnik reducible to $\mathcal{A}$, but not necessarily Medvedev reducible to $\mathcal{A}$.[51]

EXAMPLE 6.1.6. Given a ring R, $R[x] \leq_s R$.

EXAMPLE 6.1.7. Given a structure $\mathcal{A}$, there exists a graph $\mathcal{G}_{\mathcal{A}}$ such that $\mathcal{A} \equiv_s \mathcal{G}_{\mathcal{A}}$. We will develop this example later in Section 6.3.2.

EXAMPLE 6.1.8. For a group $\mathcal{G}$, $\mathcal{G}^{\omega} \leq_s \mathcal{G}$, but not necessarily $\mathcal{G} \not\leq_w \mathcal{G}^{\omega}$, where $\mathcal{G}^{\omega}$ is the sum of ω many copies of $\mathcal{G}$. Take $\mathcal{G} = \bigoplus_{n \in \mathbb{N}} \mathbb{Z}_{p_n} \oplus \bigoplus_{n \in 0'^c} \mathbb{Z}_{p_n}$.[52] We then get that $\mathcal{G}^{\omega} = \bigoplus_{n \in \mathbb{N}} \mathbb{Z}_{p_n}^{\omega}$ which has a computable copy, while $\mathcal{G}$ computably codes $0'$.

These reducibilities form upper-semi-lattices; that is, given structures $\mathcal{A}$ and $\mathcal{B}$, if we define $\mathcal{A} \oplus \mathcal{B}$ by putting together disjoint copies of $\mathcal{A}$ and $\mathcal{B}$ and adding a unary relation A that holds only of the elements in the copy of $\mathcal{A}$, then $\mathcal{A} \oplus \mathcal{B}$ is the least upper bound of $\mathcal{A}$ and $\mathcal{B}$ according to both Muchnik and Medvedev reducibilities. In both cases there is a least degree: If a structure has a computable copy, it reduces to every other structure. Another interesting observation is that there is a least non-computable structure:

---

[51] Such examples are not easy to build. Schweber showed there are ordinals which have initial segments which are not Medvedev reducible to them (cf. [Sch16, Section 8.3]).

[52] $0'^c = \mathbb{N} \setminus 0'$ and p_n is the n-th prime number.

Observation 6.1.9. The Slaman–Wehner structure $\mathcal{W}$ from Theorem 5.4.3 has no computable copies and is Medvedev reducible to all other structures without computable copies. All we have to observe is that the construction in 5.4.3 is uniform in X, i.e., that it produces a computable operator Φ such that, for every non-computable X, Φ^X is the atomic diagram of a copy of $\mathcal{W}$.

The following lemma shows how we can obtain structural information from knowing that a structure is Muchnik or Medvedev reducible to another.

LEMMA 6.1.10. *If $\mathcal{A} \leq_w \mathcal{B}$, then for every tuple $\bar{a} \in A^{<\mathbb{N}}$, there is a tuple $\bar{b} \in B^{<\mathbb{N}}$ such that $\exists\text{-}tp_\mathcal{A}(\bar{a}) \leq_e \exists\text{-}tp_\mathcal{B}(\bar{b})$. If also $\mathcal{A} \leq_s \mathcal{B}$, then $\exists\text{-}Th(\mathcal{A}) \leq_e \exists\text{-}Th(\mathcal{B})$.*

PROOF. For the first part, suppose that $\mathcal{A} \leq_w \mathcal{B}$ and take $\bar{a} \in A^{<\mathbb{N}}$. Since $\exists\text{-}tp_\mathcal{A}(\bar{a})$ is c.e. in every copy of $\mathcal{A}$, it is also c.e. in every copy of $\mathcal{B}$, and hence it is c.e.-coded by $\mathcal{B}$. By Knight's Lemma 2.1.24, $\exists\text{-}tp_\mathcal{A}(\bar{a}) \leq_e \exists\text{-}tp_\mathcal{B}(\bar{b})$ for some tuple $\bar{b} \in B^{<\mathbb{N}}$.

Suppose now that $\mathcal{A} \leq_s \mathcal{B}$. Since $\exists\text{-}Th(\mathcal{A})$ is uniformly c.e.-coded in $\mathcal{A}$, it is also uniformly c.e.-coded in $\mathcal{B}$: That is, from the diagram of a copy of $\mathcal{B}$ we can uniformly produce the diagram of a copy of $\mathcal{A}$, from which we can uniformly enumerate $\exists\text{-}Th(\mathcal{A})$. Then, using Exercise 2.1.26, we get that $\exists\text{-}Th(\mathcal{A}) \leq_e \exists\text{-}Th(\mathcal{B})$. □

So far, Muchnik and Medvedev reducibilities seem to behave in a similar way. However, one of the main differences is that adding constants to the structures does not affect Muchnik reducibility, while the following lemma shows that it does affect Medvedev reducibility.

LEMMA 6.1.11. *There are structures $\mathcal{B}$ and $\mathcal{C}$ and an element $c \in C$ with $\mathcal{B} \leq_s (\mathcal{C}, c)$, but $\mathcal{B} \not\leq_s \mathcal{C}$.*

Notice that $\mathcal{B} \leq_s (\mathcal{C}, c)$ implies $\mathcal{B} \leq_w \mathcal{C}$, and hence this is an example where the Muchnik and Medvedev reducibilities differ.

PROOF. Let Z be a non-c.e. set. Recall from Observation 5.4.2 and Lemma 5.2.3 that to each family of sets $\mathcal{S}$ we can assign a bouquet graph $\mathcal{G}_\mathcal{S}^\infty$ such that $\mathcal{G}_\mathcal{S}^\infty$ has an X-computable copy if and only if X can enumerate $\mathcal{S}$. We consider the following families of sets and their respective bouquet graphs:

- $\mathcal{S}_0 = \{F : F \subset \mathbb{N} \text{ finite}\}$ and $\mathcal{A} = \mathcal{G}_{\mathcal{S}_0}^\infty$.
- $\mathcal{S}_1 = \{Z\}$ and $\mathcal{B} = \mathcal{G}_{\mathcal{S}_1}^\infty$.
- $\mathcal{S}_2 = \mathcal{S}_0 \cup \mathcal{S}_1$ and $\mathcal{C} = \mathcal{G}_{\mathcal{S}_2}^\infty$.

The family $\mathcal{S}_0$ has a c.e. enumeration. Thus, $\mathcal{A}$ has a computable copy and $\exists\text{-}Th(\mathcal{A})$ is c.e. The family $\mathcal{S}_1$ does not have a c.e. enumeration. Furthermore, an oracle X can compute an enumeration of $\mathcal{S}_1$ if and only if X can enumerate Z. Thus, $DgSp(\mathcal{B}) = \{X \in 2^\mathbb{N} : Z \text{ is c.e. in } X\}$ is the e-cone above Z. The same is true for $\mathcal{C}$: clearly, from a copy of $\mathcal{B}$, we can produce one of C by attaching a computable copy of $\mathcal{A}$, and conversely, given a copy of C, we can produce a copy of $\mathcal{B}$ if we can identify the component of C that corresponds to $\mathcal{G}_Z$. This implies that if c is the center of the flower corresponding to the component $\mathcal{G}_Z$, we get that $\mathcal{B} \equiv_s (C, c)$.

However, every finite substructure of C is isomorphic to some finite substructure of $\mathcal{A}$, and vice versa. Since an $\exists$-formula is true of $\mathcal{A}$ if and only if it is true of some finite substructure of $\mathcal{A}$, this implies that $\exists\text{-}Th(C) = \exists\text{-}Th(\mathcal{A})$, which is c.e. On the other hand, $\exists\text{-}Th(\mathcal{B})$ can enumerate Z, and hence is not c.e. It follows from Lemma 6.1.10 that $\mathcal{B} \nleq_s C$.   □

EXERCISE 6.1.12 (Stuckachev [Stu07]). Prove that if a structure $\mathcal{A}$ has Turing degree and $\mathcal{B} \leq_w \mathcal{A}$, then for some tuple $\bar{a} \in A^{<\mathbb{N}}$, $\mathcal{B} \leq_s (\mathcal{A}, \bar{a})$.

Kalimullin [Kal09] showed that this is not true if we only assume that $\mathcal{A}$ has e-degree.

The difference between Muchnik and Medvedev reducibility is more than just adding constants, as shown in the corollary below. The following theorem gives a version of the Slaman–Wehner structure which is computable from every non-computable oracle, but not in a uniform way.

THEOREM 6.1.13 (Faizrahmanov and Kalimullin [FK15]). *There is a structure $\mathcal{A}$ that has an X-computable copy for every non-computable X, but not uniformly. That is, there is no single computable operator Φ such that Φ^X is copy of $\mathcal{A}$ for each non-computable X.*

COROLLARY 6.1.14 (Kalimullin [Kal09]). *There are structures $\mathcal{A}$ and $\mathcal{W}$ such that $\mathcal{A} \equiv_w \mathcal{W}$, but $\mathcal{A} \nleq_s (\mathcal{W}, \bar{w})$ for any tuple $\bar{w} \in W^{<\mathbb{N}}$.*

PROOF OF COROLLARY 6.1.14. The structure $\mathcal{W}$ is the Slaman–Wehner structure from Theorem 5.4.3 whose degree spectrum is the non-computable sets and for which there exists a Turing operator that outputs a copy of $\mathcal{W}$ whenever a non-computable set is used as an oracle. Moreover, for any $\bar{w} \in W^{<\mathbb{N}}$, we can produce such an operator that outputs a copy of $(\mathcal{W}, \bar{w})$: Recall that $\mathcal{W} = \bigcup_{n \in \mathbb{N}} \mathcal{W}_n$, where $\mathcal{W}_n$ is the disjoint union of the flower graphs $\mathcal{G}_{F \oplus \{n\}}$ for $F \subset \mathbb{N}$ finite with $F \neq W_n$, each appearing infinitely often. There are finitely many components $\mathcal{W}_n$ which contain an element of $\bar{w}$, so we can fix a computable enumeration of them. The rest of $\mathcal{W}$ is isomorphic to $\mathcal{W}$, so we can use the construction of Theorem 5.4.3.

The structure $\mathcal{A}$ is the one from Theorem 6.1.13. It is Muchnik equivalent to $\mathcal{W}$: On the one hand it is computable from any non-computable oracle. On the other hand, it has no computable copies, as otherwise we could produce a computable operator that always outputs the same computable ω-presentation of $\mathcal{A}$ ignoring the oracle. $\mathcal{A}$ is not Medvedev reducible to $(\mathcal{W}, \bar{w})$ for any $\bar{w} \in W^{<\mathbb{N}}$ because the set of presentations of $(\mathcal{W}, \bar{w})$ is Medvedev reducible to the set $\{X \in \mathbb{N}^{\mathbb{N}} : X \text{ non-computable}\}$, but the set of presentations of $\mathcal{A}$ is not.   □

PROOF OF THEOREM 6.1.13.   We modify Wehner's construction from Theorem 5.4.3. We still consider a family of finite sets of the form $F \oplus \{n\}$, but the difference with Wehner's construction is that we think of F as a finite subset of $\mathbb{Q}$ instead of $\mathbb{N}$, and instead of requiring F to be different from the n-th c.e. set, we just require its maximum to be different from the maximum of the n-th c.e. subset of $\mathbb{Q}$. It works.

Let $\{Q_n : n \in \mathbb{N}\}$ be an effective enumeration of the c.e. subsets of $\mathbb{Q}$. (For example, given an effective Gödel numbering $q \mapsto \ulcorner q \urcorner \colon \mathbb{Q} \to \mathbb{N}$, let $Q_n = \{q \in \mathbb{Q} : \ulcorner q \urcorner \in W_n\}$.) Consider the family of sets

$$\mathcal{F} = \{F \oplus \{n\} : F \subseteq \mathbb{Q} \text{ finite}, n \in \mathbb{N}, \max(F) \neq \max(Q_n)\},$$

where the formula $\max(F) \neq \max(Q_n)$ is assumed to be vacuously true when Q_n does not have a greatest element. Let $\mathcal{A}$ be the associated bouquet graph $\mathcal{G}_{\mathcal{F}}^{\infty}$ as in Observation 5.4.2. Recall that the existence of an X-computable presentation of $\mathcal{G}_{\mathcal{F}}^{\infty}$ is equivalent to the existence of an X-c.e. enumeration of $\mathcal{F}$, that is, an X-c.e. set V with $\mathcal{F} = \{V^{[n]} : n \in \mathbb{N}\}$.

First, let us show that $\mathcal{F}$ is computably enumerable in every non-computable set X. A real is said to be *left c.e.* if it is of the form $\sup(Q_e)$ for some c.e. set $Q_e \subseteq \mathbb{Q}$. Let α be X-left c.e., but not left c.e. To see that such an α exists, consider $\beta_0 = \sum_{i \in X} 2^{-i}$ and $\beta_1 = \sum_{i \notin X} 2^{-i}$. They cannot be both left c.e., as otherwise X would be computable.[53] Let α be whichever of β_0 or β_1 is not left c.e.; this is the only step in the construction that is not uniform in X. Let $\{\alpha_i : i \in \mathbb{N}\} \subseteq \mathbb{Q}$ be an X-computable increasing sequence with limit α.

Fix n. We want to enumerate the family

$$\mathcal{F}_n = \{F : F \subseteq \mathbb{Q} \text{ finite}, \max(F) \neq \max(Q_n)\}$$

uniformly in n. The idea is to enumerate a new component of the form F for each finite set $F \subseteq \mathbb{Q}$ at each stage, and if, at a certain stage t, we are threatened by having $\max(F) = \max(Q_{n,t})$, we add $\max(F) + \alpha_t$ to that component changing its maximum value. A component cannot be threatened infinitely often because we would end up having $\sup(Q_n) = \max(F) + \alpha$, which is not left-c.e. Let us explain this in more detail. For each finite set $F \in \mathbb{Q}$ and $s \in \mathbb{N}$, we will enumerate a set $R_{F,s}$ uniformly in X, with the objective of getting

$$\mathcal{F}_n = \{R_{F,s} : F \subseteq_{\text{fin}} \mathbb{Q}, s \in \mathbb{N}\},$$

where $\subseteq_{\text{fin}}$ means that F is a *finite* subset of $\mathbb{Q}$. The idea is that $R_{F,s}$ starts by being F at stage s and then every time it is threaten, we add a new

---

[53]If β_0 and β_1 are both left c.e. with effective approximations $\beta_0[s]$ and $\beta_1[s]$, we can compute $X \upharpoonright n$ by searching for some $F \subseteq \{0, \ldots, n-1\}$ and stage $s \in \mathbb{N}$ so that both $\beta_0[s] \geq \sum_{j \in F} 2^{-i}$ and $\beta_1[s] \geq \sum_{j \in n \setminus F} 2^{-j}$; we then know $X \upharpoonright n = F$.

element to $R_{F,s}$ so as to change its maximum value. To define $R_{F,s}$, we will define a non-decreasing sequence $\{r_{F,s}[t] : t \in \mathbb{N}\} \subset \mathbb{Q}$ and then let

$$R_{F,s} = F \cup \{r_{F,s}[t] : t \in \omega\}.$$

Let $r_{F,s}[t] = \max(F)$ for all $t \leq s$. At stage $t + 1 > s$, if $r_{F,s}[t] = \max(Q_{n,t})$, let $r_{F,s}[t + 1] = \max(F) + \alpha_t$, where $Q_{n,t}$ and α_t are the stage-t approximation to Q_n and α. We claim that this sequence eventually stabilizes. Otherwise, we would have that

$$\sup(Q_n) = \lim_t \max(Q_{n,t}) = \lim_t r_{F,s}[t] = \max(F) + \alpha,$$

contradicting that $\max(F) + \alpha$ is not left c.e. Let $r_{F,s} = \lim_t r_{F,s}[t]$. Then $r_{F,s} \neq \max(Q_n)$ and $R_{F,s} \in \mathcal{F}_n$. On the other hand, for every finite $F \subseteq \mathbb{Q}$ for which $\max(F) \neq \max(Q_n)$, we have that $R_{F,s} = F$ for large enough s: Take s so that $\max(Q_{n,t}) \neq \max(F)$ for all $t > s$, and hence so that $r_{F,s}[t] = \max(F)$ for all $t \in \mathbb{N}$.

For the second part of the theorem, let us assume that V is a c.e. operator such that V^X is an enumeration of $\mathcal{F}$ for every non-computable X, and let us try to get a contradiction. For this, we will define a uniformly c.e. sequence $\{M_n : n \in \mathbb{N}\}$ of finite subsets of $\mathbb{Q}$ with $\max(M_n) \neq \max(Q_n)$. This will give us a contradiction because, if f is a computable function such that $Q_{f(n)} = M_n$, then the recursion theorem (see page xv) gives an n_0 with $W_{f(n_0)} = W_{n_0}$, and hence with $M_{n_0} = Q_{f(n_0)} = Q_{n_0}$.

Using the operator V, we can easily produce a uniform family of c.e. operators $\{U_n : n \in \mathbb{N}\}$ such that $U_n^X \subseteq \mathbb{Q}$ is finite and $\max(U_n^X) \neq \max(Q_n)$ for all non-computable X and $n \in \mathbb{N}$: Search for a column of V^X of the form $F \oplus \{n\}$ for some F (i.e., a column that contains the number $2n + 1$), and let U_n^X be the left side of that column, namely F.

For $X \in 2^{\mathbb{N}}$, let

$$m_n^X = \sup(U_n^X) \in \mathbb{R} \cup \{\infty\},$$

which we know is actually a maximum in $\mathbb{Q}$ when X is non-computable. For $\sigma \in 2^{<\mathbb{N}}$, let $m_n^\sigma = \max(U_n^\sigma) \in \mathbb{Q} \cup \{-\infty\}$, where U_n^σ is the step-$|\sigma|$ approximation to U_n^X for $X \supset \sigma$, and where $m_n^\sigma = -\infty$ if $U_n^\sigma = \emptyset$. The map $\sigma \mapsto m_n^\sigma$ has the following properties:

- $\sigma \subseteq \tau \Longrightarrow m_n^\sigma \leq m_n^\tau$.
- If $X \in 2^{\mathbb{N}}$ is non-computable, then $m_n^X = m_n^{\sigma_X}$ for some finite $\sigma_X \subset X$.
- If $X \in 2^{\mathbb{N}}$ is non-computable, $m_n^X \neq \max(Q_n)$.

Let $T \subseteq 2^{<\mathbb{N}}$ be a computable tree with no computable paths.[54] The idea is to use T to define $\{M_n : n \in \mathbb{N}\}$ so that M_n's maximum element is the

---

[54]For instance, let $T = \{\sigma \in 2^{<\mathbb{N}} : \forall e < |\sigma| \, (\sigma(e) \neq \Phi_{e,|\sigma|}(e))\}$ whose paths are called *2-diagonally non-computable*.

minimum value of m_n^X among all the $X \in [T]$. Since such $X \in [T]$ would be non-computable, we would have that $\max(M_n) = m_n^X \neq \max(Q_n)$. Let

$$\gamma = \inf\{m_n^X : X \in [T]\} \in \mathbb{R} \cup \{-\infty\};$$

we will show that γ is actually a minimum. Consider the following sequence approximating γ:

$$\gamma[k] = \min(m_n^\sigma : \sigma \in T \cap 2^k).$$

Since $m_n^X \geq m_n^{X \restriction k}$ for all X and k, we get that $\gamma \geq \gamma[k]$. First, we claim that this sequence becomes constant and equal to γ from some point on. To see this, let us observe that the sub-tree $\{\sigma \in T : m_n^\sigma < \gamma\}$ must be finite: Otherwise, by König's lemma, it would have a path $Y \in [T]$. But then $m_n^Y = m_n^{\sigma_Y} < \gamma$ contradicting the definition of γ. So if k_0 bounds the lengths of all the strings in that tree, $\gamma[k] = \gamma$ for all $k \geq k_0$.

Second, we claim that $\gamma = m_n^X$ for some $X \in [T]$. To see this, let us observe that the tree $\{\sigma \in T : m_n^\sigma \leq \gamma\}$ must have a path: Otherwise, by König's lemma, the tree would be finite, and if k_0 bounds the lengths of all the strings in that tree, we would get $\gamma[k_0] > \gamma$, which we know does not happen. So, if X is a path through that tree, $m_n^X = m_n^{\sigma_X} \leq \gamma$ and hence m_n^X is minimum among all $X \in [T]$. It follows that $\gamma = m_n^X \neq \max(Q_n)$.

Finally, let

$$M_n = \{\gamma[k] : k \in \mathbb{N}\}.$$

Then M_n must be finite and have maximum element $\gamma \neq \max(Q_n)$. It is not hard to see that $\{M_n : n \in \omega\}$ is c.e. uniformly in n. This finishes the construction of M_n and the proof that $\mathcal{G}_{\mathcal{F}}^\infty$ cannot be uniformly computed from all non-computable sets. □

6.2. Turing-computable embeddings

Medvedev reductions can also be used to compare the complexity of classes of structures. One of the objectives of this reducibility is to reduce the problem of deciding if two structures in a class $\mathbb{K}$ are isomorphic to the problem of deciding if two structure in another class $\mathbb{S}$ are isomorphic.

DEFINITION 6.2.1. (Calvert, Cummins, Knight, and Miller [CCKM04] and Knight, Miller, and Vanden Boom [KMVB07]). A *Turing-computable embedding* from class of structures $\mathbb{K}$ to a class of structures $\mathbb{S}$ is a Turing operator Φ such that, for every ω-presentation $\mathcal{A} \in \mathbb{K}$, $\Phi^{D(\mathcal{A})}$ outputs the diagram $D(\mathcal{B})$ of some ω-presentation $\mathcal{B} \in \mathbb{S}$ and satisfies that

$$\mathcal{A} \cong \tilde{\mathcal{A}} \iff \Phi(\mathcal{A}) \cong \Phi(\tilde{\mathcal{A}})$$

for all structures $\mathcal{A}, \tilde{\mathcal{A}} \in \mathbb{K}$, where $\Phi(\mathcal{A})$ denotes the ω-presentation $\mathcal{B}$ with $\Phi^{D(\mathcal{A})} = D(\mathcal{B})$.

Turing-computable embeddings are useful to compare the complexity of the isomorphism problems of different classes of structures. Later we will see stronger notions of embeddings between structures where we can reduce more properties from one class to the other. The weaker notion that inspired the definition of Turing-computable embeddings is that of *Borel embedding*, where Φ is allowed to be Borel. It was introduced by Friedman and Stanley in [FS89], a seminal paper for what today is the field of Borel equivalence relations in Descriptive Set Theory; we will see more about this in [MonP2].

Let us see some examples.

6.2.1. Examples. Let $\mathbb{VS}$ be the class of $\mathbb{Q}$-vector spaces (as in Example 2.1.1) and let $\mathbb{LO}$ be the class of linear orderings. The following is a simple example that gives a clear picture of how the isomorphism problem for vector spaces is coded into linear orderings.

LEMMA 6.2.2. *There is a Turing-computable embedding from $\mathbb{VS}$ to $\mathbb{LO}$.*

PROOF. We build a Turing operator that, given a vector space, produces a linear ordering isomorphic to $\omega \times (1 + n)$,[55] where $n \in \mathbb{N} \cup \{\infty\}$ is the dimension of the vector space. The construction goes as follows. Let $\mathcal{V}$ be an ω-presentation of a vector space that we receive as the oracle for our Turing operator. The first step is to analyze $\mathcal{V}$ and try to guess its dimension. List the vectors in $\mathcal{V}$ as $v_0, v_1, v_2, \ldots$, and assume v_0 is the zero vector. As we discover more and more about $\mathcal{V}$, we try to pick out a basis of $\mathcal{V}$. Of course, we will make plenty of mistakes in the process, as deciding if a tuple of vectors is linearly independent requires checking an infinite amount of linear combinations. We say that a tuple of vectors is *s-linearly independent* if there is no non-trivial linear combination of them that equals zero and where all the coefficients are among the first s rational numbers. At each stage s, we define our guess $v_{t_0[s]}, v_{t_1[s]}, \ldots, v_{t_{k[s]}[s]}$ for a basis of $\mathcal{V}$ as follows: We always include v_1 in our basis; thus, let $t_0[s] = 1$. Given $t_0[s], t_1[s], \ldots, t_i[s]$, let $t_{i+1}[s]$ be the first $t > t_i[s]$ which is s-linearly independent with $v_{t_0[s]}, v_{t_1[s]}, \ldots, v_{t_i[s]}$. If there is no such $t \leq s$, we leave $t_{i+1}[s]$ undefined and let $k[s] = i$. Notice that if $\mathcal{V}$ has dimension $n \in \mathbb{N} \cup \{\infty\}$, then for each $i < n$, $v_{i[s]}$ eventually stabilizes to the first vector in the list that is linearly independent from the previous ones. These eventually stable vectors will end up forming a basis for $\mathcal{V}$.

Let us now build our linear ordering. The idea is to assign a copy of ω to each basis element, ending up with n copies of ω, plus an extra one used for garbage collection. At each stage s we will assign, to each vector that looks like it is in the basis, a linear ordering of size s which grows to the right. Thus, for the vectors that remain in the basis for ever, we end up building a copy of ω. For some vectors we will eventually stop believing they are in the basis. To them we associate a finite linear ordering that will

---

[55]$\omega \times m$ consist of m copies of ω, one after the other.

become part of the ω-chain to its right. Let us repeat this construction a bit more carefully. To each vector v_t we associate a non-trivial linear ordering $\mathcal{L}_t$ in a uniformly computable way, in such a way that $\mathcal{L}_t \cong \omega$ if v_s is a basis vector and $\mathcal{L}_t$ is finite otherwise. To do this, we build $\mathcal{L}_t$ by adding a new element to its right at every stage s that v_t is among the vectors we believe are in the basis, namely $v_{t_0[s]}, v_{t_1[s]}, \ldots, v_{t_{k[s]}[s]}$. We then define

$$\mathcal{L} = \mathcal{L}_1 + \mathcal{L}_2 + \mathcal{L}_3 + \cdots + \mathcal{L}_t + \cdots .$$

If V has finite dimension n, then n of the $\mathcal{L}_t$'s will end up being isomorphic to ω, and the rest will stay finite though non-zero, ending up with $\mathcal{L} \cong \omega \times n + \omega$. If V has infinite dimension, then infinitely many of the $\mathcal{L}_t$'s will be isomorphic to ω, ending up with a copy of $\omega \times \omega$.   □

The reversal is obviously not true, i.e., $\mathbb{LO}$ does not embed into $\mathbb{VS}$, as there are only countably many countable $\mathbb{Q}$-vector spaces while there are continuum many linear orderings.

The class of linear orderings is actually as complicated as it can be in terms of Turing computability in the following sense:

DEFINITION 6.2.3. A class $\mathbb{K}$ is said to be *on top for Turing computability* if any other class of structures Turing-computably embeds in $\mathbb{K}$.

We will see in Section 6.3.2 that the class of graphs is on top for Turing computability; actually graphs are universal in a much stronger sense. We will use that result here to prove Friedman and Stanley's result that linear orderings are top for Turing computability. They also showed that fields are on top.

LEMMA 6.2.4 (Friedman and Stanley [FS89]). *Linear orderings are on top for Turing computability.*

PROOF. As a first intermediate step, consider the class $\mathbb{LT}$ of labeled trees, where the trees are viewed as directed graphs and each node in the tree is labeled with a natural number.[56] We claim that $\mathbb{LT}$ is on top for Turing computability: Given a structure $\mathcal{A}$, let $\mathcal{A}^\star$ be the structure with domain $A^\star$, which consist of the tuples from A without repeated entries, viewed as a tree, where each $\sigma \in A^\star$ is labeled with the number $\ulcorner D_{\mathcal{A}}(\sigma) \urcorner$ coding the atomic diagram of σ within $\mathcal{A}$.[57] This transformation from $\mathcal{A}$ to $\mathcal{A}^\star$ is clearly Turing computable. It is obvious from the construction that the isomorphism type of $\mathcal{A}^\star$ is independent of the ω-presentation of $\mathcal{A}$; that is, that if $\mathcal{A} \cong \mathcal{B}$, then $\mathcal{A}^\star \cong \mathcal{B}^\star$. We need to prove the converse, that if $\mathcal{A}^\star \cong \mathcal{B}^\star$, then $\mathcal{A} \cong \mathcal{B}$.

Given a labeled tree $\mathcal{A}^\star = (A^\star; P, \ell)$, where P is the parent relation and $\ell : A^\star \to 2^{\mathbb{N}}$ is the labeling function, we will recover a copy of the

---

[56]That is, the structures in $\mathbb{LT}$ are of the form $(T; P, \ell)$ where $P : T \to T$ is the parent function of the tree and $\ell : T \to \mathbb{N}$.

[57]That is, $\mathcal{A}^\star = (A^\star; P, \ell)$ where $P(\sigma) = \sigma \upharpoonright |\sigma| - 1$ and $\ell(\sigma) = \ulcorner D_{\mathcal{A}}(\sigma) \urcorner$.

original structure $\mathcal{A}$. To each path $g \in [A^\star] \subseteq A^{\mathbb{N}}$ through the tree $A^\star$, we associate a congruence-ω-presentation $\mathcal{A}_g$ with atomic diagram $\bigcup_{\sigma \subseteq g} \ell(\sigma) = \bigcup_{\sigma \subseteq g} D_\mathcal{A}(\sigma)$. Equivalently, $\mathcal{A}_g = g^{-1}(\mathcal{A})$. If g is onto A, and hence a bijection, then g is actually an isomorphism from $\mathcal{A}_g$ to $\mathcal{A}$. The problem is that we cannot recognize which paths g through $A^\star$ are onto if all we are given is $A^\star$ as a labeled tree, and hence, in general, $\mathcal{A}_g$ need not be isomorphic to $\mathcal{A}$. However, we claim that for almost all $g \in [A^\star]$, in the sense of category, g is onto. To see this, observe that for each $a \in A$, the set of g which contain a in its image is dense and open in $[A^\star]$. The intersection of these sets among all $a \in A$ is thus comeager.[58] By the Baire category theorem, a countable intersection of dense open sets cannot be empty, thus $\mathcal{A}$ is the unique structure for which there is a co-meager set of g's such that $\mathcal{A}_g \cong \mathcal{A}$. This finishes the proof of the claim that $\mathbb{LT}$ is on top for Turing computability.

For the rest of the argument, we need a small modification of the construction above to make each branch repeat infinitely often. Define $\mathcal{A}^{\star \infty}$ to be the structure with domain

$$\{\sigma \oplus \tau \in A^\star \times \mathbb{N}^{<\mathbb{N}} : |\sigma| = |\tau|\}.$$

We view $\mathcal{A}^{\star \infty}$ as a tree where the parent relation comes form the product of the parent relations in $A^\star \times \mathbb{N}^{<\mathbb{N}}$. We label a node $\sigma \oplus \tau$ with the number $\ulcorner D_\mathcal{A}(\sigma) \urcorner$. Is easy to see that the same proof as above shows that $\mathcal{A} \cong \mathcal{B} \iff \mathcal{A}^{\star \infty} \cong \mathcal{B}^{\star \infty}$. We call these structures *infinitely repeated labeled trees*.

As a second intermediate step, let us consider the class of *labeled dense linear orderings* $\mathbb{LDLO}$. These are dense linear orderings without endpoints (i.e., copies of the rationals) where each node is labeled with a natural number. We claim that $\mathbb{LDLO}$ are also on top for Turing computability: For this, we produce a Turing-computable embedding from infinitely repeated labeled trees to label dense linear orderings. Consider an infinitely repeated labeled tree $T \subseteq \mathbb{N}^{<\mathbb{N}}$ with labeling function $\ell_T : T \to \mathbb{N}$. As domain for our labeled linear ordering we will use $\mathbb{Q}^{<\mathbb{N}}$ ordered lexicographically, which is isomorphic to $\mathbb{Q}$. We now need to add the labels. By recursion on the length of the strings, define a length-preserving map $f : \mathbb{Q}^{<\mathbb{N}} \to T$ as follows. Start by mapping the empty string to the root of T. Suppose we have already defined $f(\sigma)$. To define f on $\sigma^\frown \mathbb{Q} = \{\sigma^\frown q : q \in \mathbb{Q}\}$, let f_σ be a map from $\mathbb{Q}$ onto $\{\tau \in T : P(\tau) = \sigma\}$ such that, for each τ in the image of f_σ, the pre-image $f_\sigma^{-1}(\tau)$ is dense in $\mathbb{Q}$. Then, for each $q \in \mathbb{Q}$, define $f(\sigma^\frown q) = f_\sigma(q)$. Let us observe that, even though f is not one-to-one, (T, ℓ_T) and $(\mathbb{Q}^{<\mathbb{N}}, \ell_T \circ f)$ are isomorphic as labeled trees, because T is an *infinitely* repeated labeled tree. Label each $\sigma \in \mathbb{Q}^{<\mathbb{N}}$ with a

---

[58]A set $X \subseteq \mathbb{N}^{\mathbb{N}}$ is *comeager* if it contains a countable intersection of dense open sets. See page 51 for the connection with generic sets.

number coding the corresponding label in the tree and the length of the node, namely $\langle \ell_T(f(\sigma)), |\sigma| \rangle$.

It is easy to see that this is a computable procedure. To see that isomorphic trees yield isomorphic labeled linear orderings, start from an isomorphism between two labeled trees and build an isomorphism between their images by recursion on the length of the strings, using the fact that two dense labeled linear orderings with infinitely many dense labels are isomorphic. For the converse, suppose we have a labeled dense linear ordering $\mathcal{L} = (L; \leq_L, \ell_L)$ that is in the image of this map. We need to define a labeled tree that is isomorphic to the one used to build $\mathcal{L}$. For this, we define a tree structure on L by defining a parent relation between elements of L as follows. Recall that for $s \in L$, its label records two numbers, the second coordinate $\pi_2(\ell_L(s))$ being the length of the string σ in the tree that was associated with s. We say that $t \in L$ is the *parent* of s if $t <_L s$, $\pi_2(\ell_L(t)) = \pi_2(\ell_L(s)) - 1$, and for all $r \in L$, with $t <_L r <_L s$, $\pi_2(\ell_L(r)) \geq \pi_2(\ell_L(s))$. Now that we have a tree structure with domain L, we define a labeling function: For $s \in L$, let $\ell_T(s) = \pi_1(\ell_L(s))$. We have built the original labeled tree back.

The third step is to reduce $\mathbb{LDLO}$ to unlabeled $\mathbb{LO}$. This is the easiest step. Given a labeled dense linear ordering, replace each element $s \in L$ by a finite linear ordering with $\ell(s) + 2$ elements. $\qquad \square$

We saw above that a structure $\mathcal{A}$ can be recovered from its tree of tuples $\mathcal{A}^{\star\infty}$ but in a rather cumbersome way. Harrison-Trainor and Montalbán [HTM] recently showed that $\mathcal{A}$ cannot be recovered computably, that there is a non-computable structure $\mathcal{A}$ for which $\mathcal{A}^{\star\infty}$ has a computable copy.

When a class of structures in not on top for Turing computability, it must have a special property not all classes have. Thus, understanding why a class of structures is not on top can give valuable information about it. As we mentioned above, vector spaces are not on top because there are only countably many of them. The same reason holds for algebraically closed fields. Equivalence structures, of which there are continuum many, are not on top for a different reason: Their isomorphism problem is too simple. Deciding whether two equivalence structures are isomorphic requires four jumps, as all one needs to do is to count how many equivalence classes are of each size.[59] There are classes of structures whose isomorphism problems are much more complex than that. For instance, in [MonP2] we will see tools for showing that deciding whether a linear ordering is isomorphic to either ω^n or ω^{n+1} is Π^0_{2n+1}-complete. Furthermore, deciding if two ω-presentations of linear orderings are isomorphic is Σ^1_1-complete. Another class whose isomorphism problem is too simple is finitely branching trees: They are $\exists$-atomic and hence their isomorphism types are determined by

---

[59]The index set of the equivalence structures with infinitely many infinite equivalence classes is Π^0_4-complete.

their $\exists$-diagrams. The same applies to torsion-free abelian groups of finite rank (i.e., subgroups of $(\mathbb{Q}^n; +)$ for some $n \in \mathbb{N}$). Deciding if two such groups is isomorphic is Σ_3^0-complete. Whether torsion-free abelian groups of arbitrary rank are on top is not known and has been open since [FS89]. The class of abelian p-groups is also not on top (cf. [FS89]) because of a much more complicated reason which we will see in [MonP2].

6.3. Computable functors and effective interpretability

Let us go back to reducibilities between structures. There is a third important notion of reducibility between structures. It has more structural consequences and even has a structural characterization in terms of interpretations. The idea is to require a Medvedev reduction Φ to preserve isomorphisms effectively.

DEFINITION 6.3.1. (Miller, Poonen, Schoutens, & Shlapentokh [MPSS, Definition 3.1]). Given structures $\mathcal{A}$ and $\mathcal{B}$, a *computable functor* from $\mathcal{B}$ to $\mathcal{A}$ consists of two computable operators, Φ and Ψ, such that:

1. Φ is a Medvedev reduction witnessing $\mathcal{A} \leq_s \mathcal{B}$; that is, for every copy $\widehat{\mathcal{B}}$ of $\mathcal{B}$, $\Phi^{D(\widehat{\mathcal{B}})}$ is the atomic diagram of a copy of $\mathcal{A}$.
2. For every isomorphism f between two copies $\widehat{\mathcal{B}}$ and $\widetilde{\mathcal{B}}$ of $\mathcal{B}$, $\Psi^{D(\widehat{\mathcal{B}}),f,D(\widetilde{\mathcal{B}})}$ is an isomorphism between the copies of $\mathcal{A}$ obtained from $\Phi^{D(\widehat{\mathcal{B}})}$ and $\Phi^{D(\widetilde{\mathcal{B}})}$.

We also require that the operator Ψ preserve the identity and composition of isomorphisms:

3. $\Psi^{D(\widehat{\mathcal{B}}),id,D(\widehat{\mathcal{B}})} = id$ for every copy $\widehat{\mathcal{B}}$ of $\mathcal{B}$.
4. $\Psi^{D(\mathcal{B}_0),g\circ f,D(\mathcal{B}_2)} = \Psi^{D(\mathcal{B}_1),g,D(\mathcal{B}_2)} \circ \Psi^{D(\mathcal{B}_0),f,D(\mathcal{B}_1)}$, for copies $\mathcal{B}_0$, $\mathcal{B}_1$ and $\mathcal{B}_2$ of $\mathcal{B}$ and isomorphisms $f\colon \mathcal{B}_0 \to \mathcal{B}_1$ and $g\colon \mathcal{B}_1 \to \mathcal{B}_2$.

The pair Φ, Ψ is a functor in the sense of category theory. It is a functor from the category of ω-presentations of $\mathcal{B}$ where morphisms are the isomorphisms between the copies of $\mathcal{B}$, to the category of ω-presentations of $\mathcal{A}$.

EXAMPLE 6.3.2. Let $\mathcal{B}$ be an integral domain (i.e., a commutative ring without zero-divisors) and let $\mathcal{A}$ be the field of fractions of $\mathcal{B}$. That is, $\mathcal{A}$ consists of element of the form $\frac{p}{q}$ for $p, q \in B$, $q \neq 0$. Equivalence, addition, and multiplication of fractions is defined as in the standard construction of $\mathbb{Q}$ from $\mathbb{Z}$. One can easily build a computable functor that produces a copy of $\mathcal{A}$ out of a copy of $\mathcal{B}$ and maps isomorphisms between copies of $\mathcal{B}$ to the respective copies of $\mathcal{A}$. We let the reader check the details. We will develop this example further in Example 6.3.4 below.

We will prove that having a computable functor is equivalent to having an effective interpretation. Informally, a structure $\mathcal{A}$ is *effectively-interpretable* in a structure $\mathcal{B}$ if there is an interpretation of $\mathcal{A}$ in $\mathcal{B}$ as in model theory, but where the domain of the interpretation is allowed to be a subset of $\mathbb{N} \times B^{<\mathbb{N}}$ instead of just B^n, and where all sets in the interpretation are required to be "effectively definable" instead of elementary first-order definable.

Before giving the formal definition, we need to review one more concept. Recall that a relation R on $A^{<\mathbb{N}}$ is *uniformly r.i.c.e.* (*u.r.i.c.e.*) if there is a c.e. operator W such that $R^{\mathcal{B}} = W^{D(\mathcal{B})}$ for every copy $(\mathcal{B}, R^{\mathcal{B}})$ of $(\mathcal{A}, R)$. These are exactly the Σ_1^c-definable relations without parameters (Corollary 2.1.19). Analogously, R is *uniformly r.i. computable* if there is a computable operator Φ such that $R^{\mathcal{B}} = \Phi^{D(\mathcal{B})}$ for every copy $(\mathcal{B}, R^{\mathcal{B}})$ of $(\mathcal{A}, R)$. Recall that a relation is u.r.i. computable if and only if it is Δ_1^c-definable without parameters.

DEFINITION 6.3.3. Let $\mathcal{A}$ be a τ-structure, and $\mathcal{B}$ be any structure. Let us assume that τ is a relational vocabulary $\tau = \{P_i : i \in I\}$ where P_i has arity $a(i)$. So $\mathcal{A} = (A; P_0^{\mathcal{A}}, P_1^{\mathcal{A}}, \dots)$ and $P_i^{\mathcal{A}} \subseteq A^{a(i)}$.

We say that $\mathcal{A}$ is *effectively-interpretable* in $\mathcal{B}$ if, in $\mathcal{B}$, there are u.r.i. computable relations $A^{\mathcal{B}}$, $\sim^{\mathcal{B}}$, and $\{R_i^{\mathcal{B}} : i \in I\}$ such that

- $A^{\mathcal{B}} \subseteq \mathbb{N} \times B^{<\mathbb{N}}$ (the domain of the interpretation of $\mathcal{A}$ in $\mathcal{B}$),
- $\sim^{\mathcal{B}} \subseteq A^{\mathcal{B}} \times A^{\mathcal{B}}$ is an equivalence relation on $A^{\mathcal{B}}$ (interpreting equality),
- each $R_i^{\mathcal{B}} \subseteq (A^{\mathcal{B}})^{a(i)}$ is closed under the equivalence $\sim^{\mathcal{B}}$ (interpreting the relations P_i),

and there is a function $f_{\mathcal{A}}^{\mathcal{B}} : A^{\mathcal{B}} \to A$ which induces an isomorphism:
$$(A^{\mathcal{B}}/\sim^{\mathcal{B}}; R_0^{\mathcal{B}}, R_1^{\mathcal{B}}, \dots) \cong (A; P_0^{\mathcal{A}}, P_1^{\mathcal{A}}, \dots).$$

Let us clarify this last line. The function $f_{\mathcal{A}}^{\mathcal{B}} : A^{\mathcal{B}} \to A$ must be an onto map such that $f_{\mathcal{A}}^{\mathcal{B}}(\bar{a}) = f_{\mathcal{A}}^{\mathcal{B}}(\bar{b}) \iff \langle \bar{a}, \bar{b} \rangle \in \sim^{\mathcal{B}}$ and $f_{\mathcal{A}}^{\mathcal{B}}(\bar{a}) \in P_i^{\mathcal{A}} \iff \bar{a} \in R_i^{\mathcal{B}}$ for all $\bar{a}, \bar{b} \in (A^{\mathcal{B}})^{<\mathbb{N}}$. Notice that there is no restriction on the complexity or definability of $f_{\mathcal{A}}^{\mathcal{B}}$. We use $\mathcal{A}^{\mathcal{B}}$ to denote the structure $(A^{\mathcal{B}}/\sim^{\mathcal{B}}; R_0^{\mathcal{B}}, R_1^{\mathcal{B}}, \dots)$.

If we add parameters, this notion is equivalent to that of Σ-definability, introduced by Ershov [Ers96] and is widely studied in Russia. Ershov's definition is quite different in format: it uses $HF_{\mathcal{B}}$ instead of $\mathbb{N} \times B^{<\mathbb{N}}$ (see Section 2.4.1) and sets that are $\exists$-definable over $HF_{\mathcal{B}}$ instead of Σ_1^c-definable subsets of $\mathbb{N} \times B^{<\mathbb{N}}$ (which we know are equivalent; Theorem 2.4.3).

EXAMPLE 6.3.4. Recall Example 6.3.2 above where $\mathcal{B}$ is an integral domain and $\mathcal{A}$ its field of fractions. We claim that $\mathcal{A}$ is effectively interpretable in $\mathcal{B}$. Let $A^{\mathcal{B}} = \{\langle p, q \rangle \in B^2 : q \neq 0\}$. Let $\langle p_0, q_0 \rangle \sim^{\mathcal{B}} \langle p_1, q_1 \rangle$ if $p_0 \times^{\mathcal{B}} q_1 = p_1 \times^{\mathcal{B}} q_0$. Define the graph of addition for $\mathcal{A}^{\mathcal{B}}$ to be the set of triplets of pairs $\langle \langle p_0, q_0 \rangle, \langle p_1, q_1 \rangle, \langle p_2, q_2 \rangle \rangle \in B^{2^3}$ that satisfy

$(p_0 \times^B q_1 +^B p_1 \times^B q_0) \times^B q_2 = q_0 \times^B q_1 \times^B p_2$. Define the graph of multiplication for $\mathcal{A}^B$ to be the set of triplets of pairs $\langle \langle p_0, q_0 \rangle, \langle p_1, q_1 \rangle, \langle p_2, q_2 \rangle \rangle$ that satisfy $p_0 \times^B p_1 \times^B q_2 = q_0 \times^B q_1 \times^B p_2$.

LEMMA 6.3.5. *An effective interpretation of $\mathcal{A}$ in $\mathcal{B}$ induces a computable functor from $\mathcal{B}$ to $\mathcal{A}$.*

PROOF. Since A^B, $\sim^B$, and $\{R_i^B : i \in I\}$ are u.r.i. computable in $\mathcal{B}$, we have a computable operator that gives us those sets within any copy $\widehat{\mathcal{B}}$ of $\mathcal{B}$, using $D(\widehat{\mathcal{B}})$ as an oracle. Thus we have a computable operator Φ that, given $\widehat{\mathcal{B}} \cong \mathcal{B}$, outputs $D(\mathcal{A}^{\widehat{\mathcal{B}}})$, the atomic diagram of the congruence $(\subseteq \mathbb{N} \times \mathbb{N}^{<\mathbb{N}})$-presentation $\mathcal{A}^{\widehat{\mathcal{B}}}$ of $\mathcal{A}$ with domain $A^{\widehat{\mathcal{B}}} \subseteq \widehat{B}^{<\mathbb{N}} = \mathbb{N} \times \mathbb{N}^{<\mathbb{N}}$. Fixing a bijection between $\mathbb{N}$ and $\mathbb{N} \times \mathbb{N}^{<\mathbb{N}}$, and using Lemma 1.1.11, we get a computable operator Υ transforming congruence $(\subseteq \mathbb{N} \times \mathbb{N}^{<\mathbb{N}})$-presentations into injective ω-presentations. Both of these computable operators Φ and Υ preserve isomorphisms effectively; in other words, they can be easily made into computable functors. Composing these computable functor we get the computable functor $\Upsilon \circ \Phi$ we wanted. □

The following theorem shows the reversal. Furthermore, given a computable functor, we can get an effective interpretation that induces the original functor back, up to effective isomorphism of functors.

THEOREM 6.3.6. (Harrison-Trainor, Melnikov, Miller, and Montalbán [HTMMM]). *Let $\mathcal{A}$ and $\mathcal{B}$ be countable structures. The following are equivalent:*

1. *$\mathcal{A}$ is effectively interpretable in $\mathcal{B}$.*
2. *There is a computable functor from $\mathcal{B}$ to $\mathcal{A}$.*

We will prove this theorem in [MonP2] once we have developed more forcing techniques. The original proof from [HTMMM] does not use forcing, and the reader should be able to follow it with what we have learned so far. The proof using forcing (cf. [HTMM]) is much more informative and can be generalized to a broader setting.

6.3.1. Effective bi-interpretability. Effective interpretability and Σ-definability induce notions of equivalence between structures as usual: two structures are equivalent if they are reducible to each other. Σ-equivalence, the equivalence notion that comes from Σ-definability, has been widely studied. However, it still does not really capture the idea of two structures being "the same from a computability viewpoint." In this section, we introduce the more recent notion of effectively-bi-interpretability, which is a strengthening of Σ-equivalence. For this strengthening, we require the composition of the isomorphisms interpreting one structure inside the other and then interpreting the other back into the first one to be effective. We will show how most computability theoretic properties are preserved under this equivalence, and see some examples that show how it matches our intuitive notions of when two structures are essentially the same. Here is the formal definition:

DEFINITION 6.3.7 (Montalbán [Mon14a, Definition 5.1]).   Two structures, $\mathcal{A}$ and $\mathcal{B}$, are *effectively-bi-interpretable* if there exist effective-interpretations of each structure inside the other as in Definition 6.3.3 such that the compositions

$$f_{\mathcal{B}}^{\mathcal{A}} \circ \tilde{f}_{\mathcal{A}}^{\mathcal{B}} \colon B^{\mathcal{A}^{\mathcal{B}}} \to B \quad \text{and} \quad f_{\mathcal{A}}^{\mathcal{B}} \circ \tilde{f}_{\mathcal{B}}^{\mathcal{A}} \colon A^{\mathcal{B}^{\mathcal{A}}} \to A$$

are u.r.i. computable in $\mathcal{B}$ and $\mathcal{A}$, respectively.

Let us explain this messy notation. $B^{\mathcal{A}^{\mathcal{B}}} \subseteq \mathbb{N} \times (A^{\mathcal{B}})^{<\mathbb{N}} \subseteq \mathbb{N} \times (\mathbb{N} \times B^{<\mathbb{N}})^{<\mathbb{N}}$ is the domain of the interpretation of $\mathcal{B}$ within the interpretation of $\mathcal{A}$ within $\mathcal{B}$, and $\tilde{f}_{\mathcal{A}}^{\mathcal{B}} \colon \mathbb{N} \times (A^{\mathcal{B}})^{<\mathbb{N}} \to \mathbb{N} \times A^{<\mathbb{N}}$ is the obvious extension of $f_{\mathcal{A}}^{\mathcal{B}} \colon A^{\mathcal{B}} \to A$ from elements to tuples: $\tilde{f}_{\mathcal{A}}^{\mathcal{B}}(i, a_0, \dots, a_k) = \langle i, f_{\mathcal{A}}^{\mathcal{B}}(a_0), \dots, f_{\mathcal{A}}^{\mathcal{B}}(a_k) \rangle$. Notice that since $f_{\mathcal{B}}^{\mathcal{A}} \circ \tilde{f}_{\mathcal{A}}^{\mathcal{B}}$ is a partial function from $\mathbb{N} \times (\mathbb{N} \times B^{<\mathbb{N}})^{<\mathbb{N}}$ to B, it can be coded by a relation on $\mathbb{N} \times B^{<\mathbb{N}}$ which we require it to be u.r.i. computable.

Let us make a quick comment on non-relational vocabularies. We have defined bi-interpretability for relational vocabularies, because function symbols do not work well on congruence presentations. When the interpretations are injective, Definition 6.3.7 goes through without problems for non-relational vocabularies too.

In the next lemma, we see how effective-bi-interpretability preserves most computability theoretic properties.

LEMMA 6.3.8.   *Let $\mathcal{A}$ and $\mathcal{B}$ be effectively-bi-interpretable.*

(1) *$\mathcal{A}$ and $\mathcal{B}$ have the same degree spectrum.*
(2) *$\mathcal{A}$ is $\exists$-atomic if and only if $\mathcal{B}$ is.*
(3) *$\mathcal{A}$ is rigid if and only if $\mathcal{B}$ is.*
(4) *The automorphism groups of $\mathcal{A}$ and $\mathcal{B}$ are isomorphic.*
(5) *$\mathcal{A}$ is computably categorical if and only if $\mathcal{B}$ is.*
(6) *$\mathcal{A}$ and $\mathcal{B}$ have the same computable dimension.*
(7) *$\mathcal{A}$ has the c.e. extendibility condition if and only if $\mathcal{B}$ does.*
(8) *The index sets of $\mathcal{A}$ and $\mathcal{B}$ are Turing equivalent, provided $\mathcal{A}$ and $\mathcal{B}$ are infinite.*

(Of course, items (5), (6), and (8) assume $\mathcal{A}$ and $\mathcal{B}$ are computable.)

PROOF. Throughout this proof, assume that $\mathcal{A}$ is already the presentation $\mathcal{A}^{\mathcal{B}}$ that is coded inside $\mathbb{N} \times \mathcal{B}^{<\mathbb{N}}$, i.e., with domain $A^{\mathcal{B}}$, and $\tilde{\mathcal{B}}$ is the copy of $\mathcal{B}$ coded inside $\mathbb{N} \times \mathcal{A}^{<\mathbb{N}}$, i.e., with domain $B^{\mathcal{A}} = B^{\mathcal{A}^{\mathcal{B}}}$. We let f be the isomorphism from $\tilde{\mathcal{B}}$ to $\mathcal{B}$ obtained by $f = f_{\mathcal{B}}^{\mathcal{A}} \circ \tilde{f}_{\mathcal{A}}^{\mathcal{B}}$ which is Σ_1^c-definable.

For part (1), recall from Lemma 6.3.5 that there are computable functors between $\mathcal{A}$ and $\mathcal{B}$, and in particular, that they are Medvedev equivalent, and hence also Muchnik equivalent.

For part (2), suppose $\mathcal{A}$ is $\exists$-atomic, and hence that every automorphism orbit in $\mathcal{A}$ is $\exists$-definable. Take a tuple $\bar{b} \in \mathcal{B}^{<\mathbb{N}}$; we will show its orbit is also $\exists$-definable. Let $\bar{c} \in \widetilde{B}^{<\mathbb{N}} = (B^{\mathcal{A}^{\mathcal{B}}})^{<\mathbb{N}}$ be such that $f(\bar{c}) = \bar{b}$. The orbit of $\bar{c}$ is $\exists$-definable inside $\mathcal{A}^{\mathcal{B}}$, and since $\mathcal{A}^{\mathcal{B}}$ is Δ_1^c-definable in $\mathcal{B}$, the orbit of $\bar{c}$ is also Σ_1^c definable in $\mathcal{B}$.[60] Since f is Σ_1^c-definable in $\mathcal{B}$, the orbit of $\bar{b}$ is also Σ_1^c definable.[61] If an orbit is definable by a disjunction, it must be defined by one of its disjuncts,[62] and hence the orbit of $\bar{b}$ is $\exists$-definable in $\mathcal{B}$. It follows that $\mathcal{B}$ is $\exists$-atomic.

Part (3) is a particular case of (4), but its proof is still informative. Suppose $\mathcal{B}$ is not rigid, and let h be a nontrivial automorphism of $\mathcal{B}$. The automorphism h induces an automorphism of $\mathcal{B}^{<\mathbb{N}}$, which then induces an automorphism g_h of $\mathcal{A}^{\mathcal{B}}$, which then induces an automorphism h_{g_h} of $\mathcal{B}^{\mathcal{A}^{\mathcal{B}}}$. Since $f: \mathcal{B}^{\mathcal{A}^{\mathcal{B}}} \to \mathcal{B}$ is u.r.i. computable, it is invariant; that is, $f(\bar{a}) = b \iff f(h_{g_h}(\bar{a})) = h(b)$. In other words, $f \circ h_{g_h} = h \circ f$, and since h is nontrivial and f a bijection, h_{g_h} must be nontrivial too. It follows that the automorphism g_h of $\mathcal{A}$ cannot be trivial either.

For part (4), notice that in the previous paragraph we showed that the homomorphism $h \mapsto h_{g_h}: \operatorname{Aut}(\mathcal{B}) \to \operatorname{Aut}(\mathcal{B}^{\mathcal{A}^{\mathcal{B}}})$ has the same effect as the conjugation homomorphism induced by f, namely $h \mapsto f^{-1} \circ h \circ f: \operatorname{Aut}(\mathcal{B}) \to \operatorname{Aut}(\mathcal{B}^{\mathcal{A}^{\mathcal{B}}})$. Therefore, the composition of the following three maps is the identity on $\operatorname{Aut}(\mathcal{B})$: first the homomorphism $h \mapsto g_h: \operatorname{Aut}(\mathcal{B}) \to \operatorname{Aut}(\mathcal{A}^{\mathcal{B}})$; second the homomorphism $g \mapsto h_g: \operatorname{Aut}(\mathcal{A}^{\mathcal{B}}) \to \operatorname{Aut}(\mathcal{B}^{\mathcal{A}^{\mathcal{B}}})$; and third the inverse of the conjugation homomorphism induced by f. We thus get that they are all isomorphisms, and that $\operatorname{Aut}(\mathcal{B}) \cong \operatorname{Aut}(\mathcal{A}^{\mathcal{B}})$.

For part (5), we need the following observation. Let $\mathcal{B}_1$ and $\mathcal{B}_2$ be copies of $\mathcal{B}$. The point we need to make here is that if $\mathcal{A}^{\mathcal{B}_1}$ and $\mathcal{A}^{\mathcal{B}_2}$ are computably isomorphic, then so are $\mathcal{B}_1$ and $\mathcal{B}_2$: A computable isomorphism between $\mathcal{A}^{\mathcal{B}_1}$ and $\mathcal{A}^{\mathcal{B}_2}$ induces a computable isomorphism between $\mathcal{B}^{\mathcal{A}^{\mathcal{B}_1}}$ and $\mathcal{B}^{\mathcal{A}^{\mathcal{B}_2}}$, each of which is computably isomorphic to $\mathcal{B}_1$ and $\mathcal{B}_2$, respectively. Thus, if $\mathcal{A}$ is computably categorical, so is $\mathcal{B}$. For (6), we have that if $\mathcal{B}$ has k non-computably isomorphic copies $\mathcal{B}_1, \ldots, \mathcal{B}_k$, then the respective structures $\mathcal{A}^{\mathcal{B}_1}, \ldots, \mathcal{A}^{\mathcal{B}_k}$ cannot be computably isomorphic either. So the effective dimension of $\mathcal{A}$ is at least that of $\mathcal{B}$, and hence, by symmetry, they must be equal.

---

[60]To define the orbit of $\bar{c}$ in $\mathcal{B}$, replace, in its Σ_1^c definition inside $\mathcal{A}^{\mathcal{B}}$, each symbol in the vocabulary of $\mathcal{A}$ either by its Σ_1^c definition or its Π_1^c definition depending on whether the atomic sub-formula appears positively or negatively, and restrict the existentially quantified variables using the Σ_1^c definition of the domain of $\mathcal{A}^{\mathcal{B}}$.

[61]A tuple $\bar{y}$ is in the orbit of $\bar{b}$ if there exists a tuple $\bar{x}$ in the orbit of $\bar{c}$ such that $\langle \bar{x}, \bar{y} \rangle$ is in the graph of f.

[62]If orbit of a tuple is defined by a disjunction, one of the disjuncts must be true of the tuple, and it implies the whole disjunction, so it also defines the orbit.

For part (7), recall that we can decide if a structure has the c.e. embed-dability condition by looking at its degree spectrum, which we already proved is preserved under effective bi-interpretability.

In part (8), by the *index set* of a structure $\mathcal{A}$, we mean the set of all i's such that $\Phi_i \colon \mathbb{N} \to 2$ is the atomic diagram of a structure isomorphic to $\mathcal{A}$. Suppose we are given an index of a computable structure $\mathcal{C}$, and we want to decide if it is isomorphic to $\mathcal{B}$ using the index set of $\mathcal{A}$ as an oracle. Using the formulas in the effective interpretation of $\mathcal{A}$ in $\mathcal{B}$, we can produce a structure $\mathcal{A}^{\mathcal{C}}$ such that $\mathcal{A}^{\mathcal{C}} \cong \mathcal{A}^{\mathcal{B}}$ if $\mathcal{C} \cong \mathcal{B}$. We can then produce an index for $\mathcal{A}^{\mathcal{C}}$, and use the index set of $\mathcal{A}$ to check if it is isomorphic to $\mathcal{A}$. If it is not, then we know $\mathcal{C}$ is not isomorphic to $\mathcal{B}$. Otherwise, we need to check that the function $f_{\mathcal{B}}^{\mathcal{A}} \circ \tilde{f}_{\mathcal{A}}^{\mathcal{C}} \colon \mathcal{B}^{\mathcal{A}^{\mathcal{C}}} \to \mathcal{C}$ from the bi-interpretability does produce an isomorphism between $\mathcal{B}^{\mathcal{A}^{\mathcal{C}}}$ and $\mathcal{C}$. This would be enough because, since $\mathcal{A}^{\mathcal{C}} \cong \mathcal{A}$, we know $\mathcal{B}^{\mathcal{A}^{\mathcal{C}}} \cong \mathcal{B}$. Checking this is not computable though; it is $0''$-computable. However, all index sets compute $0''$ because we can use them to check totality of functions.[63]   □

We will see later that effectively-bi-interpretable also preserves Scott rank in [MonP2] and is preserved under taking jumps (see Remarks 9.0.12).

6.3.2. Making structures into graphs. In this section, we show how every structure is effectively-bi-interpretable with a graph. This result will allow us to reduce statements about structures in general to statements about graphs, sometimes making proofs simpler. Within the following classes of structures we can also effectively-bi-interpret all other structures: partial orderings, lattices, (cf. [HKSS02]) and fields (cf. [MPSS]); and if we add a few constants to the vocabulary also: integral domains, commutative semigroups, and 2-step nilpotent groups (cf. [HKSS02]). These classes are said to be *universal for effective-bi-interpretability*.

THEOREM 6.3.9. *For every structure $\mathcal{A}$, there is a graph $\mathcal{G}_{\mathcal{A}}$ that is effec-tively-bi-interpretable with $\mathcal{A}$.*

Furthermore, the interpretations are independent of the given structure. That is, given a vocabulary τ, the Σ_1^c formulas used to define the sets involved in the interpretations are the same for all τ-structures $\mathcal{A}$.

PROOF. We only sketch the construction and let the reader verify the details.

Similar constructions can be found in [HKSS02]. The earliest references we know of this type of coding into graphs are Rabin and Scott [RS] and Lavrov [Lav63].

Assume that τ is a relational vocabulary. The first step is to show that $\mathcal{A}$ is effectively-bi-interpretable with a structure $\mathcal{H}$ in the vocabulary

---

[63]An index set of a non-empty family of total functions is always Π_2^0-hard. To see this, let Φ be a computable function in the family. Given e, we can check if $W_e = \mathbb{N}$ which is a Π_2^0-complete question, by asking whether the index of $\Phi \upharpoonright W_e$ belongs to the index set.

$\{U, E\}$, where U is a unary relation and E a symmetric binary relation. The unary relation U picks out the elements that represent the domain of $\mathcal{A}$. The elements outside U are going to be used to code the relations in $\mathcal{A}$. Enumerate the domain of $\mathcal{A}$ as $\{a_0, a_1, \dots\}$ and let $h_0, h_1, \dots$ be the corresponding elements in $U^{\mathcal{H}}$. For each tuple $a_{i_1}, \dots, a_{i_k}$ satisfying the nth relation R_n in τ (of arity k), we attach the following configuration to $h_{i_1}, \dots, h_{i_k}$ in $\mathcal{H}$, where the top cycle has size $2n + 5$.

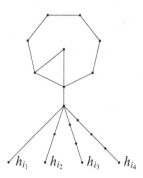

FIGURE 1. We call this configuration an *m-spider*, where m is the size of the top loop. (In this case $m = 7$ and $k = 4$.) The edges represent the pairs of elements that satisfy E. Let ℓ_m be the number of nodes in the spider ($\ell_m = 14$ in this case).

So that both the interpretation of R and that of its complement are $\exists$-definable, to each tuple $a_{i_1}, \dots, a_{i_k}$ not satisfying R_n we attach an $(2n+4)$-spider. (See Figure 1.)

It is clear that $\mathcal{A}$ can be effectively interpreted in $\mathcal{H}$: the domain of the interpretation is $U^{\mathcal{H}}$, and the interpretation of R_n is given by the set of tuples in $(U^{\mathcal{H}})^k$ that have a $(2n + 5)$-spider attached to them, which can be expressed by an $\exists$-formula. This set is also $\forall$-definable, because it is the set of tuples which do not have a $(2n + 4)$-spider attached to them.

Conversely, $\mathcal{H}$ can be interpreted in $\mathcal{A}$ as follows. Use A to interpret $U^{\mathcal{H}}$ and, for each m-spider attached to a tuple $h_{i_1}, \dots, h_{i_k}$, use the elements

$$\langle m, i, \langle a_{i_1}, \dots, a_{i_k} \rangle \rangle \in \mathbb{N} \times \mathbb{N} \times A^{<\mathbb{N}}, \quad \text{for } i < \ell_m,$$

to interpret its elements. The domain of this interpretation is u.r.i. computable because, given a tuple of the form $\langle m, i, \langle a_{i_1}, \dots, a_{i_k} \rangle \rangle$ with $i < \ell_m$, the tuple belongs to the interpretation if and only if $\langle a_{i_1}, \dots, a_{i_k} \rangle \in R_n^{\mathcal{M}}$, where $n = \lfloor (m - 3)/2 \rfloor$. Similarly, we can also decide which pairs of these elements are E-connected.[64]

---

[64]Recall from Remark 2.1.28 that it makes no difference to deal with subsets of $\mathcal{A}^{<\mathbb{N}}$ or of $\mathbb{N} \times \mathbb{N} \times \mathcal{A}^{<\mathbb{N}}$.

Checking that the compositions of the interpretations are u.r.i. computable is also straightforward: the composition of the interpretations going from $\mathcal{A}$ to $\mathcal{H}$ and back is the identity; the interpretation going from $\mathcal{H}$ to $\mathcal{A}^{<\mathbb{N}}$ and back to $\mathcal{H}^{<\mathbb{N}}$ is a bit more tedious, but not much harder to analyze.

The second step is to show that every $\{U, E\}$-structure $\mathcal{H}$ is effectively-bi-interpretable with a graph $\mathcal{G} = (G; R)$ without using an extra unary relation. Within G, we will use a subset, G_0, to interpret the domain of $\mathcal{H}$. We use the other elements of G to encode the relations U and E on G_0. Enumerate the elements of H as $\{h_0, h_1, \dots\}$ and the corresponding ones of G_0 as $g_0, g_1, \dots$. Attach to each element $g_i \in G_0$ either an A-flag or a B-flag as in Figure 2 depending on whether $h_i \in U^{\mathcal{H}}$ or not.

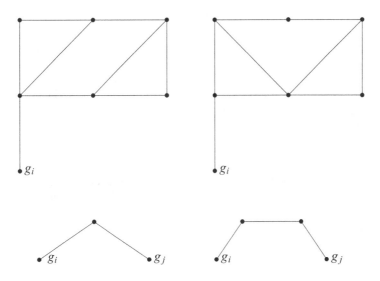

FIGURE 2. We call these configurations *A-flags, B-flags, 2-connectors,* and *3-connectors.* We attach A-flags to the elements that are in U and B-flags to the ones out of U. We use 2-connectors to encode E, and 3-connectors for the complement of E.

Connect two elements of G_0 using a 2-connector if and only if the corresponding elements in $\mathcal{H}$ are connected by E. (See Figure 2.) Connect them using a 3-connector if and only if the corresponding elements in $\mathcal{H}$ are not connected by E. The reason we cannot connect the elements of G_0 directly to code E is that we do not want to confuse the elements of G_0 with the ones used for the flags. This way, every element of G is either part of a flag (and hence out of G_0), attached to a flag (and hence in G_0), or attached to something that is attached to a flag (and hence part of either a

2-connector or a 3-connector, and out of G_0). Each of these three sets is ∃-definable, and hence G_0 is u.r.i. computable. Notice that the connectors coding the graph E among the elements of H do not get confused with these flags because since each edge in E is replaced by at least a 2-connector, the smallest cycles one could produce are 6-cycles coming from triangles in $\mathcal{H}$.

The relation E is coded by the pairs of elements of G_0 which are connected by a 2-connector or, equivalently, not connected by a 3-connector. This is u.r.i. computable.

Again, checking that the composition of the interpretations are u.r.i. computable is straightforward.   □

6.4. Reducible via effective bi-interpretability

We just defined a reduction that, for every structure $\mathcal{A}$, produces a graph $\mathcal{G}_\mathcal{A}$ effectively bi-interpretable with $\mathcal{A}$, making the class of graphs universal in any computability theoretic sense possible. Furthermore, the Σ_1^c formulas used in the effective bi-interpretation to define the domains and relations are always the same: They are independent of the τ-structure $\mathcal{A}$ and depend only on the vocabulary τ.

DEFINITION 6.4.1. A class of structures $\mathbb{K}$ is *reducible via effectively bi-interpretability* to a class $\mathbb{S}$ if there are Σ_1^c formulas defining the domains, relations, and isomorphisms of an effective bi-interpretation as in Definitions 6.3.3 and 6.3.7 so that every structure in $\mathbb{K}$ is bi-interpretable with a structure in $\mathbb{S}$.

If we also have that, under the backward direction of the bi-interpretation, every structure in $\mathbb{S}$ is interpreted by some structure in $\mathbb{K}$, we say that $\mathbb{K}$ and $\mathbb{S}$ are *effectively bi-interpretable*.

Classes that are effectively bi-interpretable are considered as the same class for computability theoretic purposes. Most, if not all, computability theoretic properties of classes of structures are invariant under effectively bi-interpretability.

Notice that $\mathbb{K}$ is reducible via effectively bi-interpretable to $\mathbb{S}$ if and only if $\mathbb{K}$ is effectively bi-interpretable with a sub-class of $\mathbb{S}$, the sub-class given by the image of the reduction. The next lemma shows that the complexity of this sub-class is similar to the complexity of $\mathbb{K}$.

LEMMA 6.4.2. *If $\mathbb{K}$ and $\mathbb{S}$ are effectively bi-interpretable and $\mathbb{K}$ is Π_2^c, then $\mathbb{S}$ is Π_2^c too.*

PROOF. Given a structure $\mathcal{S}$ that may or may not belong to $\mathbb{S}$, we can try to use the Σ_1^c formulas in the effective bi-interpretation to define a structure $\mathcal{A}^\mathcal{S}$ and then use the other direction of the interpretation to get $\mathcal{S}^{\mathcal{A}^\mathcal{S}}$. If $\mathcal{S}$ happens to be in $\mathbb{S}$, this would all work out and we would have that $\mathcal{A}^\mathcal{S}$ is in $\mathbb{K}$ and that $\mathcal{S}^{\mathcal{A}^{\mathcal{S}}}$ is isomorphic to $\mathcal{S}$. We claim that this characterizes $\mathbb{S}$. That is, that $\mathcal{S} \in \mathbb{S}$ if and only if

- the Σ_1^c formulas in the effective bi-interpretation define an actual bi-interpretation on $\mathcal{S}$,
- and the structure $\mathcal{A}^{\mathcal{S}}$ is in $\mathbb{K}$.

When we say that the Σ_1^c formulas in the effective bi-interpretation define an *actual bi-interpretation* on $\mathcal{S}$ we mean three things: First that the pairs of Σ_1^c formulas used to define the Δ_1^c relations that interpret the domain and relations of $\mathcal{A}^{\mathcal{S}}$ actually define Δ_1^c relations, that is, define complementary Σ_1^c relations; this can be checked with a Π_2^c sentence. Second, the same needs to be true about $\mathcal{S}^{\mathcal{A}^{\mathcal{S}}}$ inside $\mathcal{A}^{\mathcal{S}}$; this can also be checked with a Π_2^c about $\mathcal{A}^{\mathcal{S}}$, which can be translated into a Π_2^c sentence about $\mathcal{S}$ once we know the definition of $\mathcal{A}^{\mathcal{S}}$ within $\mathcal{S}$ is Δ_1^c. Third, that the maps defined as compositions of the interpretations are actual isomorphisms between $\mathcal{S}$ and $\mathcal{S}^{\mathcal{A}^{\mathcal{S}}}$, and between $\mathcal{A}^{\mathcal{S}}$ and $\mathcal{A}^{\mathcal{S}^{\mathcal{A}^{\mathcal{S}}}}$; this again can be stated as a Π_2^c sentence.

There might be structures $\mathcal{S}$ outside $\mathbb{S}$ where these formulas still produces a bi-interpretation. In that case, the structure $\mathcal{A}^{\mathcal{S}}$ would not be in $\mathbb{K}$, as if $\mathcal{A}^{\mathcal{S}}$ was in $\mathbb{K}$, then $\mathcal{S}^{\mathcal{A}^{\mathcal{S}}}$ would be in $\mathbb{S}$, and hence so would $\mathcal{S} \cong \mathcal{S}^{\mathcal{A}^{\mathcal{S}}}$. Since $\mathbb{K}$ is Π_2^c, one can write a Π_2^c sentence about $\mathbb{S}$ saying that $\mathcal{A}^{\mathcal{S}}$ in $\mathbb{K}$.

The conjunction of all these Π_2^c sentences gives us a Π_2^c sentence describing the structures in $\mathbb{S}$.   □

Theorem 6.3.9 above shows that every class of structures is reducible via effective bi-interpretability to the class of graphs. The lemma above shows that every Π_2^c class of structures is effective bi-interpretable with a Π_2^c class of graphs.

DEFINITION 6.4.3. A class of structure $\mathbb{S}$ is *on top for effective bi-interpretability* if every other class of structures reduces to it via effective bi-interpretability.[65]

A class that was recently shown to be on top for effective bi-interpretability is the class of fields. This was proved by Miller, Poonen, Schoutens, and Shlapentokh [MPSS] using functors. Fields were known to be on top for Turing-computable embeddings much earlier (cf. [FS89]), though that proof did not produce effective bi-interpretations.

Classes of structures that are on top for effective bi-interpretability present all the possible computability theoretic behaviors. Thus, once one proves that a class is on top for effective bi-interpretability, one knows that those structures will not have any particular computability theoretic property that was not already present on graphs. That is not to say that computability has nothing to say about them. For instance, there are

---

[65]Such classes are sometimes called *universal for effective bi-interpretability*.

sub-classes of the class of fields with various interesting properties. When a class of structures is not on top under effective bi-interpretability, it is because it has some special property not all classes have. If the class is already not on top for Turing-computable embeddings, then it is not on top for effective bi-interpretability. We mentioned various reasons why a class would not on top for Turing-computable embeddings at the end of Section 6.2.1. There are classes that are on top for Turing-computable embeddings but not for effective bi-interpretability. An example is linear orderings. Linear orderings are not top for effective bi-interpretability because they have the computable embeddability condition, a property that is preserved under effective bi-interpretability (Lemma 6.3.8(6)). The same is true for differentially closed fields.

Chapter 7

FINITE-INJURY CONSTRUCTIONS

The technique of finite-injury constructions is among the most important ones in computability theory, and is used throughout the field. It was introduced independently by Friedberg [Fri57b] and Muchnik [Muc56] to solve Post's problem, as we explain below. This technique is used to build computable objects using $0'$-computable information. On a computable construction, we can only guess at this non-computable information, so we will often be taking steps in a wrong direction based on wrong guesses. We will then need to be able to recover from those mistakes.

We will see two kinds of finite-injury constructions: priority constructions and true-stage constructions. Depending on the situation, one might be better than the other.

In a priority construction, one needs to build an object satisfying an infinite list of requirements whose actions are in conflict with one another. When we act to satisfy a requirement, we may *injure* the work done to satisfy other requirements. To control these injuries, requirements are listed in order of priority: Requirements are only allowed to injure weaker-priority requirements. In the type of constructions we will see, each requirement will be injured at most finitely many times, and hence there will be a point after which it is never injured again.

A true-stage construction works in quite a different way. It is based on a combinatorial device, the approximation of the true stages, which organizes our guesses on $0'$-computable information. One advantage of this combinatorial device is that it can be generalized to the iterates of the jump, even over the transfinite, as we will see in [MonP2].

7.1. Priority constructions

To show how priority constructions work, we give a full proof of the Friedberg–Muchnik solution to Post's problem, a seminal result in computability theory. Post [Pos44] asked whether there was a computably enumerable set that was neither computable nor Turing complete. That question was open for more than a decade, until Friedberg and Muchnik solved it independently by developing the method of finite-injury priority constructions.

We will see two other finite-injury priority constructions in Chapter 8 on computable categoricity. The reader interested in learning priority constructions should read Theorem 8.4.3 after fully understanding the proof below. The third finite-injury priority construction in the book, Lemma 8.5.8, is a beautiful construction, though is a bit more complicated.

THEOREM 7.1.1 (Friedberg [Fri57b], Muchnik [Muc56]). *There is a low, non-computable, computably enumerable set.*

A set $A \subseteq \mathbb{N}$ is *low* if its jump is as low as possible, namely $A' \equiv_T 0'$. Low sets cannot be complete. Note that for sets below $0'$, being low is equivalent to being generalized low.

PROOF. We build A as the union of a computable sequence of finite sets $A_0 \subseteq A_1 \subseteq A_2 \subseteq \cdots$ satisfying the following requirements for each $e \in \mathbb{N}$:

Negative requirements N_e: If $\Phi_{e,s}^{A_s}(e)\downarrow$ for infinitely many s's, then $\Phi_e^A(e)\downarrow$.

Satisfying the N_e requirements for all $e \in \mathbb{N}$ ensures that A is low: We would get that $e \in A'$ if and only if $\Phi_{e,s}^{A_s}(e)\downarrow$ for infinitely many s's. This makes A' a Π_2^0 set.[66] Since A' is already Σ_2^0, we get that A' is Δ_2^0.

Positive requirements P_e: If W_e is infinite, then $A \cap W_e \neq \emptyset$.

Satisfying the P_e requirements for all $e \in \mathbb{N}$ ensures that the complement of A is different from all the W_e's and hence A is not computable; well, that is unless A is co-finite. We will also make sure during the construction that A is co-infinite. Co-infinite c.e. sets which satisfy all the P_e requirements are said to be *simple sets*.

We list these requirements in decreasing order of priority as follows:

$$N_0, P_0, N_1, P_1, N_2, P_2, \ldots,$$

the ones to the left having stronger priority than those to the right. Notice that each requirement has only finitely many requirements that are stronger than it. We think of each requirement as an individual worker trying to achieve its goal. Except for possible injuries, the different requirements will work almost independently of each other. Let us look at each of these requirements individually.

Negative requirements N_e: The only way in which N_e would not be satisfied is if $\Phi_{e,s}^{A_s}(e)$ goes back and forth between converging and not converging infinitely often. What N_e needs to do if it sees that $\Phi_{e,s}^{A_s}(e)$ converges, is to try to preserve this computation forever by restraining elements from going into A below the use of this computation. Here is what N_e does at a stage s of the construction. Let r_e be the *use* of

---

[66]Because $a \in A'$ if and only if $\forall k \in \mathbb{N} \; \exists s \geq k \; (\Phi_{e,s}^{A_s}(e)\downarrow)$.

$\Phi_{e,s}^{A_s}(e){\downarrow}$, that is, the length of the initial segment of the oracle A_s used in the computation $\Phi_{e,s}^{A_s}(e){\downarrow}$. If the computation diverges, let $r_e = 0$. During the construction, N_e does not enumerate any number into A. Instead, it imposes a restraint on weaker-priority P_i requirements, not allowing them to enumerate elements below r_e into A. (This is why we call the N_i *negative* requirements.) N_e is not allowed to impose anything on stronger-priority requirements, which may enumerate elements below r_e and *injure* N_e.

Positive requirements P_e: It is the P_e requirements that enumerate elements into A. (This is why we call them *positive* requirements.) They will enumerate at most one element each. The plan to satisfy P_e is to wait until we see some number enter W_e and enumerate it into A. However, we cannot enumerate just any number, as there are a couple things we need worry about. First, P_e is not allowed to injure stronger-priority requirements. In other words, if we let $R_e = \max_{i \le e} r_i$, then P_e is not allowed to enumerate any number below R_e into A. Second, we want to make sure A is co-infinite. To do this, we only allow P_e to enumerate numbers that are greater than $2e$. The plan for P_e can now be restated as follows: At a stage $s > e$, if $W_{e,s} \cap A_s \ne \emptyset$, we consider P_e *done*, and we never do anything else for P_e again. Otherwise, if there is an $x \in W_{e,s}$ greater than $2e$ and greater than R_e, we say that P_e *requires attention*. Once P_e requires attention, it *acts* by enumerating such an x into A.

The construction: Let us now describe the full construction. At each stage s we define a finite set $A_{s+1} \supseteq A_s$, and at the end of stages we define $A = \bigcup_s A_s$. Let $A_0 = \emptyset$. At each stage $s > 0$, do the following. First, define r_e for each $e < s$; recall that r_e is the use of $\Phi_{e,s}^{A_s}(e)$. Second, check which requirements P_e, for $e < s$, require attention, and let them act; that is, for each $e < s$, if $W_{e,s} \cap A_s = \emptyset$ and there exists $x \in W_{e,s}$ with $x > \max(2e, R_e)$, add x to A_{s+1}. If no requirement requires attention, move on to the next stage without doing anything.

Verification: Each requirement P_e acts at most once. Therefore, a requirement N_e can be injured at most $e - 1$ times, and there is a stage after which it is never injured again. After this stage, if $\Phi_{e,s}^{A_s}(e)$ never converges again, N_e is satisfied. Otherwise, $\Phi_{e,t}^{A_t}(e){\downarrow}$ for some later stage t. At that stage t, N_e will define r_e to be the use of this computation. After t, no requirement of weaker priority is allowed to enumerate numbers below r_e. Since we are assuming all stronger-priority P_i requirements that ever act have acted already, we get that $A_t \restriction r_e$ is preserved forever (i.e. $A_t \restriction r_e = A \restriction r_e$), and hence so is the computation $\Phi_{e,t}^{A_t}(e){\downarrow}$, getting $\Phi_e^A(e){\downarrow}$. N_e is then satisfied. In either case, r_e is eventually constant; it is either eventually equal to zero if $\Phi_e^A(e){\uparrow}$, or eventually equal to the use of $\Phi_e^A(e){\downarrow}$. Since this is true for all e, R_e is eventually constant too.

Let us now verify that the requirements P_e are all satisfied. If a requirement P_e ever requires attention, it acts, and it is then satisfied forever. Suppose that, otherwise, there is a stage t after which P_e never requires attention again. Assume t is large enough so that R_e has reached its limit already. Either P_e does not require attention because it is done, in which case we are done, or because all the numbers in W_e are below $\max(2e, R_e)$. In that case, P_e is satisfied because W_e is finite.

Finally, let us notice that A is co-infinite, as it can have at most e elements below $2e$ for each e. This is because only the requirements P_i for $i < e$ are allowed to enumerate numbers below $2e$. $\square$

7.2. The method of true stages

Often in computability theory, we want to use Δ_2^0 information to construct computable objects. We then need to computably approximate or guess the Δ_2^0 information. This can get messy, and there are various ways to organize this guessing system. We will concentrate on the method of *true stages for the enumeration of* $0'$, introduced by Lachlan in [Lac73]. There are slightly different definitions in the literature; we use our own, which is quite flexible and applies to a large variety of situations. The reason for our choice is that, in [MonP2], we will be able to extend this notion throughout the hyperarithmetic hierarchy, obtaining a very powerful technique.

One way of approximating the halting problem $0'$ is by the sequence of finite sets

$$0'_s = \{e \in \mathbb{N} : \Phi_{e,s}(e)\!\downarrow\} \subseteq \mathbb{N}.$$

Notice $0'_s$ is finite. It is then natural to view $0'_s$ as a finite string, say by considering $0'_s \restriction m_s \in 2^{m_s+1}$, where $m_s = \max(0'_s)$.[67] A problem with $0'_s \restriction m_s$ is that it may be always wrong: It could happen that at no stage $s > 0$ is $0'_s \restriction m_s$ an initial segment of $0'$, viewed as a sequence in $2^{\mathbb{N}}$. This might be a problem for some constructions. Lachlan's idea was to consider $0'_s \restriction k_s$, where k_s is the least element enumerated into $0'$ at stage s (i.e., $k_s = \min(0'_s \smallsetminus 0'_{s-1})$). The key difference is that there are infinitely many stages where $0'_s \restriction k_s$ is correct, in the sense that $0'_s \restriction k_s$ is an initial string of $0' \in 2^{\mathbb{N}}$. Stages where our guesses for $0'$ are correct are called *true stages*.

We introduce a different approximation to the jump that enjoys better combinatorial properties. Instead of $0'$, we will use the increasing settling-time function for $0'$, which we call ∇. At each stage s, we will computably define a finite string $\nabla_s \in \mathbb{N}^{<\mathbb{N}}$ which tries to approximate $\nabla \in \mathbb{N}^{\mathbb{N}}$. A true stage will be one where ∇_s is correct; i.e., it is an initial segment of ∇. One of the main advantages of using ∇ and ∇_s is that they relativize easily, allowing us to iterate them, as we will see in [MonP2].

---

[67]Recall that $X \restriction m$ is $\{x \leq m : x \in X\}$, or, when viewed as strings, it is the initial segment of X of length $m + 1$.

7.2.1. The increasing settling-time function. The settling-time function of a c.e. set measures the speed at which its elements are enumerated. That is, the *settling time* of an enumeration $\{A_s : s \in \mathbb{N}\}$ of a c.e. set A at n is the least s such that $A_s \restriction n = A \restriction n$. The settling-time function has many uses in various constructions, and we will see a couple of examples in Subsection 7.2.3. We will deviate slightly from the standard settling-time function to consider the strictly increasing version. For now, let us fix an enumeration of the halting problem, and concentrate on it.

7.2.1.1. *The definition of* ∇. The settling-time function of a set measures the time a given enumeration takes to settle on an initial segment of the set. The increasing settling-time function is the least strictly-increasing function ∇ such that $0'_{\nabla(i)} \restriction i = 0' \restriction i$ for every $i \in \mathbb{N}$:

DEFINITION 7.2.1. The *i-th true stage* (*in the enumeration of* $0'$), denoted $\nabla(i)$, is defined by recursion on i by any of the following three equivalent definitions:[68]

$$\nabla(i) = \text{the least } t > \nabla(i-1) \text{ such that } 0'_t \restriction i = 0' \restriction i,$$
$$= \text{the least } t > \nabla(i-1) \text{ such that } \Phi_i(i)\downarrow \iff \Phi_{i,t}(i)\downarrow,$$
$$= \begin{cases} \nabla(i-1)+1 & \text{if } \Phi_i(i) \text{ diverges,} \\ \nabla(i-1)+1 & \text{if } \Phi_i(i) \text{ converges by stage } \nabla(i-1), \\ \mu t\ (\Phi_{i,t}(i)\downarrow) & \text{if } \Phi_i(i) \text{ converges after stage } \nabla(i-1). \end{cases}$$

We use the value $\nabla(-1) = -1$ as the base case for the recursion, so that $\nabla(0) \geq 0$. We call t a *true stage* if $t = \nabla(i)$ for some i. We call ∇ the *increasing settling-time function* for $0'$.

Observe that $\nabla \equiv_T 0'$: Clearly $\nabla \leq_T 0'$. For the other reduction, notice that $i \in 0' \iff i \in 0'_{\nabla(i)}$.

LEMMA 7.2.2. *The set of true stages is co-c.e., and the set*

$$\{\langle i, t\rangle \in \mathbb{N}^2 : t < \nabla(i)\}$$

is c.e.

PROOF. Let us first observe that the set of initial segments of ∇, $\{\nabla \restriction i : i \in \mathbb{N}\} \subseteq \mathbb{N}^{<\mathbb{N}}$, is Π_1^0. To see this, note that given $\sigma \in 2^{<\mathbb{N}}$, $\sigma \subseteq \nabla$ if and only if, for every $e < |\sigma|$,

- either $\Phi_e(e)\uparrow$ and $\sigma(e) = \sigma(e-1) + 1$,
- or $\Phi_{e,\sigma(e)}(e)\downarrow$ and $\sigma(e)$ is the least $t > \sigma(e-1)$ such that $\Phi_{e,t}(e)\downarrow$.

Notice the first item is Π_1^0 and the second computable. It follows that the set of true stages, i.e., the image of ∇, is Π_1^0: This is because t is a true stage if and only if there exists an increasing finite string σ whose last value is t (and the previous values are less than t) that is an initial segment of ∇.

---

[68] Recall that $\mu t\ \varphi(t)$ denotes the least t that satisfies $\varphi(t)$.

As for the second part of the statement, $t > \nabla(i)$ if and only if some $\sigma \in (t+1)^{i+1}$ is an initial segment of ∇. $\qquad \square$

7.2.2. Domination properties. One of the useful properties of ∇ is that it grows rapidly when compared to computable functions. We characterize it below as the fastest ω-c.a. function up to computable speed up. For functions $f, g \colon \mathbb{N} \to \mathbb{N}$, we say that

- f *majorizes* g if $(\forall m)\ f(m) \geq g(m)$;
- f *dominates* g if $(\exists n)\ (\forall m \geq n)\ f(m) \geq g(m)$.

The function ∇ is fast growing in this sense: It dominates all computable functions (Exercise 7.2.5), and every function that dominates ∇ computes $0'$ (Lemma 7.2.3).

LEMMA 7.2.3. *If* $g \colon \mathbb{N} \to \mathbb{N}$ *dominates* ∇, *then* g *computes* $0'$.

PROOF. First, modify the first few values of g to get a function $f \equiv_T g$ that majorizes ∇. Given $x \in \mathbb{N}$, we can decide whether $x \in 0'$ by checking if $x \in 0'_{f(x)}$. $\qquad \square$

COROLLARY 7.2.4. *Every infinite subset of the set of true stages computes* $0'$.

PROOF. If we enumerate in increasing order the elements of a subset of the set of true stages, we obtain a function that majorizes ∇. $\qquad \square$

EXERCISE 7.2.5. (Hard) Prove that ∇ dominates every computable function. Hint in footnote.[69]

We can still talk about domination in the case of partial computable functions: A function f *majorizes* a partial function g if $f(n) \geq g(n)$ for every n at which $g(n)$ is defined. The exercise above is not true for partial computable functions, although, ∇ is still faster than the partial computable functions in the following sense.

DEFINITION 7.2.6. We say that $f \colon \mathbb{N} \to \mathbb{N}$ is *faster* than $g \colon \mathbb{N} \to \mathbb{N}$ *up to computable speed up* if there is a computable function $h \colon \mathbb{N} \to \mathbb{N}$ such that $f \circ h$ majorizes g.

LEMMA 7.2.7. ∇ *is faster than every partial computable function up to a computable speed up.*

PROOF. Let g be a partial computable function. We define the computable speed-up function as follows: Let $h(i)$ be the index of a computable function that, independently of the input, converges after $g(i)$ converges; that is, if $g(i)\uparrow$, then $\Phi_{h(i)}(x)\uparrow$, and if $g(i)\downarrow$, then $\Phi_{h(i)}(x)$ converges in at least $g(i)$ steps. Since $\nabla(j)$ is larger than the time-use of $\Phi_j(j)$, we get that $\nabla(h(i)) \geq g(i)$ whenever $g(i)\downarrow$. $\qquad \square$

---

[69]Given a non-decreasing computable function g, use the recursion theorem to define a computable function s such that $\nabla(s(i)) \geq g(s(i+1))$ for each i.

Definition 7.2.8. A function $f : \mathbb{N} \to \mathbb{N}$ is said to be ω-*computably approximable* (denoted ω-c.a.) if it has a computable approximation $\{f_s : s \in \mathbb{N}\}$ for which the number of mind-changes is computably bounded: That is, there are a computable list of computable functions $\{f_s : s \in \mathbb{N}\}$ and a computable function $c : \mathbb{N} \to \mathbb{N}$ such that $|\{s : f_s(i) \neq f_{s+1}(i)\}| \leq c(i)$ for all i, and, of course, $\lim_{s \to \infty} f_s(i) = f(i)$. (Cf. limit lemma on page xxi.)

Notice that ∇ is ω-c.a. since the number of mind-changes of $\nabla_s(i)$ is at most $i + 1$, as it changes only if a number below $i + 1$ is enumerated into $0'$. Actually, the following lemma shows that ∇ is the fastest ω-c.a. function up to computable speed up.

Lemma 7.2.9. *If* $f : \mathbb{N} \to \mathbb{N}$ *is increasing and faster than every partial computable function up to a computable speed up, then it is faster than every* ω-*c.a. function up to a computable speed up.*

Proof. Let g be ω-c.a. as witnessed by the computable function c that bounds the number of mind-changes in the approximation $\{g_s : s \in \mathbb{N}\}$. Define a partial computable function p such that $p(i, j)$ is the value of $g_s(i)$ after j mind changes. That is, $p(i, j)$ is $g_s(i)$ for the least s such that $|\{t < s : g_t(i) \neq g_{t+1}(i)\}| = j$ if such an s exists, and $p(i, j)$ is undefined otherwise. Notice that since $g(i)$ changes at most $c(i)$ times, $g(i) \leq \max\{p(i, j) : j \leq c(i)\}$. Since f is faster than every partial computable function up to a computable speed up, there is a computable h such that $f(h(i, j)) \geq p(i, j)$ for all $i, j \in \mathbb{N}$.[70] The function $\tilde{h}(i) = \max\{h(i, j) : j \leq c(i)\}$ is the computable speed up witnessing that f is faster than g: Using that f is increasing,

$$f(\tilde{h}(i)) \geq \max\{f(h(i, j)) : j \leq c(i)\}$$
$$\geq \max\{p(i, j) : j \leq c(i)\} \geq g(i). \qquad \square$$

Exercise 7.2.10. Given a computable well-ordering $\alpha = (A; \leq_\alpha)$, we say that $f : \mathbb{N} \to \mathbb{N}$ is α-c.a. if there is a computable approximation $\{f_s : s \in \mathbb{N}\}$ of f, and a computable function $c : \mathbb{N}^2 \to A$ that counts the number of mind changes in f_s in the following sense: If $f_s(i) \neq f_{s+1}(i)$, then $c(i, s) >_\alpha c(i, s + 1)$.

(a) Prove that for every computable well-ordering $\alpha = (A; \leq_\alpha)$ there is an α-c.a. function f_α that is faster than any other α-c.a. function up to a computable speed up.

(b) (Hard) Prove that if $\beta < \alpha$, then f_β is not faster than f_α even after a computable speed up.

---

[70]We are using a computable bijection between $\mathbb{N}$ and $\mathbb{N}^2$.

7.2.3. A couple of examples. The facts that the set of true stages is co-c.e., and that ∇ grows so fast are enough to make ∇ useful. We give a couple of examples to illustrate its use. In Section 1.1.2, we built a copy $\mathcal{A}$ of the ordering $\omega = (\mathbb{N}; \leq)$ so that the isomorphism between $\mathcal{A}$ and $(\mathbb{N}; \leq)$ computes $0'$. We now produce another such copy using a different method.

LEMMA 7.2.11. *There is a computable ω-presentation $\mathcal{A}$ of the ordering $(\mathbb{N}; \leq)$ such that any embedding from $\mathcal{A}$ to $(\mathbb{N}; \leq)$ computes $0'$.*

PROOF. The idea is to define $\mathcal{A} = (A; \leq_A)$ together with a computable sequence $a_0 <_A a_1 <_A a_2 <_A \cdots$ such that there are at least $\nabla(i)$ elements $<_A$-below a_{i+1} for every i. This way, if $g : A \to \mathbb{N}$ is an embedding from $\mathcal{A}$ to $(\mathbb{N}; \leq)$, we would have that the function $i \mapsto g(a_{i+1})$ majorizes ∇ and hence computes $0'$. Recall that the set $\{\langle i, t \rangle : i \in \mathbb{N}, \ t < \nabla(i)\}$ is c.e. We build $\mathcal{A}$ by first laying down elements $a_0 <_A a_1 <_A a_2 <_A \cdots$ (say, using the even integers: $a_n = 2n$), and then adding elements $b_{i,t}$ $\leq_A$-in-between a_i and a_{i+1} for each i, t with $t < \nabla(i)$. More formally, if f is a computable one-to-one enumeration of $\{\langle i, t \rangle : t < \nabla(i), i \in \mathbb{N}\}$, name the odd number $2n + 1$ with the label $b_{i,t}$ if $f(n) = \langle i, t \rangle$ and then define

$$a_j <_A b_{i,t} \iff j \leq i \quad \text{and} \quad b_{j,s} <_A b_{i,t} \iff j < i \vee (j = i \wedge s < t).$$

We then get that there are $\nabla(i)$ elements $\leq_A$-between a_i and a_{i+1} as needed.   □

The following lemma answers the question of how difficult is it to find a basis on a vector space. A jump is sufficient, as we can computably enumerate a maximal linearly independent set using the linear dependence relation, which we know is r.i.c.e. The lemma below shows it is necessary.

LEMMA 7.2.12. *There is a computable copy of the infinite dimensional $\mathbb{Q}$-vector space $\mathbb{Q}^\infty$ where every basis computes $0'$.*

We will actually show that every infinite linearly independent set in this ω-presentation computes $0'$. Let $\mathbb{Q}^\infty$ denote the standard ω-presentation of the infinite dimensional $\mathbb{Q}$-vector space, which has a computable basis $\{e_i : i \in \mathbb{N}\}$.

PROOF. The idea is to define a copy $\mathcal{A}$ of $\mathbb{Q}^\infty$ by taking the quotient of $\mathbb{Q}^\infty$ over a computable subspace U with infinite co-dimension. The equivalence relation generated by a computable subspace U—namely $u \sim v \iff u - v \in U$—is computable, and hence we have a computable congruence ω-presentation $\mathcal{A}$ of $\mathbb{Q}^\infty / U \cong \mathbb{Q}^\infty$, where the projection map from $\mathbb{Q}^\infty$ to $\mathcal{A}$ is also computable (see Lemma 1.1.11).

Define U so that, for every s_1 and s_2 which are not true stages, e_{s_1} and e_{s_2} are linearly dependent in $\mathbb{Q}^\infty / U$. To get this, all we need to do is add to U a vector of the form $a e_{s_1} - e_{s_2}$ for some $a \in \mathbb{Q}$ as soon as we realize s_1 and s_2 are not true. Before showing how to define U in a computable way, let us see why having such a U is enough. Suppose $I \subseteq A$ is an infinite linearly independent set in $\mathcal{A}$; we need to show $I \geq_T 0'$. Since the projection map is computable, by choosing pre-images we can get an infinite set $J \subseteq \mathbb{Q}^\infty$ which is not just linearly independent, but also linearly independent modulo U. The subspace generated by $e_0, e_1, \ldots, e_{\nabla(n)-1}$ has dimension $n + 1$ when projected to $\mathcal{A}$, because, except for $e_{\nabla(0)}, e_{\nabla(1)}, \ldots, e_{\nabla(n-1)}$, all the other vectors are mutually linearly dependent. Therefore, if we take $n + 2$ vectors $v_0, \ldots, v_{n+1}$ from J, they cannot all belong to the subspace of $\mathbb{Q}^\infty$ generated by $e_0, e_1, \ldots, e_{\nabla(n)-1}$. Recall that in $\mathbb{Q}^\infty$, every vector is given as a linear combination of the bases of e_i's. One of the vectors v_i must then use some e_t for $t \geq \nabla(n) - 1$ in its representation. Let $g(n)$ be the largest t such that e_t appears in the representation of one of the vectors v_i for $i \leq n + 1$. The function g majorizes ∇, and hence we can use g to compute $0'$ as in Lemma 7.2.3.

We now have to show how to build U effectively. At each stage s, we define a finite subset $U_s \subseteq \mathbb{Q}^\infty$ and at the end, define $U = \bigcup_{s \in \mathbb{N}} U_s$. Consider the finite sets

$$V_s = \left\{ \sum_{i < s} \frac{p_i}{q_i} e_i : p_i, q_i \in \mathbb{Z}, |p_i| < s, 0 < q_i < s \right\},$$

whose union is $\mathbb{Q}^\infty$. To make sure U is computable, we will ensure that $U \cap V_s = U_s \cap V_s$ for every s. Therefore, after each stage s, we must ensure that no element of $V_s \setminus U_s$ ever enters U. To get U to be a subspace, we will ensure that each U_s is closed under linear combinations within V_s (i.e., $U_s \cap V_s = \langle U_s \rangle \cap V_s$).

Suppose that, at stage s, we discover that s_1 and s_2 are not true stages and we have not made e_{s_1} and e_{s_2} dependent in $\mathcal{A}$ yet. (Recall that the set of non-true stages is c.e.) We then want to add a vector of the form $ae_{s_1} - e_{s_2}$ to U so that we make e_{s_1} and e_{s_2} dependent in $\mathcal{A}$ without changing U within V_s: All we have to do is search for such an $a \in \mathbb{Q}$ such that when we add $ae_{s_1} - e_{s_2}$ to U, we keep all the vectors in $V_s \setminus U_s$ outside U. That is, we need to make sure that no vector in $V_s \setminus U_s$ belongs to the subspace generated by $U_s \cup \{ae_{s_1} - e_{s_2}\}$. Once we find such an a, we can verify this computably, and thus we just need to know that one such a exists. Using basic linear algebra, if $a_0 \neq a_1$, and e_{s_1} and e_{s_2} are independent over U_s, then the intersection of the spaces generated by $U_s \cup \{a_0 e_{s_1} - e_{s_2}\}$ and by $U_s \cup \{a_1 e_{s_1} - e_{s_2}\}$ is the subspace generated by U_s.[71] Since V_s is finite, there can be at most finitely many a's which generate elements in $V_s \setminus U_s$. In other words, for all but finitely many a's, the space generated by $U_s \cup \{ae_{s_1} - e_{s_2}\}$ adds no new vectors to V_s that were not in the subspace

---

[71]Suppose v in the intersection of the spaces generated by $U_s \cup \{a_0 e_{s_1} - e_{s_2}\}$ and $U_s \cup \{a_1 e_{s_1} - e_{s_2}\}$. Thus $v = u_0 + \lambda_0(a_0 e_{s_1} - e_{s_2}) = u_1 + \lambda_1(a_1 e_{s_1} - e_{s_2})$ for some $u_0, u_1 \in \langle U_s \rangle$ and $\lambda_0, \lambda_1 \in \mathbb{Q}$. We then get that $u_0 - u_1 = (\lambda_1 a_1 - \lambda_0 a_0) \cdot e_{s_1} + (\lambda_1 - \lambda_0) \cdot e_{s_2}$. Since e_{s_1} and e_{s_2} are independent over U_s, we get that $\lambda_1 a_1 - \lambda_0 a_0 = 0 = \lambda_1 - \lambda_0$, from which we deduce that $\lambda_0 = \lambda_1 = 0$ and that $v = u_0 = u_1 \in \langle U_s \rangle$.

generated by U_s already. Now that we know such a exist, all we have to do is look for one. When we find it, we add $ae_{s_1} - e_{s_2}$ to U_{s+1}. At stage s we might discover that various pairs s_1 and s_2 are not true, so we do this for each such pair. Finally, to get $U_{s+1} \cap V_{s+1} = \langle U_{s+1} \rangle \cap V_{s+1}$, once we have added all these vectors, we close U_{s+1} under linear combinations that are in V_{s+1}, that is, we add to U_{s+1} all vectors in V_{s+1} which are linear combination of U_s and these new vectors.

Notice that U has infinite co-dimension as whenever $t_1, \ldots, t_k$ are true stages, $e_{t_1}, \ldots, e_{t_k}$ are linearly independent modulo U, as only vectors in the subspace generated by the remaining basis vectors are ever added to U.   □

Historical Remark 7.2.13. Metakides and Nerode [MN79] prove a similar result, and Friedman, Simpson, and Smith [FSS83] prove this same result for reverse mathematics purposes.

7.3. Approximating the settling-time function

Every true stage can figure out all the previous true stages in a uniformly computable way. More precisely: Suppose $t = \nabla(i)$ is the ith true stage. Using the fact that $0'_t \upharpoonright i = 0' \upharpoonright i$, we have that, for $j \le i$, $\nabla(j)$ is the least $s > \nabla(j-1)$ such that $0'_s \upharpoonright j = 0'_t \upharpoonright j$. If t is not a true stage, we can still apply the same procedure and get the stages that t *believes* should be true.

DEFINITION 7.3.1. Given $j < t$, we define the jth *apparent true stage at* t, denoted $\nabla_t(j)$, as the least $s \le t$ such that $s > \nabla_t(j-1)$ and $0'_s \upharpoonright j = 0'_t \upharpoonright j$. Again, to match with ∇, we are using $\nabla_t(-1) = -1$ in the definition of $\nabla_t(0)$.

This definition only makes sense if $s \le t$, so once we reach a j with $\nabla_t(j) = t$, we cannot define any more apparent true stages, and we let ∇_t be the string defined up to that point. Thus, ∇_t is a finite increasing string whose last element is always t.

From the paragraph preceding the definition, we get that if t is the ith true stage, then $\nabla_t = \nabla \upharpoonright i$. Furthermore, for every $s > t$, since $0'_t \upharpoonright i = 0'_s \upharpoonright i = 0' \upharpoonright i$, we get that $\nabla_s \upharpoonright i$ is also correct and equal to $\nabla \upharpoonright i$. On the other hand, if t is not a true stage, since t is the last entry of ∇_t, we have that $\nabla_t \not\subset \nabla$. For the same reason, if $s > t$ is a true stage, then $\nabla_t \not\subseteq \nabla_s$. In short, for $t \in \mathbb{N}$,

$$t \text{ is a true stage } \iff \nabla_t \subset \nabla \iff \forall s > t \ (\nabla_t \subseteq \nabla_s).$$

By essentially the same argument, we get the following property:

♣ For every $r < s < t$, if $\nabla_r \subseteq \nabla_t$, then $\nabla_r \subseteq \nabla_s$.

The reason is that if $\nabla_r \subseteq \nabla_t$, then no number below $|\nabla_r|$ is enumerated into $0'$ between the stages r and t. That would then also be true between the stages r and s, and hence $\nabla_r \subseteq \nabla_s$.

The following two lemmas are intended to give us a feeling for how the sequence $\{\nabla_s : s \in \mathbb{N}\}$ behaves. Let $\mathcal{T}$ be the image of the function $s \mapsto \nabla_s$. To gain some intuition, we recommend the reader see how the sequence $\{\nabla_s : s \in \mathbb{N}\}$ moves around $\mathcal{T}$ in Figure 1 below.

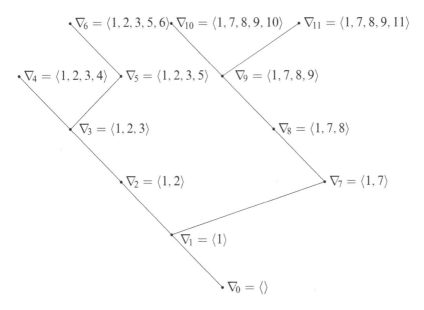

FIGURE 1. Example where 3 is enumerated into $0'$ at stage 5, 1 at stage 7, and 4 at stage 11.

EXERCISE 7.3.2. Draw a tree like the one in Figure 1 in the following situation: 4 is enumerated at stage 5, 3 at 7, 5 at 9, and 0 at 11.

LEMMA 7.3.3. *The set $\mathcal{T} = \{\nabla_s : s \in \mathbb{N}\} \subseteq \mathbb{N}^{<\mathbb{N}}$ is a computable tree whose only path is ∇.*

PROOF. $\mathcal{T}$ is computable because given $\sigma \in \mathbb{N}^{<\mathbb{N}}$, we can calculate ∇_t, where t is the last entry of σ, and then check if $\sigma = \nabla_t$.

To show that $\mathcal{T}$ is a tree, we need to show that it is closed downward. To do this, all we have to observe is that if $\nabla_s(i) = t$, then $\nabla_s \restriction i = \nabla_t$. This is because $0'_s \restriction i = 0'_t \restriction i$, and hence the computations of $\nabla_t \restriction i$ and $\nabla_s \restriction i$ are the same.

About the paths of $\mathcal{T}$, clearly ∇ is one of them. We claim that if $\nabla_s \not\subseteq \nabla$, the set of extensions of ∇_s in $\mathcal{T}$ is finite, and hence there is no path extending ∇_s: Let $t > s$ be a true stage. Then $\nabla_s \not\subseteq \nabla_t$. By ($\clubsuit$), for all $u \geq t$, $\nabla_s \not\subseteq \nabla_u$. □

The *Kleene-Brower ordering*, $\leq_{KB}$, on $\mathbb{N}^{<\mathbb{N}}$ is defined as follows: $\sigma \leq_{KB} \tau$ if either $\sigma \supseteq \tau$ or σ and τ are incomparable and, for the least i with $\sigma(i) \neq \tau(i)$, we have $\sigma(i) < \tau(i)$.

LEMMA 7.3.4. *The Kleene-Brower ordering, $\leq_{KB}$, on $\mathcal{T}$ produces a computable ordering of order type $\omega + \omega^*$ on which every descending sequence computes $0'$.*

PROOF. To prove that $(\mathcal{T}; \leq_{KB}) \cong \omega + \omega^*$, we prove that if s is a true stage, then there are only finitely many strings in $\mathcal{T}$ that are $\geq_{KB} \nabla_s$; and if s is not a true stage, then there are only finitely many strings in $\mathcal{T}$ that are $\leq_{KB} \nabla_s$. For the former claim, if s is a true stage, then for every $t \geq s$, we have $\nabla_t \supseteq \nabla_s$, and hence $\nabla_t \leq_{KB} \nabla_s$. For the latter claim, if s is not a true stage and $t > s$ is a true stage, then there is a least i such that $\nabla_s(i) \neq \nabla_t(i)$. The reason for this difference must be that $i \notin 0'_s$ while $i \in 0'_t$, and hence $\nabla_t(i) > s \geq \nabla_s(i)$. Since t is true, we have that, for every $u \geq t$, $\nabla_u \supseteq \nabla_t$, and hence $\nabla_u \upharpoonright i = \nabla_s \upharpoonright i$ and $\nabla_u(i) = \nabla_t(i) > \nabla_s(i)$. Thus, $\nabla_u \geq_{KB} \nabla_s$.

Every descending sequence must be a subsequence of $\{\nabla_t : t \text{ is a true stage}\}$, and hence computes $0'$ by Corollary 7.2.4. □

EXERCISE 7.3.5. Show that $(\mathcal{T}; \leq_{KB})$ has a computable ascending sequence.

EXERCISE 7.3.6. (Hard) Use a priority argument to show that there is an ω-presentation of $\omega + \omega^*$ which has no computable ascending sequence and no computable descending sequence.

Remark 7.3.7. Hirschfeldt and Shore [HS07, Theorem 2.11] showed that every computable ω-presentation of $\omega + \omega^*$ must have either an ascending sequence or a descending sequence that is low.

EXERCISE 7.3.8. (Hard) A small modification of the proof of Theorem 5.4.3 can produce another interesting spectrum. Let us view a set $\Gamma \subseteq 2^{<\mathbb{N}}$ as an operator by letting $\Gamma^X = \{|\tau| : \tau \subseteq X, \tau \in \Gamma\}$. Given a finite set $F \subseteq \mathbb{N}$, let

$$\Gamma_F = \{\nabla \upharpoonright i : i \in F\} \cup \{\tau \in 2^{<\mathbb{N}} : \tau \not\subseteq \nabla\}.$$

Notice that $\Gamma_F^\nabla = F$. Consider the family of sets:

$$\mathcal{F} = \{\Gamma_F \oplus \{n\} : F \subseteq \mathbb{N} \text{ finite } \& F \neq W_n^\nabla\}.$$

Prove that

$$DgSp(\mathcal{G}_\mathcal{F}) = \{X \in 2^\mathbb{N} : X \text{ not } \Delta_2^0\}.$$

(The first one to construct a structure with this spectrum was Kalimullin [Kal08]. The construction above is due to Montalbán [ACK⁺16, Theorem 2].) Hint in footnote.[72]

7.4. A construction of linear orderings

In this section, we prove a well-known result that is best proved using the method of true stages we just developed. Given linear orderings $\mathcal{A}$ and $\mathcal{B}$, we let $\mathcal{A} \cdot \mathcal{B}$ be the ordering on $A \times B$ given by $\langle a_0, b_0 \rangle \leq_{\mathcal{A} \cdot \mathcal{B}} \langle a_1, b_1 \rangle$ if either $b_0 <_{\mathcal{B}} b_1$, or $b_0 = b_1$ and $a_0 \leq_{\mathcal{A}} a_1$. Notice that the coordinates are compared from right to left, and not as in the lexicographic ordering; it is the tradition. Then, for instance $\mathcal{A} + \mathcal{A} = \mathcal{A} \cdot 2$, and $\mathbb{Z} \cdot \mathcal{A}$ is the linear ordering obtained by replacing each element in $\mathcal{A}$ with a copy of $\mathbb{Z}$.

THEOREM 7.4.1 (Fellner [Fel76]). *Let $\mathcal{L}$ be a linear ordering. Then $\mathbb{Z} \cdot \mathcal{L}$ has a computable copy if and only if $\mathcal{L}$ has a $0''$-computable copy.*

The left-to-right direction is the easy one. On a computable copy of $\mathbb{Z} \cdot \mathcal{L}$, the equivalence relation $\sim$, given by $a \sim b$ if and only if they are finitely apart, is $0''$ computable, and hence we can make the copy of $\mathbb{Z} \cdot \mathcal{L}$ into a $0''$-computable congruence ω-presentation of $\mathcal{L}$.

The proof of the other direction is divided into a few steps which we prove in separate lemmas. The first lemma is a general one that will be useful in other settings too. It gives a way of approximating $0'$-computable structures in a way that correct approximations to the structure happen at the same stages where we have correct approximations to ∇.

LEMMA 7.4.2. *Let $\mathcal{B}$ be a $0'$-computable ω-presentation of a structure in a relational vocabulary τ. There is a computable sequence of finite $\tau_{|\cdot|}$-structures $\{\mathcal{B}_s : s \in \mathbb{N}\}$ such that*

$$(\forall s < t) \quad \nabla_s \subseteq \nabla_t \Longrightarrow \mathcal{B}_s \text{ is a substructure of } \mathcal{B}_t,$$

and

$$\mathcal{B} = \bigcup \{\mathcal{B}_s : s \text{ a true stage}\}.$$

Moreover, if φ is a $\forall$-formula true of $\mathcal{B}$, we can make the $\mathcal{B}_s$'s satisfy φ too.

PROOF. Let $\mathcal{A}_t$ be the τ_t-substructure of $\mathcal{B}$ with domain $\{0, \ldots, t-1\}$. The sequence $\{\mathcal{A}_t : t \in \mathbb{N}\}$ is $0'$ computable. Let Φ be a computable function such that $\Phi^{\nabla}(t)$ is an index for the finite structure $\mathcal{A}_t$. If at a stage s we believe ∇_s is an initial segment of ∇, we also believe that Φ^{∇_s} outputs the indices of the first few structures in the sequence $\{\mathcal{A}_t : t \in \mathbb{N}\}$.

---

[72]For the construction, use as an oracle a set U with no Δ_2^0 computable subsets, and when there is a threat $\Gamma^{\nabla_s} = W_n^{\nabla_s}$ add to Γ the extensions of ∇_s with a certain length in U.

For each s, let t_s be the largest t so that, for every $i \leq t$, $\Phi^{\nabla_s}(i)$ converges and outputs an index for a finite structure $\tilde{\mathcal{A}}_i$ satisfying φ and so that

$$\tilde{\mathcal{A}}_0 \subseteq \tilde{\mathcal{A}}_1 \subseteq \cdots \subseteq \tilde{\mathcal{A}}_t.$$

Let $\mathcal{B}_s = \tilde{\mathcal{A}}_{t_s}$. We then have that if $\nabla_s \subseteq \nabla_r$, $\Phi^{\nabla_s}(i) = \Phi^{\nabla_r}(i)$ for all $i \leq t_s$, and hence $\mathcal{B}_s \subseteq \mathcal{B}_r$. If $\nabla_s \subseteq \nabla$, then $\tilde{\mathcal{A}}_{t_s}$ is actually one of the $\mathcal{A}_t$'s, and hence $\mathcal{B}_s \subset \mathcal{B}$.   □

LEMMA 7.4.3. *If a linear ordering $\mathcal{L}$ has a $0'$-computable copy, then the adjacency linear ordering $(\mathbb{Z} \times \mathcal{L}; \leq_{\mathbb{Z} \times \mathcal{L}}, \mathrm{Adj})$ has a computable copy.*

PROOF. Let $\{\mathcal{L}_s : s \in \mathbb{N}\}$ be a sequence of finite linear orderings approximating $\mathcal{L}$ as in Lemma 7.4.2. Notice that being a linear ordering can be described by a $\forall$-sentence.

At each stage s, we build a finite linear ordering $\mathcal{A}_s = (\{0, \ldots, k_s\}; \leq_{\mathcal{A}_s}, \mathrm{Adj}_s)$ and an onto, order-preserving map $g_s \colon \mathcal{A}_s \to \mathcal{L}_s$ such that $g_s(a) = g_s(b)$ if and only if there is a finite sequence of Adj_s-adjacent elements in between a and b in $\mathcal{A}_s$. The binary relations Adj_s satisfy that if $\mathcal{A}_s \models \mathrm{Adj}_s(a, b)$, then there is no element in between a and b in $\mathcal{A}_s$, but there could be elements $a, b \in \mathcal{A}_s$ without elements in between for which Adj_s does not hold. Thus, $\mathcal{A}_s$ is partitioned into *adjacency chains*, where an adjacency chain is a maximal string of elements $a_0 <_{\mathcal{A}_s} \cdots <_{\mathcal{A}_s} a_k$ with $\mathrm{Adj}_s(a_i, a_{i+1})$ for all $i < k$. The condition on g above implies that for each $\ell \in \mathcal{L}_s$, $g_s^{-1}(\ell)$ is an adjacency chain.

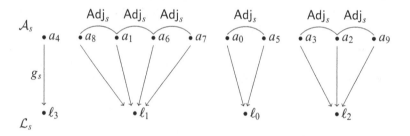

FIGURE 2. The top row are the points in $\mathcal{A}_s$ ordered by $\leq_{\mathcal{A}_s}$ from left to right. The bottom row are the points in $\mathcal{L}_s$ ordered by $\leq_{\mathcal{L}_s}$ from left to right.

At each stage s, we need to satisfy the following two properties:

1. If $t \leq s$, then $\mathcal{A}_t \subseteq \mathcal{A}_s$ (as structures, i.e., preserving $\leq$ and Adj).
2. If $\nabla_t \subset \nabla_s$, then $g_t \subseteq g_s$, and for every $\ell \in \mathcal{L}_t$, $1 + g_t^{-1}(\ell) + 1 \subseteq g_s^{-1}(\ell)$.[73]

---

[73] That is, $g_t^{-1}(\ell) \subseteq g_s^{-1}(\ell)$, and there is an element in $g_s^{-1}(\ell)$ that is less than all the elements in $g_t^{-1}(\ell)$ and another one that is greater.

Let us first note that these conditions are enough to build the desired structure $\mathcal{A}$. Condition (1) allows us to define a computable adjacency linear ordering $\mathcal{A} = \bigcup_s \mathcal{A}_s$. Condition (2) allows us to define an onto, order-preserving map $g = \bigcup\{g_s : s \text{ is a true stage}\}: \mathcal{A} \to \mathcal{L}$. Furthermore, for every $\ell \in \mathcal{L}$, $g^{-1}(\ell)$ must be infinite in both directions and satisfy that any two elements in it are linked by a finite sequence of adjacencies. Therefore, $g^{-1}(\ell)$ is isomorphic to $\mathbb{Z}$, and we get that $\mathcal{A}$ is isomorphic to $\mathbb{Z} \cdot \mathcal{L}$.

Last, we need to show that, at each stage $s + 1$, we can define $\mathcal{A}_{s+1}$ and g_{s+1} so that they satisfy (1) and (2). Let $t \leq s$ be the largest such that $\nabla_t \subseteq \nabla_{s+1}$. Thus, we know that $\mathcal{L}_t \subseteq \mathcal{L}_{s+1}$, and we need to define $\mathcal{A}_{s+1}$ extending $\mathcal{A}_s$ and g_{s+1} extending g_t. The rest of the proof is just a brute-force combinatorial argument proving that such an $\mathcal{A}_{s+1}$ and g_{s+1} exist. We recommend the reader to try to prove it and to draw pictures like Figures 2 and 3 before reading it.

First, define $\widetilde{\mathcal{A}}_{s+1}$ by adding a new element at the end of each adjacency chain in $\mathcal{A}_s$, and by attaching each new adjacency chain to one that existed in $\mathcal{A}_t$. (To *attach* two adjacency chains, we add a new element in between the chains and make it satisfy Adj_{s+1} with the ends of the two chains.) Thus, we end up with $\widetilde{\mathcal{A}}_{s+1}$ having the same adjacency chains as $\mathcal{A}_t$, though these chains are longer in $\widetilde{\mathcal{A}}_{s+1}$. Extend $g_t: \mathcal{A}_t \to \mathcal{L}_t$ to $\tilde{g}_{s+1}: \widetilde{\mathcal{A}}_{s+1} \to \mathcal{L}_t$ so that, for each $\ell \in \mathcal{L}_t$, $\tilde{g}_{s+1}^{-1}(\ell)$ is an adjacency chain in $\widetilde{\mathcal{A}}_{s+1}$. We have now fixed the mess done at stage s.

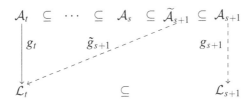

FIGURE 3. The diagram above commutes.

Second, define $\mathcal{A}_{s+1} \supseteq \widetilde{\mathcal{A}}_{s+1}$ by adding a new element a_ℓ in between chains for each new $\ell \in \mathcal{L}_{s+1} \setminus \mathcal{L}_t$. Of course, if $\ell_0 < \ell < \ell_1$ with $\ell_0, \ell_1 \in \mathcal{L}_t$, then the a_ℓ must be in between the chains corresponding to $g^{-1}(\ell_0)$ and $g^{-1}(\ell_1)$. Finally, extend $\tilde{g}_{s+1}: \widetilde{\mathcal{A}}_{s+1} \to \mathcal{L}_t$ to $g_{s+1}: \mathcal{A}_{s+1} \to \mathcal{L}_{s+1}$ by mapping each a_ℓ to ℓ.   □

LEMMA 7.4.4. *If* $(\mathbb{Z} \cdot \mathcal{L}; \leq, \text{Adj})$ *has a* $0'$ *computable copy, then* $\mathbb{Z} \cdot \mathcal{L}$ *has a computable copy.*

PROOF. Let $\mathcal{B}$ be the $0'$-computable copy of $(\mathbb{Z} \cdot \mathcal{L}; \leq, \text{Adj})$. Let $\{\mathcal{B}_s : s \in \mathbb{N}\}$ be a sequence of finite structures approximating $\mathcal{B}$ as in Lemma 7.4.2.

We assume each $\mathcal{B}_s$ satisfies the $\forall$-sentence saying that they are linear orderings and that if $\mathcal{B}_s \models \mathrm{Adj}(a,b)$, there is no element between a and b. However, as for the structures $\mathcal{A}_s$ in the previous lemma, there will be elements a and b not satisfying $\mathrm{Adj}(a,b)$ in $\mathcal{B}_s$ and without anything in $\mathcal{B}_s$ between them.

At each stage s, we build a finite linear ordering

$$\mathcal{A}_s = (\{0,\dots,k_s\}; \leq_{A_s}, \mathrm{Adj}_s)$$

and an order-preserving, one-to-one map $h_s \colon \mathcal{B}_s \to \mathcal{A}_s$. Again, as with the structures $\mathcal{A}_s$ from the previous lemma, Adj_s satisfies $\forall a,b \leq k_s(\mathrm{Adj}_s(a,b) \wedge a <_{A_s} b \to \nexists c(a <_{A_s} c <_{A_s} b))$, and hence $\mathcal{A}_s$ is partitioned into *adjacency chains*. We do not require h_s to be onto, not even in the limit. Instead, all we require is that every adjacency chain in $\mathcal{A}_s$ has an element in the image of h_s. Also, we require that two elements of $\mathcal{B}_s$ are in the same adjacency chain if and only if their images are. Thus, h_s induces a bijection between the adjacency chains in $\mathcal{B}_s$ and the adjacency chains in $\mathcal{A}_s$. Notice that we do not require h_s to preserve Adj, but only to preserve the property of being in the same adjacency chain.

At each stage s, we need to satisfy the following two properties:

1. If $t \leq s$, then $(\{0,\dots,k_t\}; \leq_{A_t}) \subseteq (\{0,\dots,k_s\}; \leq_{A_s})$.
2. If $\nabla_t \subseteq \nabla_s$, then $\mathcal{A}_t \subseteq \mathcal{A}_s$ preserving order and adjacency, and $h_t \subseteq h_s$.

Condition (1) allows us to define a computable linear ordering

$$\mathcal{A} = (\mathbb{N}; \leq_A) = \bigcup_s (\{0,\dots,k_s\}; \leq_{A_s}).$$

Notice that we lost the adjacency relation, which may not be computable. Condition (2) allows us to define an embedding

$$h = \bigcup\{h_s : s \text{ a true stage}\} \colon \mathcal{B} \to \mathcal{A},$$

which preserves ordering and adjacency chains. The embedding h produces a bijection between the adjacency chains in $\mathcal{B}$ and those in $\mathcal{A}$, and an embedding of each adjacency chain in $\mathcal{B}$ to the corresponding one in $\mathcal{A}$. Since the adjacency chains in $\mathcal{B}$ are isomorphic to $\mathbb{Z}$, the ones in $\mathcal{A}$ must also be isomorphic to $\mathbb{Z}$, and we get that $\mathcal{A}$ and $\mathcal{B}$ are isomorphic.

Last, we need to show that, at each stage $s+1$, we can define $\mathcal{A}_{s+1}$ and h_{s+1} so they satisfy (1) and (2). Let $t \leq s$ be the largest such that $\nabla_t \subseteq \nabla_{s+1}$. We need to define $(\mathcal{A}_{s+1}; \leq_{s+1})$ extending $(\mathcal{A}_s; \leq_s)$ and Adj_{s+1} and h_{s+1} extending Adj_t and h_t. The rest of the proof is just a brute-force combinatorial argument proving that such $\mathcal{A}_{s+1}$, Adj_{s+1}, and h_{s+1} exist. Again, we recommend the reader to try to prove and to draw pictures before reading it.

Ignoring Adj_s, define $\widetilde{\mathrm{Adj}}_{s+1}$ on $\mathcal{A}_s$ so that it is compatible with Adj_t and so that every element belongs to an adjacency chain that existed in

$\mathcal{A}_t$. We can do this because, since $\nabla_t \subseteq \nabla_s$ (which follows from $\nabla_t \subseteq \nabla_{s+1}$ and ($\clubsuit$)), Adj_t is preserved in $\mathcal{A}_s$, and hence if two elements satisfy Adj_t in $\mathcal{A}_t$, they is still nothing in between them in $\mathcal{A}_s$. Extend $(A_s; \leq_s)$ to $(\widetilde{A}_{s+1}; \leq_{s+1})$ by adding one new element a_ℓ for each $\ell \in B_{s+1} \smallsetminus B_t$ so that we can extend $h_t \colon \mathcal{B}_t \to \mathcal{A}_t$ to $h_{s+1} \colon \mathcal{B}_{s+1} \to A_{s+1}$ (recall that $\mathcal{B}_t \subseteq \mathcal{B}_{s+1}$). Also, if two adjacency chains in $\mathcal{B}_t$ have collapsed to one in $\mathcal{B}_{s+1}$, we need to collapse the respective chains in $\mathcal{A}_{s+1}$: Thus, if two consecutive elements $\ell_0, \ell_1 \in \mathcal{B}_{s+1}$ belong to adjacency chains that were part of separate chains in $\mathcal{B}_t$, but are part of a single chain in $\mathcal{B}_{s+1}$, we add a new element a_{ℓ_0, ℓ_1} to A_{s+1} in between the adjacency chains corresponding to $h_t(\ell_0)$ and $h_t(\ell_1)$ so that we can attach those chains. Define Adj_{s+1} on A_{s+1} so that h_{s+1} produces a bijection between the adjacency chains in $\mathcal{B}_{s+1}$ and those in $\mathcal{A}_{s+1}$.   □

Finally, the right-to-left direction of Theorem 7.4.1 follows from first applying Lemma 7.4.3 relativized to $0'$, and then Lemma 7.4.4.

EXERCISE 7.4.5 (Downey [DK92]). Prove that $\mathcal{L}$ has a $0'$ computable copy if and only if $(\mathbb{Q} + 2 + \mathbb{Q}) \cdot \mathcal{L}$ has a computable copy.

Chapter 8

COMPUTABLE CATEGORICITY

Computably categorical structures are the ones for which all computable ω-presentations have the same computational properties. This is a desirable property on a structure, of course, but the structures which have it are rather few. The notion was originally introduced by Mal'cev [Mal62] in 1962 for groups, and has been intensively studied over the past few decades.

A second objective of this chapter is to get the reader acquainted with finite-injury priority constructions.

8.1. The basics

Most of the properties one considers in computable structure theory are invariant under computable isomorphisms, though not necessarily under all isomorphisms: Two computable ω-presentations may be isomorphic and still have different computational properties. For instance, there are computable ω-presentations of the countable, infinite-dimensional $\mathbb{Q}$-vector space $\mathbb{Q}^\infty$ where all the finite-dimensional subspaces are computable, and there are computable ω-presentations of $\mathbb{Q}^\infty$ where no non-trivial finite-dimensional subspace is computable (see [DHK$^+$07]).

DEFINITION 8.1.1. A computable structure $\mathcal{A}$ is *computably categorical* if there is a computable isomorphism between any two computable copies of $\mathcal{A}$.

The following somewhat trivial lemma shows how computably categorical structures are exactly the ones that avoid the behavior of the example above, that is, the ones where all computable copies have the same computable relations.

LEMMA 8.1.2. *Let $\mathcal{A}$ be a computable structure. The following are equivalent*:

1. *$\mathcal{A}$ is computably categorical.*
2. *For every computable $R \subseteq A^n$ and every computable copy $\mathcal{B}$ of $\mathcal{A}$, there is a computable $R^{\mathcal{B}} \subseteq B^n$ such that $(\mathcal{B}, R^{\mathcal{B}}) \cong (\mathcal{A}, R)$.*

PROOF. To show that (1) implies (2), consider a computable isomorphism $g : \mathcal{B} \to \mathcal{A}$, and define $R^{\mathcal{B}} = g^{-1}(R)$. For the other direction, consider a computable copy $\mathcal{B}$ of $\mathcal{A}$; we need to build a computable isomorphism between them. Of course, we are assuming $\mathcal{A}$ is infinite, and hence we may assume its domain is $\mathbb{N}$. Let

$$R = \{ \langle n, n + 1 \rangle : n \in \mathbb{N} \} \subseteq \mathbb{N}^2 = A^2.$$

Since R is computable, there is a computable $R^{\mathcal{B}}$ such that $(\mathcal{A}, R) \cong (\mathcal{B}, R^{\mathcal{B}})$. Once we know what element of B corresponds to $0 \in A$ under this isomorphism, we can use $R^{\mathcal{B}}$ to computably find the element of B that corresponds to $1 \in A$, and then the one that corresponds to $2 \in A$, etc. Continuing this process, we get the desired computable isomorphism between $\mathcal{A}$ and $\mathcal{B}$.   □

Now that we are convinced that computable categoricity is a desirable property, the next question that is "what makes a structure computable categorical?" This question is currently being investigated, and there has been a lot of work characterizing the computably categorical structures within certain classes of structures. See Table 2.

Such clean characterizations as in Table 2 are not always possible. Downey, Kach, Lempp, Lewis-Pye, Montalbán, and Turetsky [DKL$^+$15] showed that there is no structural characterization of the notion of computable categoricity. They did this by showing that the index set of the computably categorical structures[74] is Π_1^1-complete (defined in [MonP2]). There are, however, structural characterizations of variations of the notion of computable categoricity. For instance, we already proved in Section 3.4 that the *uniformly computably categorical structures* coincide with the effectively $\exists$-atomic ones. This chapter is dedicated to the non-uniform notions which are, arguably, more natural. In particular, it is dedicated to the notion of relative computable categoricity and its connections to plain computable categoricity.

8.2. Relative computable categoricity

In this section, we give a purely structural characterization of the computational notion of relative computable categoricity.

DEFINITION 8.2.1. (Ash, Knight, Manasse, and Slaman [AKMS89, Section 4]; Chisholm [Chi90, Definition V.9]). Given $X \in 2^{\mathbb{N}}$, an X-computable structure $\mathcal{A}$ is X-*computably categorical* if there is an X-computable isomorphism between any two X-computable copies of $\mathcal{A}$.

---

[74]The *index set* of a class of structures is the set of indices for computable functions that are the diagrams of ω-presentations of structures in the class.

Class	Condition for computable categoricity	Reference
Linear orderings	Finitely many pairs of adjacent elements	Goncharov and Dzgoev [DG80], Remmel [Rem81]
Boolean algebras	Finitely many atoms	Goncharov [Gon75b], La Roche [LR78]
$\mathbb{Q}$-vector spaces	Finite dimension	
Algebraically closed fields	Finite transcendence degree over prime subfield	Ershov [Ers77]
Ordered abelian groups	Finite rank	Goncharov, Lempp, and Solomon [GLS03]
Trees of finite height	Finite type	Lempp, McCoy, Miller, and Solomon [LMMS05]
Torsion-free abelian groups	Finite rank	Nurtazin [Nur74]
Abelian p-groups	Either (i) $(\mathbb{Z}(p^\infty))^\ell \oplus G$ for $\ell \in \mathbb{N} \cup \{\infty\}$ and G finite, or (ii) $(\mathbb{Z}(p^\infty))^n \oplus (\mathbb{Z}_{p^k})^\infty \oplus G$ where G is finite, and $n, k \in \mathbb{N}$	Goncharov [Gon80], Smith [Smi81]

TABLE 2. The middle column describes a necessary and sufficient condition for a structure within the given class to be computably categorical. For the definitions of the relevant terms and the proofs, we refer the reader to the references given in the third column. Each case requires a different priority argument to show that structures that do not satisfy the condition are not computably categorical.

A computable structure $\mathcal{A}$ is *relatively computably categorical* if it is X-computably categorical for all $X \in 2^{\mathbb{N}}$.

Equivalently, $\mathcal{A}$ is relatively computably categorical if, for every copy $\mathcal{B}$ (computable or not) of $\mathcal{A}$, there is an isomorphism between $\mathcal{B}$ and $\mathcal{A}$ that is computable in $D(\mathcal{B})$.

THEOREM 8.2.2. (Ash, Knight, Manasse, and Slaman [AKMS89, Theorem 4]; Chisholm [Chi90, Theorem V.10]). *Let $\mathcal{A}$ be a computable structure. The following are equivalent*:

1. $\mathcal{A}$ *is relatively computably categorical.*

2. $(\mathcal{A}, \bar{a})$ is uniformly computably categorical for some $\bar{a} \in A^{<\mathbb{N}}$.

3. $(\mathcal{A}, \bar{a})$ is effectively $\exists$-atomic for some $\bar{a} \in A^{<\mathbb{N}}$.

PROOF. The equivalence between (2) and (3) was proved in Theorem 3.4.2. To see that (2) implies (1), just notice that for any copy $\mathcal{B}$ of $\mathcal{A}$, one can non-uniformly pick the corresponding tuple $\bar{a}^{\mathcal{B}}$ so that $(\mathcal{B}, \bar{a}^{\mathcal{B}}) \cong (\mathcal{A}, \bar{a})$, and then use part (2) of Theorem 3.4.2 to get a $D(\mathcal{B})$-computable isomorphism between them.

The interesting direction is the implication from (1) to (3), which shares some ideas with the proof of Theorem 3.4.2; we recommend the reader studies it first. Assume $\mathcal{A}$ is relatively computably categorical. Out of this computational assumption we need to build a syntactical object, namely a c.e. Scott family of $\exists$-definitions for the automorphism orbits of the tuples in $A^{<\mathbb{N}}$, over some parameters.

Let $g : \mathbb{N} \to \mathcal{A}$ be an enumeration of $\mathcal{A}$ that is 2-*generic relative to the presentation of* $\mathcal{A}$ as in Definition 5.3.7. Let $\mathcal{B}$ be the generic presentation obtained as the pull-back of $\mathcal{A}$ through g (as in Definition 4.3.3). Since $\mathcal{A}$ is relatively computably categorical, and $\mathcal{B} \cong \mathcal{A}$, there is a computable operator Γ such that $\Gamma^{D(\mathcal{B})}$ is an isomorphism from $\mathcal{B}$ to $\mathcal{A}$.

The first step is to get a tuple $\bar{p} \subseteq g$ which *forces* that $\Gamma^{D(\mathcal{B})}$ is an isomorphism as follows:

CLAIM 8.2.3. *There is a tuple $\bar{p} \subseteq g$ such that any tuple $\check{q} \supseteq \bar{p}$ can be extended to an enumeration $\check{g}$ with pull-back $\check{\mathcal{B}} = \check{g}^{-1}(\mathcal{A})$ so that $\Gamma^{D(\check{\mathcal{B}})}$ is an isomorphism from $\check{\mathcal{B}}$ to $\mathcal{A}$.*

Let us leave the proof of the claim for later, and start by proving the theorem from it.

Given tuples $\bar{q} = \langle q_0, q_1, \ldots \rangle \in A^{\leq \mathbb{N}}$ and $\bar{n} = \langle n_0, \ldots, n_\ell \rangle \in \mathbb{N}^{<\mathbb{N}}$, we use $\bar{q} \upharpoonright \bar{n}$ to denote $\langle q_{n_0}, \ldots, q_{n_\ell} \rangle \in A^{|\bar{n}|}$. Since g and $\Gamma^{D(\mathcal{B})}$ are isomorphisms from $\mathcal{B}$ to $\mathcal{A}$, for every $\bar{n} \in \mathbb{N}^{<\mathbb{N}}$,

$$(\mathcal{A}, g \upharpoonright \bar{n}) \cong (\mathcal{B}, \bar{n}) \cong (\mathcal{A}, \Gamma^{D(\mathcal{B})} \upharpoonright \bar{n}).$$

Recall that if $\bar{q} \subseteq g$, then $D_{\mathcal{A}}(\bar{q}) \subseteq D(\mathcal{B})$ (Observation 1.1.15). Therefore, if we have $\bar{q} \subseteq g$ so that $\Gamma^{D_{\mathcal{A}}(\bar{q})} \upharpoonright \bar{n}$ converges (i.e., if $\Gamma^{D_{\mathcal{A}}(\bar{q})}(n_i)\!\downarrow$ for all $i \leq \ell$), then $\Gamma^{D_{\mathcal{A}}(\bar{q})} \upharpoonright \bar{n}$ is automorphic to $\bar{q} \upharpoonright \bar{n}$ as in the diagram below.

$$
\begin{array}{ccccc}
\mathcal{A} & \xleftarrow{\;\;\cong\;\;} & \mathcal{B} & \xrightarrow{\;\;\cong\;\;} & \mathcal{A} \\
 & {\scriptstyle g} & & {\scriptstyle \Gamma^{D(\mathcal{B})}} & \\
\vdots & & \vdots & & \vdots \\
\bar{q} \upharpoonright \bar{n} & \longleftarrow\!\!\!\mid & \bar{n} & \mid\!\!\!\longrightarrow & \Gamma^{D_{\mathcal{A}}(\bar{q})} \upharpoonright \bar{n}
\end{array}
$$

Here comes the key observation: the value of $\Gamma^{D_{\mathcal{A}}(\bar{q})} \upharpoonright \bar{n}$ depends only on $D_{\mathcal{A}}(\bar{q}) \in 2^{<\mathbb{N}}$, and it determines the automorphism orbit of $\bar{q} \upharpoonright \bar{n}$. Thus, informally: for $\bar{a} = \bar{q} \upharpoonright \bar{n}$, the existential formula that says that $\bar{a}$ is part of a tuple $\bar{q}$ with this particular diagram defines the automorphism orbit of

$\bar{a}$. Let us explain this in more detail. The key observation above can be formally stated as follows:

CLAIM 8.2.4. *If $\bar{q}, \check{q} \supseteq \bar{p}$ and $\Gamma^{D_{\mathcal{A}}(\bar{q})} \upharpoonright \bar{n} \downarrow$, then*

$$D_{\mathcal{A}}(\bar{q}) = D_{\mathcal{A}}(\check{q}) \Longrightarrow (\mathcal{A}, \bar{q} \upharpoonright \bar{n}) \cong (\mathcal{A}, \check{q} \upharpoonright \bar{n}).$$

To see this, from the previous claim we get an enumeration $\check{g} \supset \check{q}$ such that if $\check{\mathcal{B}} = \check{g}^{-1}(\mathcal{A})$, then $\Gamma^{D(\check{\mathcal{B}})}$ is an isomorphism. Then, using the observation from the diagram above and that $\Gamma^{D_{\mathcal{A}}(\bar{q})} \upharpoonright \bar{n} = \Gamma^{D_{\mathcal{A}}(\check{q})} \upharpoonright \bar{n}$, we get that

$$(\mathcal{A}, \bar{q} \upharpoonright \bar{n}) \cong (\mathcal{B}, \bar{n}) \cong (\mathcal{A}, \Gamma^{D_{\mathcal{A}}(\bar{q})} \upharpoonright \bar{n}) = (\mathcal{A}, \Gamma^{D_{\mathcal{A}}(\check{q})} \upharpoonright \bar{n})$$

$$\cong (\check{\mathcal{B}}, \bar{n}) \cong (\mathcal{A}, \check{q} \upharpoonright \bar{n}),$$

as needed for the claim.

Fix a tuple $\bar{a}$; let us find a $\exists$-definition for the orbit of $\bar{a}$ under automorphisms of $\mathcal{A}$ that fix $\bar{p}$. Computably, search for a tuple $\bar{q}_{\bar{a}} \in A^{<\mathbb{N}}$ and a tuple $\bar{n}_{\bar{a}} \in \mathbb{N}^{<\mathbb{N}}$ such that

$$\bar{q}_{\bar{a}} \supseteq \bar{p}, \quad \bar{q}_{\bar{a}} \upharpoonright \bar{n}_{\bar{a}} = \bar{p}\bar{a} \quad \text{and} \quad \Gamma^{D_{\mathcal{A}}(\bar{q}_{\bar{a}})} \upharpoonright \bar{n}_{\bar{a}} \downarrow.$$

We will eventually find such tuples because one can always take $\bar{q}_{\bar{a}}$ to be a long enough initial segment of g and take $\bar{n}_{\bar{a}}$ so that $g \upharpoonright \bar{n}_{\bar{a}} = \bar{p}\bar{a}$. We claim that, for any tuple $\bar{b}$,

$$(\mathcal{A}, \bar{p}\bar{a}) \cong (\mathcal{A}, \bar{p}\bar{b}) \iff \exists \check{q}(\check{q} \supseteq \bar{p} \wedge \check{q} \upharpoonright \bar{n}_{\bar{a}} = \bar{p}\bar{b} \wedge D_{\mathcal{A}}(\check{q}) = D_{\mathcal{A}}(\bar{q}_{\bar{a}})).$$

For the right-to-left direction, consider such a tuple $\check{q}$, and observe that $\bar{p}\bar{a}$ and $\bar{p}\bar{b}$ are automorphic by Claim 8.2.4. For the left-to-right direction, let $\check{q}$ be the tuple that corresponds to $\bar{q}_{\bar{a}}$ through the automorphism mapping $\bar{p}\bar{a}$ to $\bar{p}\bar{b}$.

We can rewrite the right-hand side as an existential formula about $\mathcal{A}$ with parameters $\bar{p}$:

$$\varphi_{\bar{a}}(\bar{p}, \bar{x}) \equiv \exists \bar{y} \left(\bar{y} \supseteq \bar{p} \wedge \bar{y} \upharpoonright \bar{n}_{\bar{a}} = \bar{p}\bar{x} \wedge D(\bar{y}) = D_{\mathcal{A}}(\bar{q}_{\bar{a}}) \right),$$

where $\bar{x}$ and $\bar{y}$ are replacing $\bar{b}$ and $\check{q}$, and where "$D(\bar{y}) = \sigma$" is shorthand for $\varphi_{\sigma}^{\text{at}}(\bar{y})$, as defined in 1.1.10. The formula $\varphi_{\bar{a}}$ defines the orbit of $\bar{a}$ under automorphisms that fix $\bar{p}$. The set $\{\varphi_{\bar{a}} : \bar{a} \in A^{<\mathbb{N}}\}$ is thus the desired c.e. Scott family of $\exists$-formulas over $\bar{p}$.

We still have to prove Claim 8.2.3, that there is a $\bar{p}$ that forces $\Gamma^{D(\mathcal{B})}$ to be an isomorphism from $\mathcal{B}$ to $\mathcal{A}$.

PROOF OF CLAIM 8.2.3. This is a standard forcing proof as we will see in [MonP2]. For this particular forcing application, the techniques we have developed so far in Chapter 4 are enough, as we did in Theorem 5.3.6.

Recall that g is a 2-generic enumeration of $\mathcal{A}$, and $\mathcal{B} = g^{-1}(\mathcal{A})$. Let us start by forcing $\Gamma^{D(\mathcal{B})}$ to behave correctly wherever it converges. For this, consider the set of strings which force it not to:

$$Q_1 = \{\bar{q} \in A^\star : \exists n < |\bar{q}| (\Gamma^{D_A(\bar{q})} \restriction n\downarrow \ \& \ D_A(\Gamma^{D_A(\bar{q})} \restriction n) \neq D_A(\bar{q} \restriction n))\}.$$

(Recall that $A^\star$ is the set of tuples of different elements from A.) The set Q_1 is r.i. computable in $\mathcal{A}$ (using that $D(\mathcal{A})$ is computable), and hence decided by some initial segment of the enumeration g.[75] No initial segment of g is in Q_1 because $\Gamma^{D_A(\mathcal{B})}$ is an isomorphism, so there must be an initial segment $\bar{p}_1 \in A^\star$ of g such that no extension of $\bar{p}_1$ is in Q_1. This means that whenever $\bar{q} \in A^\star$ extends $\bar{p}_1$, if $\Gamma^{D_A(\bar{p})}(n)\downarrow$, then $D_A(\Gamma^{D_A(\bar{q})} \restriction n) = D_A(\bar{q} \restriction n)$.

Second, we force that $\Gamma^{D(\mathcal{B})}$ is total: For this, consider the set of strings which force $\Gamma^{D(\mathcal{B})}$ to be undefined at some $n \in \mathbb{N}$:

$$Q_2 = \{\bar{q} \in A^\star : \exists n \in \mathbb{N} \ \forall \bar{r} \in A^\star (\bar{r} \supseteq \bar{q} \rightarrow \Gamma^{D_A(\bar{r})}(n)\uparrow)\}.$$

The set Q_2 is Σ^c_2 in $\mathcal{A}$, and hence r.i.c.e. in $(\mathcal{A}, \vec{K}^{\mathcal{A}})$ and decided by an initial segment of g.[76] We cannot have an initial segment of g in Q_2 because we would have that $\Gamma^{D(\mathcal{B})}(n)\uparrow$ for some n. So, for some initial segment $\bar{p}$ of g, we have that, for every $\bar{q} \in A^\star$ extending $\bar{p}$ and every n, there is a $\bar{r} \in A^\star$ extending $\bar{q}$ for which $\Gamma^{D_A(\bar{r})}(n)\downarrow$. We may assume $\bar{p} \supseteq \bar{p}_1$.

We claim that $\bar{p}$ is as wanted in Claim 8.2.3. Since $\bar{p}$ forces out of Q_2, for any $\breve{q} \supseteq \bar{p}$, we can build a sequence $\breve{q} \subseteq \bar{r}_1 \subseteq \bar{r}_2 \subseteq \bar{r}_3 \subseteq \cdots \in A^\star$ so that $\Gamma^{D_A(\bar{r}_n)}(n)\downarrow$ for each n. If we also make sure that n is in the range of $\bar{r}_n$, we get an onto enumeration $\tilde{g} = \bigcup_{n \in \mathbb{N}} \bar{r}_n \colon \mathbb{N} \to A$, which satisfies that $\Gamma^{D(\tilde{\mathcal{B}})}$ is total, where $\tilde{\mathcal{B}} = \tilde{g}^{-1}(\mathcal{A})$. Since $\bar{p}$ forces out of Q_1 (meaning that no extension of $\bar{p}$ is in Q_1), $\Gamma^{D(\tilde{\mathcal{B}})} \circ \tilde{g}^{-1} \colon \mathcal{A} \to \mathcal{A}$ must preserve diagrams and hence be an isomorphism. It follows that $\Gamma^{D(\tilde{\mathcal{B}})} \colon \tilde{\mathcal{B}} \to \mathcal{A}$ must be an isomorphism too.   □

EXERCISE 8.2.5. (Hard) (Originated after conversations between Harrison-Trainor, Hirschfeldt, Kalimullin, Melnikov, Montalbán, and Solomon.) The proof above uses the fact that $\mathcal{A}$ has a computable ω-presentation. We can still have relatively computably categorical structures that do not have computable ω-presentations: between any two copies $\mathcal{B}$ and $\mathcal{C}$ of $\mathcal{A}$ there is an isomorphism computable from $D(\mathcal{B}) \oplus D(\mathcal{C})$.

(a) Prove that Theorem 8.2.2 is still true when $\mathcal{A}$ does not have computable copies. (In this case, the Scott family will have extra formulas that are not satisfied by any tuple in the structure.) Hint in footnote.[77]

---

[75] Notice that $\Gamma^{D_A(\bar{q})} \restriction n$ is a tuple in $\mathbb{N}^{<\mathbb{N}}$, and we need to use $D(\mathcal{A})$ to figure out $D_A(\Gamma^{D_A(\bar{q})} \restriction n) \in 2^{<\mathbb{N}}$ in this particular presentation of $\mathcal{A}$. When we wrote $D_A(\Gamma^{D_A(\bar{q})} \restriction n) \neq D_A(\bar{q} \restriction n)$, it was a shorthand for $\neg\varphi^{at}_\sigma(\bar{q} \restriction n)$ for $\sigma = D_A(\Gamma^{D_A(\bar{q})} \restriction n)$, where φ^{at}_σ is as in Observation 1.1.10.

[76] To see that Q_2 is Σ^c_2, observe that $\{\langle \bar{q}, n \rangle : \forall \bar{r} \in A^\star (\bar{r} \supseteq \bar{q} \rightarrow \Gamma^{D_A(\bar{r})}(n)\uparrow)\}$ is co-r.i.c.e. in $\mathcal{A}$ and hence Π^c_1-definable.

[77] You need to consider a generic presentation of $\mathcal{A} \sqcup \mathcal{A}$.

(b) Show that in this setting, if the $\exists$-type of the parameters is c.e. in an oracle X, then $\mathcal{A}$ has a Π_2^{cX} Scott sentence.

(c) Show that $\mathcal{A}$ has enumeration degree given by the $\exists$-type of the parameters.

8.3. Categoricity on a cone

Recall that by the *Turing cone* above X, we mean the set $\{ Y \in 2^{\mathbb{N}} : Y \geq_T X \}$. Sometimes, we will just call it a *cone*. A set $\mathcal{R} \subseteq 2^{\mathbb{N}}$ is said to be *degree invariant* if, for every $X, Y \in 2^{\mathbb{N}}$, if $X \in \mathcal{R}$ and $Y \equiv_T X$, then $Y \in \mathcal{R}$ too. Martin showed that every degree-invariant set of reals either contains a cone or is disjoint from a cone—if one assumes enough determinacy, whatever that means. This prompts us to view degree-invariant sets that contain cones as *large*, and the ones disjoint from cones as *small*. It is not hard to show that countable intersections of large sets are still large, and countable unions of small sets are still small.

THEOREM 8.3.1 (Martin [Mar68]). *If $\mathcal{R} \subseteq 2^{\mathbb{N}}$ is Borel and degree-invariant, it either contains a cone or is disjoint from a cone.*

We sketch this proof for the readers familiar with infinite games. The theorem is not relevant for the rest of the text, other than as a motivation for Definition 8.3.2. The reader not familiar with determinacy may freely skip it.

PROOF. Consider a game where player I and player II alternatively play binary bits $x_0, y_0, x_1, y_1, \cdots \in \{0, 1\}$ for infinitely many steps.

Player I	x_0		x_1		x_2	$\cdots$	$\cdots$	$\bar{x} \in 2^{\mathbb{N}}$
Player II		y_0		y_1		$\cdots$	$\cdots$	$\bar{y} \in 2^{\mathbb{N}}$

Let player I win the game if the sequence $\bar{x} \oplus \bar{y}$ belongs to $\mathcal{R}$, and let player II win if it does not. By Borel determinacy (cf. Martin [Mar75]) one of the two players must have a winning strategy $s: 2^{<\mathbb{N}} \to 2$.

We claim that if player I has a winning strategy, the cone above s is included in $\mathcal{R}$; while if player II has a winning strategy, the cone above s is disjoint from $\mathcal{R}$. Suppose s is a winning strategy for player I, and let $\bar{y}$ be any real in the cone above s; we want to show that $\bar{y} \in \mathcal{R}$. Assume player II plays $\bar{y}$, and let $\bar{x}$ be the response to $\bar{y}$ by a player I following the strategy s. Since s is a winning strategy, we have that $\bar{x} \oplus \bar{y} \in \mathcal{R}$ and $\bar{x} \leq_T s \oplus \bar{y}$. Since $\bar{y} \geq_T s$, we get that $\bar{y} \equiv_T \bar{x} \oplus \bar{y}$. Since $\mathcal{R}$ is degree invariant, this implies that $\bar{y} \in \mathcal{R}$, as needed. The proof of the case where II has a winning strategy is completely parallel. □

If instead of assuming $\mathcal{R}$ is Borel, we have that it is analytic, the theorem is still true, but does not follow from ZFC. It follows from the existence of sharps, which a weak large-cardinal hypothesis (cf. Harrington [Har78]). If we do not want to impose any complexity assumption on $\mathcal{R}$, we would need omit the axiom of choice and assume the full axiom of determinacy.

Suppose now we have a property of reals that is invariant under Turing equivalence. For instance, consider the set of $X \in 2^{\mathbb{N}}$ such that a given structure $\mathcal{A}$ is X-computably categorical. By Martin's theorem, this set must be either large or small, assuming analytic determinacy. In other words, either, relative to almost all oracles $\mathcal{A}$ is computably categorical; or, relative to almost all oracles $\mathcal{A}$ is not computably categorical.

DEFINITION 8.3.2. A structure $\mathcal{A}$ is *computably categorical on a cone* if there is a $Y \in 2^{\mathbb{N}}$ such that $\mathcal{A}$ is X-computably categorical for all $X \geq_T Y$.

In Section 8.5, we will construct a computable categorical structure which is not relatively so. That structure is far from being natural, and it was purposely build diagonalizing against lists of computable functions. However, if $\mathcal{A}$ is a *natural* structure, a proof that it has a property like categoricity, or a proof that it does not have it, would typically relativize. Thus, for natural $\mathcal{A}$, the three notions of computable categoricity—plain, relative, and on a cone—should coincide. If we want to understand how computable categoricity works on "natural" structures, our best bet is to look at it on a cone. The reason is that on-a-cone properties avoid counterexamples one can build by diagonalizing against all computable functions. This is because one would have to diagonalize against all X-computable functions for almost all X, and there are continuum many of those. This is why it is often the case that on-a-cone properties have cleaner structural characterizations, as is the case for computable categoricity:

THEOREM 8.3.3. *Let $\mathcal{A}$ be a countable structure. The following are equivalent*:

1. $\mathcal{A}$ *is computably categorical on a cone.*
2. $\mathcal{A}$ *is $\exists$-atomic over a finite set of parameters.*
3. $\mathcal{A}$ *has an Σ_3^{in} Scott sentence.*

PROOF. The equivalence between the top two statements follows from the relativized version of Theorem 8.2.2: Notice that $\mathcal{A}$ is computably categorical on a cone if and only if it is "relatively computably categorical" relative to some oracle X. The equivalence between the bottom two statements was proved in Lemma 3.7.4.   □

8.4. When relative and plain computable categoricity coincide

We saw in Table 2 that computable categoricity can be completely understood within certain classes of structures, despite being Π_1^1-complete in the general case. Something that is special about the classes from Table 2 is that, for them, plain and relative computable categoricity coincide. As we argued above, for "natural" structures within any class, the two notions should also coincide. Goncharov proved that, under certain effectiveness conditions, computably categoricity is indeed well-behaved. His result is based on a theorem by Nurtazin that deals with yet another variation of the notion of computable categoricity.

DEFINITION 8.4.1. Given an ω-presentation $\mathcal{A}$ of a τ-structure, we define $ED(\mathcal{A}) \in 2^{\mathbb{N}}$, the *elementary diagram* of $\mathcal{A}$, the same way we defined its atomic diagram in 1.1.2, but now considering all elementary first-order formulas instead of just the atomic ones.[78] For $i \in \mathbb{N}$,

$$ED(\mathcal{A})(i) = \begin{cases} 1 & \text{if } \mathcal{A} \models \varphi_i^{\text{el}}[x_j \mapsto j : j \in \mathbb{N}], \\ 0 & \text{otherwise,} \end{cases}$$

where $\{\varphi_i^{\text{el}} : i \in \mathbb{N}\}$ is an effective listing of the elementary first-order τ-formulas.

An ω-presentation $\mathcal{A}$ is said to be *decidable* if $ED(\mathcal{A})$ is computable.

The notion of decidable structure is quite important in computable structure theory. If one were interested in studying theorems from model theory from a computational perspective, dealing with decidable structures may be more appropriate than with computable ones. The notions of computable categoricity and effective $\exists$-atomicity translate as follows:

DEFINITION 8.4.2. $\mathcal{A}$ is *computably categorical for decidable copies* if there is a computable isomorphism between any two decidable copies of $\mathcal{A}$. $\mathcal{A}$ is *effectively atomic* if it has a c.e. Scott family of elementary first-order formulas (see Definition 3.1.2).

Atomic structures are quite important in model theory, as $\exists$-atomic structure are relevant in computable structure theory. Exactly as in Theorem 3.5.2, a structure is atomic if and only if every elementary type realized in the structure is supported by an elementary formula, and its Scott family consists of these supporting formulas. (In the case of full types, supported types are called *principal types*, and the supporting formulas are called *generating formulas*.)

THEOREM 8.4.3 (Nurtazin [Nur74]). *Let $\mathcal{A}$ be a decidable structure. The following are equivalent*:

1. $\mathcal{A}$ *is computably categorical for decidable copies*.
2. $\mathcal{A}$ *is effectively atomic over a finite set of parameters*.

Let us highlight that, while in Theorem 8.2.2 we could build a non-computable (generic) copy of $\mathcal{A}$ to apply relatively computable categoricity, we now need to build a decidable copy of $\mathcal{A}$ to apply the assumptions. Thus, generics will not be useful here, and the proof will have to be quite different.

---

[78]The *elementary* formulas are the finitary first-order formulas.

PROOF. An easy back-and-forth argument shows that effective atomicity implies computable categoricity for decidable copies as in Theorem 3.4.2.

The other implication, from (1) to (2), requires a finite-injury priority construction. The reader not familiar with priority construction should read Section 7.1 first. This is a long and elaborated proof, so brace for it.

The idea is to build a decidable copy $\mathcal{B}$ of $\mathcal{A}$ so that we can deduce that either there are no computable isomorphisms between $\mathcal{B}$ and $\mathcal{A}$, or there is a c.e. Scott family for $\mathcal{A}$. Thus, either part (1) fails or part (2) holds. There are two sets of requirements. First, for each e, we have:

> *Requirement R_e:* Either Φ_e is not an isomorphism from $\mathcal{B}$ to $\mathcal{A}$, or $\mathcal{A}$ has a c.e. Scott family over parameters.

If all these requirements are satisfied, then either one of them succeeds in building a Scott family and we get that $\mathcal{A}$ is effectively atomic over parameters, or all of them succeed in making sure no Φ_e is an isomorphism, and hence showing that $\mathcal{A}$ is not computably categorical for decidable copies.

As usual, we will build $\mathcal{B}$ by building a one-to-one enumeration $g \colon \mathbb{N} \to A$ and defining $\mathcal{B}$ as the pull-back $g^{-1}(\mathcal{A})$. The other set of requirements will guarantee that g is onto.

> *Requirement P_e:* The eth element of the ω-presentation $\mathcal{A}$ is in the range of g.

The requirements are listed in order of priority as usual: $P_0, R_0, P_1, R_1, \ldots$.

We need to ensure that $\mathcal{B}$ is decidable despite g not being computable. To be able to speak in precise terms, we need to define the elementary diagram of finite tuples the same way we did for atomic diagrams in Definition 1.1.9. Given a tuple $\bar{a} = \langle a_0, \ldots, a_s \rangle \in A^{<\mathbb{N}}$, we define the *elementary diagram of $\bar{a}$ in $\mathcal{A}$*, denoted $ED_{\mathcal{A}}(\bar{a})$, as the string in $2^{|\bar{a}|}$ such that, for $i < |\bar{a}|$,[79]

$$ED_{\mathcal{A}}(\bar{a})(i) = \begin{cases} 1 & \text{if } \mathcal{A} \models \varphi_i^{\text{el}}[x_j \mapsto a_j, j < s], \\ 0 & \text{otherwise.} \end{cases}$$

As in Observation 1.1.15, we have that if g is an enumeration of $\mathcal{A}$, then

$$ED(g^{-1}(\mathcal{A})) = \bigcup_{k \in \mathbb{N}} ED_{\mathcal{A}}(g \restriction k).$$

---

[79]We are choosing to make $ED_{\mathcal{A}}(\bar{a})$ have length $|\bar{a}|$, but we could have chosen many other finite bounds for it. What matters is that it is a finite string, that it only involves formulas that use the first $|\bar{a}|$ variables, and that all formulas are eventually taken into account as $|\bar{a}|$ goes to infinity.

At each stage s of the construction, we will build an injective finite tuple $g_s \in A^{<\mathbb{N}}$. The g_s's will not form a nested sequence, so we will not be able to define g as their union. But the sequence will have a pointwise limit, and we will be able to define $g(i) = \lim_s g_s(i)$. We still need $\mathcal{B}$ to be decidable, though. So even if the g_s's are not nested, we require that the strings $ED_A(g_s) \in 2^{<\mathbb{N}}$ are nested; that is, for all $s < t$, $ED_A(g_s) \subseteq ED_A(g_t)$. We will then have that

$$ED(\mathcal{B}) = \bigcup_{s \in \mathbb{N}} ED_A(g_s) \in 2^{\mathbb{N}}$$

is computable.

Informally, the idea for satisfying R_e is as follows. R_e will try to define g_s so that, for some tuple $\bar{n} \in \mathbb{N}^{<\mathbb{N}}$, $\Phi_{e,s} \restriction \bar{n}$ converges and disagrees with $g_s \restriction \bar{n}$ on some elementary formula. This way, if R_e manages to preserve this tuple g_s so that it ends up being an initial segment of g, since g will be an isomorphism from $\mathcal{B}$ to $\mathcal{A}$, Φ_e will not. To do this, for every tuple $\bar{b} \in A^{<\mathbb{N}}$, once we see $\Phi_{e,s} \restriction \bar{n}{\downarrow} = \bar{b}$ for some $\bar{n}$ and s, we enlist $\bar{b}$ as a possible candidate for diagonalization. From that point on, we will be looking for another tuple $\bar{c}$ disagreeing with $\bar{b}$ on some elementary formula, so we can try to define $g \restriction \bar{n} = \bar{c}$ while preserving $ED_A(g_s)$. If we find it, R_e will require attention, and if attention is given to it at some stage t, it will define g_t so that $g_t \restriction \bar{n} = \bar{c}$ and then try to preserve this initial segment of g. If we do not find such a disagreeing tuple $\bar{c}$, the reason is that whatever commitment we made at stage s about $\bar{n}$—namely that we must preserve $ED_A(g_s)$—had to imply all other formulas about $\bar{b}$, and hence be a principal formula for the type of $\bar{b}$. If this happens for all tuples $\bar{b}$, we can build a Scott family for $\mathcal{A}$. To make sure this works, we will be monitoring that everything we commit to regarding $\bar{b}$ later on—namely that we must preserve $ED_A(g_t)$ for some new g_t—is implied by the potentially principal formula. If it is, then we are not really committing anything new; if it is not, we have found an opportunity to diagonalize.

What makes this more difficult is that R_e must respect the work done by weaker priority requirements. The same way R_e would like to preserve the initial segment of g it defined, higher-priority requirements will like to preserve their initial segments. At the beginning of stage $s + 1$, we will define $\bar{p}_e[s] \subseteq g_s$ to be the initial segment of g_s that has been defined by higher-priority requirements R_i for $i < e$ and P_i for $i \le e$. R_e must preserve $\bar{p}_e[s]$; that is, it is only allowed to define g_{s+1} extending $\bar{p}_e[s]$. R_e must also preserve $ED(g_s)$; that is, it is only allowed to define g_{s+1} satisfying $ED(g_{s+1}) \supseteq ED(g_s)$.

The construction: At any given stage, the first few requirements will be *active* and the rest *inactive*. At each stage, the highest-priority inactive requirement will be *initialized* and become active. During the construction,

requirements may be *canceled*, making them inactive again. At each
stage, each active P_e requirement will have an *output* string $\bar{p}_e \in A^{<\mathbb{N}}$,
and each active R_e requirement an *output* string $\bar{r}_e$. These strings will
be nested, $\bar{p}_0 \subseteq \bar{r}_0 \subseteq \bar{p}_1 \subseteq \bar{r}_1 \subseteq \cdots$, and g_s will be the union of the
output strings of the active requirements at stage s. These are not fixed
strings, and the value of $\bar{p}_e$ or $\bar{r}_e$ may change throughout the stages. We
write $\bar{p}_e[s]$ or $\bar{r}_e[s]$ if we want to highlight that we are referring to their
values at stage s. We will show they will eventually reach a limit and stop
changing.

Requirement P_e only acts the first time it is active after being initialized.
If it is ever canceled, it will act again once it gets initialized again. When
it acts at stage $s + 1$, its action consists of defining $g_{s+1} = g_s{}^\frown e$ (where
e refers to the eth element of the ω-presentation $\mathcal{A}$). Well, that is if e
is not in the range of g_s already, in which case we just define $g_{s+1} = g_s$.
Once P_e acts, stage $s + 1$ is over, and we move on directly to the next
stage, $s + 2$. We define the *output* of P_e to be $\bar{p}_e = g_{s+1}$, and this will
stay this way unless P_e is later canceled. Since P_e will only act at a
stage when no other requirement acts, we will have that $\bar{r}_{e-1}$, the output
of R_{e-1}, is included in g_s. Thus, P_e indeed respects higher-priority
requirements.

Requirement R_e works as follows. At each stage that is active, R_e may
go through four *phases*:

- *waiting*,
- *internal calculations*,
- *requiring attention*, or
- *acting*.

We need to describe what R_e does in each of these phases. We leave the
internal calculations phase for last.

Recall that $\bar{p}_e$ is the initial segment of g_s given by the output of the
requirement of immediately higher priority, namely P_e. Once R_e has been
activated, it will stay in the *waiting* phase until we reach a stage s at which
$\Phi_{e,s} \upharpoonright |\bar{p}_e|$ converges. At stages where $\Phi_{e,s} \upharpoonright |\bar{p}_e|$ does not converge, R_e does
not do anything, and we move on to consider the next active requirement.
During these waiting stages, and until the requirement acts (if ever), its
output is $\bar{r}_e = \bar{p}_e$. When we reach a stage s where $\Phi_{e,s} \upharpoonright |\bar{p}_e|$ converges, we
let

$$\bar{a} = \Phi_{e,s} \upharpoonright |\bar{p}_e|$$

and move to the next phases of internal calculations to decide if we require
attention.

For a tuple $\bar{p} \subseteq g_s$ of elements and a tuple $\bar{n} \in \mathbb{N}^{<\mathbb{N}}$ of numbers be-
tween $|\bar{p}|$ and $|g_s| - 1$, we let $\psi_{\bar{n},g_s}(\bar{p}, \bar{x})$ be the elementary formula
describing the commitments we have made about $\bar{n}$ relative to $\bar{p}$ in

$ED(g_s)$:[80]

$$\psi_{\bar{n},g_s}(\bar{p},\bar{x}) \equiv \exists \bar{y}\,(\bar{y} \supseteq \bar{p} \wedge \bar{y} \restriction \bar{n} = \bar{x} \wedge ED(\bar{y}) = \sigma),$$

$$\text{where } \sigma = ED_A(g_s) \in 2^{<\mathbb{N}}.$$

Notice that $A \models \psi_{\bar{n},g_s}(\bar{p}, g_s \restriction \bar{n})$ with witness $\bar{y} = g_s$.

R_e *requires attention* if it finds an opportunity to diagonalize, that is, if it finds a tuple $\bar{n} \in \mathbb{N}^{<\mathbb{N}}$ of numbers greater than $|\bar{p}_e|$, a tuple $\bar{c} \in A^{<\mathbb{N}}$, and an elementary formula φ such that:

1. $\Phi_{e,s} \restriction \bar{n}$ converges,
2. the tuples $\bar{p}_e{}^\frown \bar{c}$ and $\bar{a}{}^\frown \Phi_{e,s} \restriction \bar{n}$ disagree on φ, and
3. $A \models \psi_{\bar{n},g_s}(\bar{p}_e, \bar{c})$.

After R_e requires attention, it may be allowed to *act*. Let $\bar{q}$ be the witness to $A \models \psi_{\bar{n},g_s}(\bar{p}_e, \bar{c})$. That is,

$$\bar{q} \supseteq \bar{p}_e \wedge \bar{q} \restriction \bar{n} = \bar{c} \wedge ED_A(\bar{q}) = ED_A(g_s).$$

The *action* of R_e is to define $g_{s+1} = \bar{q}$ and re-define $\bar{r}_e$, the *outcome* of R_e, to be $\bar{q}$ too. If R_e is never canceled again, and g ends up being an isomorphism from B to A extending $\bar{r}_e$, R_e would have succeeded in diagonalizing against Φ_e, ensuring that Φ_e is not an isomorphism from B to A. This is because, if Φ_e was an isomorphism, the automorphism $\Phi_e \circ g^{-1}$ should map $\bar{p}_e{}^\frown \bar{c}$ to $\bar{a}{}^\frown \Phi_{e,s} \restriction \bar{n}$, contradicting the fact that they disagree on φ. After this action, we cancel all the weaker-priority requirements making them inactive and finish stage $s + 1$. R_e will not act again, and $\bar{r}_e$ will not change anymore, unless R_e is later canceled and re-initialized, in which case it will start all over again.

The *initial calculations of R_e* are as follows. While R_e waits for a chance to require attention, it enumerates a set S of formulas hoping it ends up being a Scott family for A over $\bar{p}_e$. Every time $\Phi_{e,s}$ converges on some new tuple $\bar{n}$ of numbers between $|\bar{p}_e|$ and $|g_s|$,

- define $\varphi_{\bar{n}}(\bar{x})$ to be the formula $\psi_{\bar{n},g_s}(\bar{p}_e, \bar{x})$, and
- enumerate $\varphi_{\bar{n}}$ into S.

By doing this, R_e is betting $\varphi_{\bar{n}}(\bar{x})$ generates the type of $\bar{b} = g \restriction \bar{n}$ within A over $\bar{p}_e$. To secure its bet, R_e will verify at each later stage u that

$$A \models \forall \bar{x}(\varphi_{\bar{n}}(\bar{x}) \rightarrow \psi_{\bar{n},g_u}(\bar{p}_e, \bar{x})).$$

---

[80]"$ED(\bar{y}) = \sigma$" is shorthand for what one would expect:

$$\left(\bigwedge_{i:\sigma(i)=1} \varphi_i^{\text{el}}(\bar{y})\right) \wedge \left(\bigwedge_{i:\sigma(i)=0} \neg\varphi_i^{\text{el}}(\bar{y})\right)$$

It does this as follows: at each stage $u + 1$ where a weaker-priority require-ment R_i of P_i for $i > e$ requires attention and wants to extend g_u to some tuple $\bar{h}$, we first check that

$$\mathcal{A} \models \forall \bar{x}(\varphi_{\bar{n}}(\bar{x}) \rightarrow \psi_{\bar{n},\bar{h}}(\bar{p}_e, \bar{x})). \tag{1}$$

If it does, we let the weaker-priority requirement do its thing and define $g_{u+1} = \bar{h}$. If it does not, R_e does not allow the weaker-priority requirement to act, because, instead, R_e is in a position to require attention itself: We know there is a tuple $\bar{c}_1$ satisfying $\varphi_{\bar{n}}(\bar{c}_1) \wedge \psi_{\bar{n},\bar{h}}(\bar{p}_e, \bar{c}_1)$, namely $\bar{h} \upharpoonright \bar{n}$,[81] and we know there is another tuple $\bar{c}_2$ that satisfies $\varphi_{\bar{n}}(\bar{c}_2) \wedge \neg\psi_{\bar{n},\bar{h}}(\bar{p}_e, \bar{c}_2)$ because the implication (1) does not hold. Let $\bar{c}$ be whichever of these two tuples disagrees with $\Phi_{e,s} \upharpoonright \bar{n}$ on $\psi_{\bar{n},\bar{h}}(\bar{p}_e, \bar{x})$. Since, at the previous stage u, we verified that $\mathcal{A} \models \forall \bar{x}(\varphi_{\bar{n}}(\bar{x}) \rightarrow \psi_{\bar{n},g_u}(\bar{p}_e, \bar{x}))$, we have that $\mathcal{A} \models \psi_{\bar{n},g_u}(\bar{p}_e, \bar{c})$. R_e has now found the witnesses $\bar{n}$, $\bar{c}$, and $\varphi \equiv \psi_{\bar{n},\bar{h}}(\bar{p}_e, \bar{x})$ necessary to require attention at stage $u + 1$.

Verifications: After a requirement is initialized, it will act at most once before it is re-initialized again, if ever. One can then prove, by induction on the list of requirements, that each requirement will eventually stop being canceled and will then eventually stop acting, and hence the next requirement will stop being canceled and then eventually stop acting, and so on. Since the outputs of the requirements only change when they act, we get that each $\bar{p}_e$ and $\bar{r}_e$ reaches a limit, and that g is the union of all these limits. Since each requirement P_e is eventually given the chance to act without being canceled again, we get that g is onto. Notice that g is one-to-one because each g_s is.

Let us now verify that each R_e is satisfied. Let s_e be the last stage in which P_e acted, so that R_e is never canceled after s_e. Suppose Φ_e is a computable isomorphism from $\mathcal{B}$ to $\mathcal{A}$. It must then be the case that R_e never requires attention after s_e, as otherwise, R_e would have acted and diagonalized against Φ_e, as we argued before. We claim that this implies that R_e is successful in making S a Scott family. For each tuple $\bar{b} \in A^{<\mathbb{N}}$ disjoint from $\bar{p}_e$, there will be some $\bar{n}$ such that $g \upharpoonright \bar{n} = \bar{b}$, and there will be a first stage $s_{\bar{b}} > s_e$ at which $\Phi_{e,s_{\bar{b}}} \upharpoonright \bar{n}\downarrow$. At that stage, we enumerate $\varphi_{\bar{n}}(\bar{x})$ $(= \psi_{\bar{n},g_{s_{\bar{b}}}}(\bar{p}_e, \bar{x}))$ into S. We need to show that $\varphi_{\bar{n}}$ is indeed a generating formula for the elementary type of $\bar{b}$ over $\bar{p}_e$. First, notice that even if $g_s \upharpoonright \bar{n} \neq \bar{b}$, we still have that $\mathcal{A} \models \varphi_{\bar{n}}(\bar{b})$, because, for every $t \geq s$, since $ED(g_t) \supseteq ED(g_s)$, we have that $\mathcal{A} \models \varphi_{\bar{n}}(g_t \upharpoonright \bar{n})$ as witnessed by $\bar{y} = g_t \upharpoonright |g_s|$. Since R_e never requires attention again, at every later stage $u > s_{\bar{b}}$, we have that

$$\mathcal{A} \models \forall \bar{x}(\varphi_{\bar{n}}(\bar{x}) \rightarrow \psi_{\bar{n},g_u}(\bar{p}_e, \bar{x})).$$

---

[81] We know $\bar{h} \upharpoonright \bar{n}$ satisfies $\varphi_{\bar{n}}$ because $ED(g_s) \subseteq ED(g_u) \subseteq ED(\bar{h})$.

Every elementary formula $\theta(\bar{p}_e, \bar{x})$ that is true of $\bar{b}$ in $\mathcal{A}$ will eventually be part of $ED_{\mathcal{A}}(g_u)$ for large enough u. Thus, θ is implied by $\psi_{\bar{n}, g_u}(\bar{p}_e, \bar{x})$, and hence implied by $\varphi_{\bar{n}}(\bar{x})$. □

If we want to go back to the notion of computable categoricity (for computable copies), we can modify the proof above so long as we assume the two-quantifier theory of $\mathcal{A}$ is computable.

DEFINITION 8.4.4. A $\forall\exists$-*formula* is one of the form

$$\forall x_0 \forall x_1 \ldots \forall x_n \exists y_0 \exists y_1 \ldots \exists y_k \ \psi(\bar{x}, \bar{y}, \bar{z})$$

where ψ is finitary and quantifier-free. An ω-presentation $\mathcal{A}$ is $\forall\exists$-*decidable* if we can effectively decide all $\forall\exists$-formulas about the tuples of $\mathcal{A}$, i.e., if there exists a computable function that, given an index for a $\forall\exists$-formula $\varphi(\bar{z})$ and a tuple $\bar{a} \in \mathcal{A}^{<\mathbb{N}}$, returns 1 or 0 depending on whether $\mathcal{A} \models \varphi(\bar{a})$.

THEOREM 8.4.5 (Goncharov [Gon75a]). *If $\mathcal{A}$ is $\forall\exists$-decidable, then $\mathcal{A}$ is computably categorical if and only if it is effectively $\exists$-atomic over a finite set of parameters.*

SKETCH OF THE PROOF. The proof is very similar to the proof above, but it requires being extra careful with the complexity of certain formulas at various steps of the construction. For this proof, we only need to preserve our usual atomic diagrams $D(g_s)$ instead of the elementary diagrams $ED(g_s)$. This will get us a computable ω-presentation $\mathcal{B}$. The formulas $\psi_{\bar{n}, g_s}$ are now defined using $D(g_s)$ instead of $ED(g_s)$. Notice that $\psi_{\bar{n}, g_s}$ is now an $\exists$-formula. When R_e is deciding if it requires attention, it now wants the tuples $\bar{p}_e{}^\frown \bar{c}$ and $\bar{a}^\frown \Phi_{e,s} \upharpoonright \bar{n}$ to disagree on some $\forall\exists$-formula, as that is what we can check computably. The key point where we used the decidability of $\mathcal{A}$ was during the initial-calculations phase to check whether

$$\mathcal{A} \models \forall\bar{x}(\varphi_{\bar{n}}(\bar{x}) \to \psi_{\bar{n}, \bar{h}}(\bar{p}, \bar{x})).$$

This formula is now $\forall\exists$, which we can decide by the assumption on $\mathcal{A}$. However, we need to check a bit more. Let $\psi_{\bar{n}, g_s}^{\forall}(\bar{p}_e, \bar{x})$ be the conjunction of all the $\forall$-formulas with indices less than $|g_s|$ that are true of $g_s(\bar{n})$ over $\bar{p}_e$. We also check that

$$\mathcal{A} \models \forall\bar{x}(\varphi_{\bar{n}}(\bar{x}) \to \psi_{\bar{n}, \bar{h}}^{\forall}(\bar{p}, \bar{x})),$$

as this also gives us an opportunity to diagonalize. When we are verifying that R_e works, we only need to show that $\varphi_{\bar{n}}$ supports the $\forall$-type of $g \upharpoonright \bar{n}$ over $\bar{p}_e$. All these formulas are implied by $\psi_{\bar{n}, g_u}^{\forall}(\bar{p}_e, \bar{x})$ for large enough u, so the proof is the same. □

Kudinov [Kud96a] showed this result is sharp by building a $\forall$-decidable computably categorical structure that is not effectively $\exists$-atomic. It is still true that $\forall$-decidable computably categorical structures are effectively

Σ_2^c-atomic, as proved by Downey, Kach, Lempp, and Turetksy [DKLT13, Theorem 1.13].

8.5. When relative and plain computable categoricity diverge

This section is dedicated to proving the following theorem.

THEOREM 8.5.1 (Goncharov [Gon77, Theorem 4]). *There is a structure which is computably categorical, but not relatively so.*

This is an important theorem, and its proof illustrates a couple of techniques that are useful throughout the field. One is the use of families of sets to build structures with particular properties, a very common technique in the Russian school. The other one is the use of a finite-injury priority argument that is a bit more elaborate than the two we have seen so far.

To prove Theorem 8.5.1, we will build a c.e. family of sets $\mathcal{F} \subseteq \mathcal{P}(\mathbb{N})$, and then take the graph

$$\mathcal{G}_{\mathcal{F}}^1 = \bigsqcup_{X \in \mathcal{F}} \mathcal{G}_X,$$

where $\mathcal{G}_X$ is the flower graph that consists of loops of size $n + 3$, one for each $n \in X$, all with a common node. This is almost the same as the graph $\mathcal{G}_{\mathcal{F}}^\infty$ we considered in Observation 5.4.2 and Lemma 6.1.11, with the difference that, in $\mathcal{G}_{\mathcal{F}}^1$, each $X \in \mathcal{F}$ is associated to exactly one flower graph $\mathcal{G}_X$ instead of infinitely many as in $\mathcal{G}_{\mathcal{F}}^\infty$. Let us see how the relevant properties about structures translate to families.

DEFINITION 8.5.2. A *computable Friedberg enumeration* of a family $\mathcal{F}$ is a c.e. set W whose columns exactly are the sets in $\mathcal{F}$ without repetition, i.e., not only $\mathcal{F} = \{ W^{[i]} : i \in \mathbb{N} \}$, but also $W^{[i]} \neq W^{[j]}$ for all $i \neq j$.

Recall from Definition 5.4.1 that a computable enumeration for a family $\mathcal{F}$ is a c.e. set W with $\mathcal{F} = \{ W^{[i]} : i \in \mathbb{N} \}$, allowing for repeating columns. In a Friedberg enumeration, every set in $\mathcal{F}$ corresponds to exactly one column. In Observation 5.4.2, we showed that $\mathcal{F}$ has a computable enumeration if and only if $\mathcal{G}_{\mathcal{F}}^\infty$ has a computable copy. As in Observation 5.4.2, one can easily produce a computable Friedberg enumeration of $\mathcal{F}$ out of a computable ω-presentation of $\mathcal{G}_{\mathcal{F}}^1$, and vice versa.

DEFINITION 8.5.3. A family $\mathcal{F} \subseteq \mathcal{P}(\mathbb{N})$ is *discrete* if there is a family S of finite sets such that, for each $A \in \mathcal{F}$, there is an $F \in S$ with $F \subseteq A$, and for each $F \in S$, there is a unique $A \in \mathcal{F}$ with $F \subseteq A$. We call such a set S a *separating family* for $\mathcal{F}$. We say that $\mathcal{F}$ is *effectively discrete* if $\mathcal{F}$ has a c.e. separating family.

LEMMA 8.5.4. *Let $\mathcal{F} \subseteq \mathcal{P}(\mathbb{N})$ be a family with a c.e. enumeration. Then $\mathcal{G}^1_{\mathcal{F}}$ is effectively $\exists$-atomic if and only if $\mathcal{F}$ is effectively discrete.*

PROOF. Suppose $\mathcal{F}$ has a separating set S. We need to find $\exists$-formulas defining each node of $\mathcal{G}^1_{\mathcal{F}}$. Notice that each center of a flower graphs $\mathcal{G}_X$ is alone in its own automorphism orbit because each $\mathcal{G}_X$ appears only once in $\mathcal{G}^1_{\mathcal{F}}$. Also notice that if we have an $\exists$-formula defining the center of $\mathcal{G}_X$, we can find $\exists$-definitions for all the nodes in $\mathcal{G}_X$.[82] Thus, we will concentrate on enumerating $\exists$-definitions for the centers of the flower graphs. For each $X \in \mathcal{F}$, there is a finite set $A \in S$ such that X is the only set in $\mathcal{F}$ that contains A. Let $\varphi_X(x)$ be the formula that says that x is part of a loop of size $n + 3$ for each $n \in A$. The center of G_X would be the only element of $\mathcal{G}^1_{\mathcal{F}}$ satisfying that formula. Notice that if S is c.e., this produces a c.e. Scott family.

Suppose now that $\mathcal{G}^1_{\mathcal{F}}$ is $\exists$-atomic. For each X, let φ_X be the $\exists$-formula in the Scott family satisfied by the center of $\mathcal{G}_X$. Let A_X be a finite subset of X such that the center of a flower graph $\mathcal{G}_A$ also satisfies φ_X. Such an A_X must exist because if an $\exists$-formula is true of a relational structure, it is also true of a finite substructure (Observation 1.1.8). We claim that $\{A_X : X \in \mathcal{F}\}$ is a separating family for $\mathcal{F}$. We already argued that such an A_X exists for each X. If $A_X \subseteq Y$ for $Y \in \mathcal{F}$, then, since $\exists$-formulas are preserved under embeddings and $\mathcal{G}_{A_X}$ embeds into G_Y, we would have that φ_X holds of the center of G_Y too. Since φ_X defines the orbit of the center of G_X, we must have $X = Y$.

Notice that if we have a c.e. enumeration of $\mathcal{F}$, for each column X of the enumeration, we can effectively find φ_X within the given c.e. Scott family, and we then effectively find some A_X. □

Recall that a structure is relatively computably categorical if and only if it is effectively $\exists$-atomic over some parameters. So, we need to add the parameters to the previous lemma. We only need one direction.

COROLLARY 8.5.5. *Let $\mathcal{F} \subseteq \mathcal{P}(\mathbb{N})$ be a discrete family of computable sets with a c.e. enumeration. Then if $\mathcal{G}^1_{\mathcal{F}}$ is effectively $\exists$-atomic over parameters, $\mathcal{F}$ is effectively discrete.*

PROOF. Let $\bar{p}$ be the parameters over which $\mathcal{G}^1_{\mathcal{F}}$ is effectively $\exists$-atomic. We can assume the elements of $\bar{p}$ are the centers of flowers, as from each $p \in \mathcal{G}^1_{\mathcal{F}}$ we can effectively find the center of the flower it belongs to and, vice-versa, we can effectively find p from the center of its flower. Since all flowers are completely independent, if we remove the flowers that contain $\bar{p}$ from $\mathcal{G}^1_{\mathcal{F}}$, we get a bouquet graph $\mathcal{G}^1_{\widetilde{\mathcal{F}}}$ that is effectively $\exists$-atomic over no parameters. By the previous lemma, the corresponding family $\widetilde{\mathcal{F}}$ is effectively discrete, and has a c.e. separating family $\widetilde{S}$. So, for each $F \in \widetilde{S}$

---

[82]We need to say that the node belongs to a loop of a certain size and that the loop also contains the center of $\mathcal{G}_X$.

is included in a unique $X \in \widetilde{\mathcal{F}}$, but it might also be included in some $Y \in \mathcal{F} \setminus \widetilde{\mathcal{F}}$. Since $\mathcal{F}$ was discrete to begin with, there is an extension of F which is still included in X, but not included in any of the finitely many sets in $Y \in \mathcal{F} \setminus \widetilde{\mathcal{F}}$. We can find such extension as we can find X using the c.e. enumeration of $\mathcal{F}$ and then find the extension using that the sets Y are computable. Let $\check{S}$ consist of the set of all the extensions of all the F's in S. Also using that $\mathcal{F}$ is discrete, there is a finite set of finite sets S_0, such each $F \in S_0$ is included in a unique set in $\mathcal{F}$ and that set is one of the finitely many ones in $\mathcal{F} \setminus \widetilde{\mathcal{F}}$. We then get that $\check{S} \cup S_0$ is a c.e. separating family for $\mathcal{F}$.   □

DEFINITION 8.5.6. A *computable equivalence* between two computable enumerations V and W of a family $\mathcal{F}$ is a computable permutation f of $\mathbb{N}$ such that $V^{[n]} = W^{[f(n)]}$ for every n. When such a computable equivalence exists, we say that V and W are *computably equivalent*.

LEMMA 8.5.7. $\mathcal{G}_{\mathcal{F}}^1$ *is computably categorical if and only if $\mathcal{F}$ has only one Friedberg enumeration up to computable equivalence.*

PROOF. We already know that computable ω-presentations of $\mathcal{G}_{\mathcal{F}}^1$ are in correspondence with c.e. Friedberg enumerations of $\mathcal{F}$. It is not hard to see that computable isomorphisms between ω-presentations of $\mathcal{G}_{\mathcal{F}}^1$ are then in correspondence with computable equivalences between c.e. Friedberg enumerations of $\mathcal{F}$.   □

Theorem 8.5.1 now follows from the following lemma which contains the bulk of the proof.

LEMMA 8.5.8 (Badaev [Bad77]). *There is a family $\mathcal{F} \subseteq \mathcal{P}(\mathbb{N})$ that is not effectively discrete and has only one computable Friedberg enumeration up to computable equivalence.*

PROOF. Let

$$ E = \{0, 2, 4, 6, 8, \dots\} \quad \text{and} \quad E_k = \{0, 2, 4, \dots, 2k\} \cup \{2k+1\}. $$

For each $n \in \mathbb{N}$, the family $\mathcal{F}$ will contain one set of the form $E \oplus \{n\}$, and at most one set of the form $E_k \oplus \{n\}$. There will be no other sets in $\mathcal{F}$. We will build a computable Friedberg enumeration U of $\mathcal{F}$.

To make sure $\mathcal{F}$ is not effectively discrete, we have the following requirements:

Positive Requirement P_e: W_e is not a separating family for $\mathcal{F}$.

To make sure $\mathcal{F}$ has a unique Friedberg enumeration, we have the following requirements:

Negative Requirement N_e: If W_e is an Friedberg enumeration of $\mathcal{F}$, then W_e is computably equivalent to U.

The requirements are listed in decreasing order of priority as usual: $N_0, P_0, N_1, P_1, \dots$. All the sets $E \oplus \{n\}$, for $n \in \mathbb{N}$, are enumerated into

U from the beginning, say on the even columns of U. The sets $E_k \oplus \{n\}$ will be enumerated later on by the positive requirements P_e. Each P_e will act at most once, enumerating at most one such set. At each stage, each negative requirement N_i will impose a restraint on the P_e requirements of weaker priority by not allowing them to enumerate any set of the form $E_k \oplus \{n\}$ with $n < M_{i,s} \le k$, where $M_{i,s}$ is a number defined by N_i at stage s of the construction. Each stage s of the construction starts with all the requirements N_i, for $i < s$, independently doing their own calculations and defining $M_{i,s}$. Then, all the requirements P_e for $e < s$ will independently do their thing as we describe below.

What makes these requirements "positive" and "negative," is that P_e enumerates elements into U, while N_e prevents elements from being enumerated.

The requirement P_e works as follows. Let $\{C_e : e \in \mathbb{N}\}$ be a computable partition of $\mathbb{N}$; for instance, let $C_e = \{\langle e, m \rangle : m \in \mathbb{N}\}$. The set C_e is reserved for requirement P_e. Suppose P_e has not been declared done yet. If we see a finite subset G with $\ulcorner G \urcorner \in W_e$ such that, for some $n \in C_e$ and some $k \in \mathbb{N}$, we have

- $G \subseteq \{0, 2, \ldots, 2k\} \oplus \{n\}$, and
- for each $i \le e$, either $M_{i,s} \le n$ or $k < M_{i,s}$,

then we add $E_k \oplus \{n\}$ to $\mathcal{F}$ (i.e., we enumerate it as a column in U), getting $G \subseteq E \oplus \{n\}$ and $G \subseteq E_k \oplus \{n\}$ which are both in $\mathcal{F}$. We declare P_e done, and we re-initialize all lower-priority N_i requirements. Recall that $M_{i,s}$ will be defined by N_i below. All we need to know for now about the sequence $M_{i,s}$ is that it is non-decreasing in s, and therefore that it converges to a limit—either to a number or to ∞. If W_e were indeed a separating family for F, then for every n, since $E \oplus \{n\} \in \mathcal{F}$, W_e would contain some set of the form $G = F \oplus \{n\}$ with $F \subseteq \{0, 2, \ldots, 2k\}$ for some k. Consider some $n \in C_e$ which is above $\lim_s M_{i,s}$ for all the $i \le e$ for which the limit is finite. The corresponding k would eventually be below all the $M_{i,s}$ for all the $i \le e$ for which the limit is infinite. P_e would then be allowed to act and enumerate $E_k \oplus \{n\}$ into $\mathcal{F}$. This contradicts that W_e is a separating family because G would be included in both $E \oplus \{n\}$ and $E_k \oplus \{n\}$; P_e succeeds.

The requirement N_e works as follows. It will be *initialized* at stage $s + 1 = e$ and then will be re-initialized every time a higher-priority P_i requirement acts. Since each P_i acts at most once in the whole construction, there will be a point after which N_e will never be re-initialized again. Every time N_e is initialized, it starts building a computable matching g_e between the columns of W_e and those of U by finite approximations $g_{e,0} \subseteq g_{e,1} \subseteq g_{e,2} \subseteq \cdots \to g_e$, with $g_{e,s} \in \mathbb{N}^{<\mathbb{N}}$. If it turns out that W_e is a Friedberg enumeration of $\mathcal{F}$ and that N_e is never re-initialized again, we have to make sure g_e is a computable equivalence between W_e and

U. The rough idea is as follows: At each stage s, we will look at the columns of $W_{e,s}$ and $U[s]$, and hope there is an obvious way to match them. Whenever we see a set of the form $E_k \oplus \{n\}$ in both $W_{e,s}$ and in $U[s]$, we can safely match these columns through $g_{e,s}$. The problem arises when we need to match columns of the form $\{0, 2, \ldots, 2m\} \oplus \{n\}$: These apparently matching columns may later grow in different ways and become $E_k \oplus \{n\}$ for some $k \geq m$ in W_e and become $E \oplus \{n\}$ in U. To deal with this, N_e will impose a restraint not allowing sets of the form $E_k \oplus \{n\}$ for any $k \geq m$ to be enumerated into U by lower-priority requirements.

Let us start by defining an enumeration $\{V_{e,s} : s \in \mathbb{N}\}$ of $\mathcal{F}$ that is tidier than W_e. We do this by delaying the enumeration of certain elements, but in a way that if W_e is actually an enumeration of $\mathcal{F}$, then all elements of W_e eventually enter some $V_{e,s}$, so that $W_e = \bigcup_{s \in \mathbb{N}} V_{e,s}$. We want $V_{e,s}$ to satisfy the following properties for every $s \in \mathbb{N}$:

- $V_{e,s} \subseteq W_{e,s}$.
- Every non-empty column of $V_{e,s}$ is of the form $F \oplus \{n\}$ for some F and n.
- For every n, there are at most two such columns, one included in $E \oplus \{n\}$, and if there is a second one, it must be of the form $E_k \oplus \{n\}$.
- If $V_{e,s}$ contains a column of the form $E_k \oplus \{n\}$, then so does $U[s]$.

We can easily get such an enumeration $\{V_{e,s} : s \in \mathbb{N}\}$ by slowing down the enumeration of $W_{e,s}$ and enumerating the elements of a column of $W_{e,s}$ into $V_{e,s}$ only once the properties above are satisfied.

Let $M_{e,s}$ be the largest m such that, for every $n < m$, there is a column in $V_{e,s}$ containing $\{0, 2, 4, \ldots, 2m\} \oplus \{n\}$. (See Figure 1 below.) Notice that $M_{e,s}$ is non-decreasing with s, and that if W_e is indeed an enumeration of $\mathcal{F}$, then $M_{e,s}$ converges to ∞. N_e imposes the following restraint on the lower-priority requirements:

> No set of the form $E_k \oplus \{n\}$ with $n < M_{e,s} \leq k$ can be enumerated into $\mathcal{F}$ at stage s.

At each stage s, we define a finite partial map $g_{e,s}$ matching columns in $V_{e,s}$ with columns in $U[s]$. We let $g_{e,s}(i) = j$ if and only if $V_{e,s}^{[i]}$ and $U^{[j]}[s]$ are of the forms $A \oplus \{n\}$ and $B \oplus \{n\}$ for the same n and one the following holds:

1. A and B are equal and of the form E_k for some k.
2. $n < M_{e,s}$, $A \subseteq E$, $B = E$, and there are no columns in $U[s]$ of the form $E_k \oplus \{n\}$ with $A \subseteq E_k$ (and hence none in $V_{e,s}$ either).

We claim that, unless N_e is re-initialized, $g_{e,s} \subseteq g_{e,s+1}$ for all s: If $g_{e,s}$ matches two columns of the form $E_k \oplus \{n\}$, those columns will still be matched in $g_{e,s+1}$. Suppose now $g_{e,s}$ matches two columns of the form $A \oplus \{n\}$ and $B \oplus \{n\}$ with $A, B \subseteq E$. We then must have that $n < M_{e,s}$, which implies that $\{0, \ldots, 2M_{e,s}\} \subseteq A$, and there is no column in $U[s]$

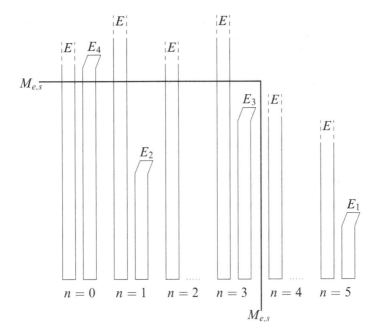

FIGURE 1. These are the columns of $V_{e,s}$. In this example, the restrain forbids us to enumerate a column of the form $E_k \oplus \{2\}$ for $k \geq M_{e,s}$ at stage s. That is, we cannot cross the horizontal $M_{e,s}$-line. So, for instance, the column that currently looks like $E \oplus \{2\}$ is not allowed to become of the form $E_k \oplus \{2\}$. The column $E_4 \oplus \{0\}$ crossing the line in the picture was enumerated before the current stage.

of the form $E_k \oplus \{n\}$ with $A \subseteq E_k$. Suppose, toward a contradiction, $g_{e,s+1}$ does not match those two columns. There could be only two possible reasons: (1) that the column in $V_{e,s+1}$ that contains A is not a subset of E anymore, and (2) that there is a column in $U[s+1]$ of the form $E_k \oplus \{n\}$ with $A \subseteq E_k$. Since columns of this form can only enter V_e after entering U, in either case we have a new column in U_{s+1} of the form $E_k \oplus \{n\}$ with $A \subseteq E_k$. Because of the restraint imposed by N_e, only columns of the form $E_k \oplus \{n\}$ with $k < M_{e,s}$ are allowed to be enumerated into $V_{e,s+1}$. But then we could not have $A \subseteq E_k$ as $A \supseteq \{0, \ldots, 2M_{e,s}\} \not\subseteq E_k$. This proves our claim, and we get that, if N_e is never re-initialized again, $g_e = \bigcup_s g_{e,s}$ is a computable equivalence between W_e and U. □

Notice that the family $\mathcal{F}$ is discrete, even if it is not effectively discrete. We thus get that $\mathcal{G}_{\mathcal{F}}^1$ is $\exists$-atomic and hence computably categorical on a cone.

After Goncharov's result, there have been various other constructions of computably categorical structures which are not relatively so. For instance, Khoussainov, Semukhin, and Stephan [KSS07] built one without using a priority argument using effective randomness instead. Their structure is not $\exists$-atomic over any finite set of parameters, so it is not computably categorical on a cone. Another example is due to Khoussainov and Shore [KS98, Theorem 4.2]. They built a computably categorical structure $\mathcal{A}$ such that, for each element $a \in \mathcal{A}$, the structure $(\mathcal{A}, a)$ is not computably categorical. The Khoussainov–Shore structure is not relatively computably categorical, as otherwise, it would remain relatively computably categorical if one added parameters.

EXERCISE 8.5.9. Let $\mathcal{G}_{\mathcal{F}}^1$ be a bouquet graph as in Section 8.5. Show that if the degree spectrum of $\mathcal{G}_{\mathcal{F}}^1$ has measure 1, then $\mathcal{G}_{\mathcal{F}}^1$ has a $0''$-computable copy. Hint in footnote.[83]

---

[83] Use Sacks's theorem that the measure of every non-trivial cone is 0.

Chapter 9

THE JUMP OF A STRUCTURE

In Definition 2.2.3, we defined Kleene's complete r.i.c.e. relation $\vec{K}^{\mathcal{A}}$ on a structure $\mathcal{A}$ by putting together all Σ_1^c-definable relations:

$$\vec{K}^{\mathcal{A}} = \{\langle i, \bar{b}\rangle : \mathcal{A} \models \varphi_{i,|\bar{b}|}^{\Sigma_1^c}(\bar{b})\} \subseteq \mathbb{N} \times A^{<\mathbb{N}},$$

where $\varphi_{i,j}^{\Sigma_1^c}(\bar{x})$ is the ith Σ_1^c τ-formula with j free variables. We then used this construction to define the jump of a relation $Q \subseteq \mathbb{N} \times A^{<\mathbb{N}}$ to be the relation $Q' = \vec{K}^{(\mathcal{A},Q)}$ (Definition 2.2.7), and proved that this is an actual jump, that is, that $Q <_{rT} Q'$ for all $Q \subseteq \mathbb{N} \times A^{<\mathbb{N}}$ (Corollary 2.2.10). In this chapter, we consider this same construction, but view it as an operation that maps structures to structures.

DEFINITION 9.0.10. Given a τ-structure $\mathcal{A}$, we define its jump to be the new structure obtained by adding the complete r.i.c.e. relation to it. That is, we let

$$\mathcal{A}' = (\mathcal{A}, \vec{K}^{\mathcal{A}}).$$

Thus, $\mathcal{A}'$ has the same domain as $\mathcal{A}$, but a larger vocabulary. It is a τ'-structure, where τ' consists of τ together with infinitely many new symbols naming the relations $K_{i,j} = \{\bar{b} \in A^j : \mathcal{A} \models \varphi_{i,j}^{\Sigma_1^c}(\bar{b})\}$.

Notice that this definition is independent of the presentation of $\mathcal{A}$. The isomorphism type of $\mathcal{A}'$ depends only on the isomorphism type of $\mathcal{A}$. We should mention that the isomorphism type of $\mathcal{A}'$ also depends—in an totally unessential way—on the Gödel numbering of the Σ_1^c τ-formulas, the same way the Turing jump of a set depends on the Gödel numbering of the partial computable functions. Also notice that the extended vocabulary τ' is still a computable relational vocabulary.

Historical Remark 9.0.11. The jump of structures has been introduced on various independent occasions over the last few years. Other definitions can be found in [Mor04, Bal06, Sos07, SS09, Puz09, Mon09, Stu09]. The history of the different definitions is explained in more detail in [Mon12]. The definition we give here comes from [Mon12, Definition 5.1].

Remark 9.0.12. Let us remark that the jump preserves effective bi-interpretability. That is, if $\mathcal{A}$ and $\mathcal{B}$ are effectively bi-interpretable, then so are $\mathcal{A}'$ and $\mathcal{B}'$. The interpretation maps are the same. All one has to observe is that the relation $\vec{K}^{\mathcal{A}^{\mathcal{B}}}$ within the copy $\mathcal{A}^{\mathcal{B}}$ of $\mathcal{A}$ interpreted in $\mathcal{B}$ is r.i.c.e. in $\mathcal{B}$ and therefore r.i. computable in $\mathcal{B}'$.

9.1. The jump-inversion theorems

Friedberg's jump-inversion theorem (Theorem 4.1.6) says that every Turing degree above $0'$ is the jump of some degree. There are a couple of different ways in which one could generalize Friedberg theorem to the jump of structures. We call them the first and second jump-inversion theorems.

9.1.1. The first jump-inversion theorem. This theorem is a generalization of the Friedberg jump-inversion theorem to the semi-lattice of structures ordered by effective interpretability.

THEOREM 9.1.1 (Soskova, Stukachev). *For every structure $\mathcal{A}$ which computably codes $0'$, there is a structure $\mathcal{C}$ whose jump is effectively bi-interpretable with $\mathcal{A}$.*

PROOF. We proved in Theorem 6.3.9 that every structure is effectively bi-interpretable with a graph. Therefore, we may assume $\mathcal{A}$ is a graph $(A; E)$ with domain A and edge relation E. The key idea behind this proof is the following: If we are given a linear ordering isomorphic to either ω or ω^*, deciding which one is the case is a Δ^0_2-complete question. We will thus define $\mathcal{C}$ by removing the edge relation E and instead attaching to each pair of elements of A one of these two linear orderings, depending on whether there is an edge between the two elements or not.

We define $\mathcal{C}$ as $(C; A, R)$, where A is a unary relation and R a 4-ary relation. The domain C of $\mathcal{C}$ consists of the disjoint union of the domain A of $\mathcal{A}$ and another set B, and we use the unary relation A to identify the elements of A. We define the 4-ary relation

$$R \subseteq A \times A \times B \times B$$

so that: If we let $B_{a,b} = \{c \in B : R(a, b, c, c)\}$, and $R_{a,b} = \{\langle c, d \rangle \in B^2 : R(a, b, c, d)\}$, then $(B_{a,b}; R_{a,b})$ is a linear ordering isomorphic to either ω or ω^*, and it is isomorphic to ω if and only if $\langle a, b \rangle \in E$. We also assume the sets $B_{a,b}$ for $a, b \in A$ partition B.

$\mathcal{C}$ can be easily effectively interpreted in $\mathcal{A}$ as follows. Let $B = \mathbb{N} \times A^2$ and let $C = A \cup B$. Then define R as follows:

$$R = \{\langle a, b, \langle n, a, b \rangle, \langle m, a, b \rangle \rangle \in A^2 \times B^2 : \text{for } \langle a, b \rangle \in E \ \& \ n \leq m\}$$
$$\cup \{\langle a, b, \langle n, a, b \rangle, \langle m, a, b \rangle \rangle \in A^2 \times B^2 : \text{for } \langle a, b \rangle \in A^2 \smallsetminus E \ \& \ n \geq m\}.$$

To show that this is actually an effective interpretation of $\mathcal{C}'$, and not just of $\mathcal{C}$, we need to show that $\vec{K}^{\mathcal{C}}$ (viewed as a relation in $\mathbb{N} \times A^{<\mathbb{N}}$) is r.i. computable in $\mathcal{A}$. To see this, fix an ω-presentation of $\mathcal{A}$. The construction above then gives us an ω-presentation of $\mathcal{C}$. Use Friedberg's jump-inversion theorem to get an oracle $X \in 2^{\mathbb{N}}$ such that $X' \equiv_T D(\mathcal{A})$ (using that $\mathcal{A}$ computably codes $0'$). We will now construct a second copy, $\widetilde{\mathcal{C}}$, of $\mathcal{C}$ that is computable in X. For each $\langle a, b \rangle \in A^2$, X' knows whether or not $\langle a, b \rangle \in E$, and hence computably in X, we can uniformly build a linear ordering $\widetilde{\mathcal{B}}_{a,b}$ such that

$$\widetilde{\mathcal{B}}_{a,b} \cong \begin{cases} (\mathbb{N}; \leq) & \text{if } \langle a, b \rangle \in E, \\ (\mathbb{N}; \geq) & \text{if } \langle a, b \rangle \notin E. \end{cases}$$

To do this, if $f(a, b, s)$ is an X-computable function such that
- $\lim_{s \in \mathbb{N}} f(a, b, s) = 1$ if $\langle a, b \rangle \in E$ and
- $\lim_{s \in \mathbb{N}} f(a, b, s) = 0$ if $\langle a, b \rangle \notin E$,

then we can define $\widetilde{\mathcal{B}}_{a,b} = (\mathbb{N}; \leq_{\widetilde{\mathcal{B}}_{a,b}})$ by

$$s \leq_{\widetilde{\mathcal{B}}_{a,b}} r \iff \left(s \leq_{\mathbb{N}} r \ \& \ f(a, b, r) = 1 \right) \vee \left(r \leq_{\mathbb{N}} s \ \& \ f(a, b, s) = 0 \right).$$

In other words, for each $s \in \mathbb{N}$ we have that, if $f(a, b, s) = 1$, s is $\geq_{\widetilde{\mathcal{B}}_{a,b}}$-above all $r <_{\mathbb{N}} s$, and, if $f(a, b, s) = 0$, s is $\leq_{\widetilde{\mathcal{B}}_{a,b}}$-below all $r <_{\mathbb{N}} s$. We let the reader verify this ordering is as needed. We then define $\widetilde{\mathcal{C}}$ by putting together $\mathcal{A}$ and disjoint copies of all the $\widetilde{\mathcal{B}}_{a,b}$ for $\langle a, b \rangle \in A^2$ and defining $\widetilde{R}(a, b, n, m) \iff n \leq_{\widetilde{\mathcal{B}}_{a,b}} m$. An important point is that $D(\mathcal{A})$ can compute an isomorphism between $\widetilde{\mathcal{C}}$ and $\mathcal{C}$. This is because X' can compute isomorphisms between $\widetilde{\mathcal{B}}_{a,b}$ and $\mathcal{B}_{a,b}$ for all $\langle a, b \rangle \in A^2$. Since $D(\widetilde{\mathcal{C}}) \leq_T X$, we have that $\vec{K}^{\widetilde{\mathcal{C}}}$ is computable in X', and hence in $D(\mathcal{A})$. Going through the isomorphism between $\widetilde{\mathcal{C}}$ and $\mathcal{C}$, we get that $\vec{K}^{\mathcal{C}}$ is also computable in $D(\mathcal{A})$. Since this worked for every ω-presentation of $\mathcal{A}$, we have that $\vec{K}^{\mathcal{C}}$ is r.i. computable in $\mathcal{A}$. This proves that we have an effective interpretation of $\mathcal{C}'$ in $\mathcal{A}$.

The effective interpretation of $\mathcal{A}$ within $\mathcal{C}'$ is more direct. The domain of the interpretation is, of course, A itself, as identified by the relation A within $\mathcal{C}$. Notice that E is now r.i. Δ_2^0 in $\mathcal{C}$. This is because, to decide if $\langle a, b \rangle \in A^2$, we need to decide whether $\mathcal{B}_{a,b} \cong \omega$ or $\mathcal{B}_{a,b} \cong \omega^*$. For this, we need to decide whether there exists an element in $\mathcal{B}_{a,b}$ without predecessors, or one without successors; both are Σ_2^c questions.

The last step is to check that these two effective interpretations form an effective bi-interpretation; i.e., that the composition of the isomorphisms are r.i. computable in the respective structures. First, notice that the interpretation of $\mathcal{A}$ inside $\mathcal{C}$ inside $\mathcal{A}$ is the identity, and hence obviously r.i. computable in $\mathcal{A}$. Second, for the interpretation of $\mathcal{C}$ inside $\mathcal{A}$ inside $\mathcal{C}$, the A-part stays the same. The copies of $\mathcal{B}_{a,b}$ are not the same, but since they are isomorphic to either ω or ω^*, the isomorphism between them can be computed within a jump of $\mathcal{C}$. $\square$

Historical Remark 9.1.2. For the case of Muchnik equivalence, this theorem was proved independently on two occasions. One is due to Goncharov, Harizanov, Knight, McCoy, R. Miller, and Solomon by essentially the same proof we gave above (cf. [GHK$^+$05, Lemma 5.5 for $\alpha = 2$]), although they were not considering jumps of structures. Their objective was to prove various result like Theorem 9.1.5 below and their transfinite versions. The other is due to Alexandra Soskova [Sos07, SS09]. Her construction is quite different and uses Marker extensions. Stukachev [Stu10, Stu] proved that Soskova's constructions actually gives effective interpretations instead of just Muchnik reductions.

9.1.2. An application of the first jump-inversion theorem.

DEFINITION 9.1.3. A computable structure $\mathcal{A}$ is Δ_2^0-*categorical* if there is a $0'$-computable isomorphism between it and any computable copy. It is *relatively* Δ_2^0-*categorical* if every copy $\mathcal{B}$ is isomorphic to $\mathcal{A}$ via a $D(\mathcal{B})'$-computable isomorphism.

Notice that being Δ_2^0-categorical is not the same as being $0'$-computably categorical (i.e., computably categorical relative to $0'$). The latter means that every $0'$-computable copy $\mathcal{B}$ is $0'$-computably isomorphic to $\mathcal{A}$, while the former only considers computable copies $\mathcal{B}$.

EXERCISE 9.1.4. Show that $(\omega; \leq)$ is relatively Δ_2^0-categorical.

THEOREM 9.1.5. (Goncharov, Harizanov, Knight, McCoy, Miller, and Solomon [GHK$^+$05]). *There is a structure that is* Δ_2^0-*categorical but not relatively so.*

PROOF. Relativizing Theorem 8.5.1 to $0'$, let $\mathcal{A}$ be a $0'$-computable structure that is $0'$-computably categorical, but not $0'$-relatively computably categorical. Let $\mathcal{C}$ be the structure built from $\mathcal{A}$ in the proof of the first jump-inversion theorem. From the proof of the theorem we get that if $\mathcal{A}$ has an X'-computable ω-presentation for some $X \in 2^{\mathbb{N}}$, then $\mathcal{C}$ has an X-computable presentation that is X'-computably isomorphic to $\mathcal{C}$. Thus, we may assume that $\mathcal{C}$ is computable and $\mathcal{A}$ is obtained from the effective bi-interpretation with $\mathcal{C}'$. We claim that $\mathcal{C}$ is Δ_2^0-categorical but not relatively so.

To prove that $\mathcal{C}$ is Δ_2^0-categorical, let $\widehat{\mathcal{C}}$ be a computable copy of $\mathcal{C}$. Then $\widehat{\mathcal{C}}'$ is associated via the effectively bi-interpretation with a copy $\widehat{\mathcal{A}}$ of $\mathcal{A}$. Notice the ω-presentation $\widehat{\mathcal{A}}$ is computable in $0'$. Since $\mathcal{A}$ is $0'$-computably categorical, $0'$ can compute an isomorphism between $\mathcal{A}$ and $\widehat{\mathcal{A}}$. Using the effective bi-interpretations, $0'$ can then compute an isomorphism from $\mathcal{C}$ to $\widehat{\mathcal{C}}$.

Let us now prove that C is not relatively Δ_2^0-categorical. Since $\mathcal{A}$ is not $0'$-relatively computably categorical, there is a copy $\widehat{\mathcal{A}}$ of $\mathcal{A}$ computable in some oracle $Y \geq_T 0'$ that is not Y-computably isomorphic to $\mathcal{A}$. Let $\widehat{C}$ be the copy of C associated via the effectively bi-interpretation with $\widehat{\mathcal{A}}$. Use the Friedberg's jump-inversion theorem to get $X \in 2^{<\mathbb{N}}$ with $X' \equiv_T Y$. The oracle X might not compute the ω-presentation $\widehat{C}$, but as in the proof of the theorem, it computes a copy $\widetilde{C}$ of $\widehat{C}$ that is X'-computably isomorphic to $\widehat{C}$. We claim that there is no X'-computable isomorphism between C and $\widetilde{C}$. That would prove that C is not relatively Δ_2^0-categorical. As for the claim, if there was an X'-computable isomorphism between C and $\widetilde{C}$, there would be one between C and $\widehat{C}$, and using the effective bi-interpretations, we would get an X'-computable isomorphism between A and $\widehat{A}$, which we assumed does not exist. □

One can of course iterate this proof and produce, for each $n \in \mathbb{N}$, a computable structure that is Δ_n^0-categorical but not relatively so.

9.1.3. The second jump-inversion theorem. This jump-inversion theorem is not a generalization of the usual jump-inversion theorem to a more general class of degrees, but a generalization in the sense that, given $X \in 2^{\mathbb{N}}$, it yields $Y \in 2^{\mathbb{N}}$ with $Y' \equiv_T X$ and some extra properties.

THEOREM 9.1.6 (Soskov). *If $X \in 2^{\mathbb{N}}$ computes a copy of $\mathcal{B}'$, then there is a $Y \in 2^{\mathbb{N}}$ satisfying $Y' \equiv_T X$ that computes a copy C of $\mathcal{B}$.*

PROOF. By Lemma 4.3.2, there is a 1-generic enumeration g of $\mathcal{B}$ computable in $\vec{K}^{\mathcal{B}}$, and hence in X. Let $C = g^{-1}(\mathcal{B})$ and $Z = D(C)$. Since $\vec{K}^C = g^{-1}(\vec{K}^{\mathcal{B}})$, we have that

$$\vec{K}^C \leq_T \vec{K}^{\mathcal{B}} \leq_T X.$$

Since C is 1-generic,

$$\vec{K}^C \equiv_T D(C)' = Z',$$

as proved in Lemma 4.4.4. Thus, $Z' \leq_T X$. By the relativized Friedberg's theorem, there is a $Y \in 2^{\mathbb{N}}$ such that $Y \geq_T Z$ and $Y' \equiv_T X$. This Y computes C, a copy of $\mathcal{B}$. □

For future reference, let us remark that X can compute the isomorphism g between C and $\mathcal{B}$.

As a corollary, we get that the degree spectrum of the jump of a structure is what it should be: the set of jumps of the degrees in the spectrum of the original structure.

COROLLARY 9.1.7. *For every structure $\mathcal{B}$,*

$$DgSp(\mathcal{B}') = \{X \in 2^{\mathbb{N}} : X \geq_T Y' \text{ for some } Y \in DgSp(\mathcal{B})\}.$$

PROOF. For the right-to-left inclusion, it is clear that if $X \geq_T Y'$ for some $Y \in DgSp(\mathcal{B})$, then X computes a copy of $\mathcal{B}'$. For the left-to-right inclusion, if X computes a copy of $\mathcal{B}'$, then by the theorem, there is a $Y \in DgSp(\mathcal{B})$ such that $X \geq_T Y'$.   □

Historical Remark 9.1.8. Theorem 9.1.6 was first introduced by Soskov at a talk at the LC'02 in Münster; a full proof then appeared in [SS09]. It was also independently proved in [Mon09].

9.1.4. Application of the second jump-inversion theorem. First, let us note how second jump-inversion theorem can be applied to structures for which we understand their jump. For instance, we know from Lemma 2.3.2 and Definition 2.3.6, that the jump of a linear ordering $\mathcal{L} = (L; \leq_L)$ is effectively bi-interpretable with $(L; \leq_L, \bar{\mathsf{A}}\mathsf{dj}, 0')$. It follows from the second jump-inversion theorem that if $(\mathcal{L}, \bar{\mathsf{A}}\mathsf{dj})$ has a $0'$-computable copy, then $\mathcal{L}$ must have a low copy (cf. Lemma 7.4.4).

Anther application is the generalization of Theorem 2.1.16 from r.i.c.e. relations to r.i. Σ_n^0 relations for $n \in \mathbb{N}$.

DEFINITION 9.1.9. A relation $R \subseteq \mathbb{N} \times A^{<\mathbb{N}}$ is *relatively intrinsically Σ_2^0* if $R^{\mathcal{B}}$ is c.e. in $D(\mathcal{B})'$ for every copy $(\mathcal{B}, R^{\mathcal{B}})$ of $(\mathcal{A}, R)$.

THEOREM 9.1.10. (Ash, Knight, Manasse, and Slaman [AKMS89] and Chisholm [1990]). *Let $\mathcal{A}$ be a structure, and $R \subseteq \mathbb{N} \times A^{<\mathbb{N}}$ a relation on it. The following are equivalent*:

(A1) *R is relatively intrinsically Σ_2^0.*
(A2) *R is Σ_2^c definable in $\mathcal{A}$ with parameters.*

PROOF. The upward direction follows from the observation that a Σ_2^c relation $R \subseteq \mathbb{N} \times A^{<\mathbb{N}}$ is always Σ_2^0 relative to the diagram of $D(\mathcal{A})$. We concentrate on the downward direction.

First, we claim that if R is relatively intrinsically Σ_2^0 in $\mathcal{A}$, it is r.i.c.e. in $\mathcal{A}'$. For this, let $\mathcal{B}'$ be a copy of $\mathcal{A}'$ computable in some oracle X; We need to show that $R^{\mathcal{B}}$ is c.e. in X. This would follow if we knew that $D(\mathcal{B})' \leq_T X$, but that might not be the case. Use the second jump-inversion theorem with $X = D(\mathcal{B}')$ to get an oracle Y with $Y' \equiv_T X$ and a Y-computable copy $\mathcal{C}$ of $\mathcal{B}$. Recall from the remark after the proof of the theorem that X can compute an isomorphism between $\mathcal{C}$ and $\mathcal{B}$. Since R is relatively intrinsically Σ_2^0, $R^{\mathcal{C}}$ is c.e. in $D(\mathcal{C})' \equiv_T Y' \equiv_T X$. Pushing $R^{\mathcal{C}}$ through the isomorphisms, we get that $R^{\mathcal{B}}$ is c.e. in X as needed for our claim that R is r.i.c.e. in $\mathcal{A}'$.

Now, by Theorem 2.1.16, R is Σ_1^c-definable in $\mathcal{A}'$ with parameters. It follows that R is Σ_2^c-definable in $\mathcal{A}$ with parameters.   □

One can, of course, iterate the proof of this theorem and prove that r.i. Σ_n^0 relations are Σ_n^c definable with parameters.

A third application is the generalization of Theorem 8.2.2 from relative computable categoricity to relative Δ_n^0 categoricity for $n \in \mathbb{N}$. We prove the case $n = 2$ for simplicity. For the definitions of relative Δ_2^0 categoricity and Scott families see definitions 9.1.3 and 3.1.2.

THEOREM 9.1.11. (Ash, Knight, Manasse, and Slaman [AKMS89, Theorem 4] and Chisholm [Chi90, Theorem V.10]). *Let $\mathcal{A}$ be a computable structure. The following are equivalent*:

1. $\mathcal{A}$ *is relatively Δ_2^0-categorical.*
2. $(\mathcal{A}, \bar{a})$ *has a c.e. Scott family of Σ_2^c formulas using a finite tuple of parameters.*

PROOF. As in the Δ_1^0 case, the implication from (2) to (1) follows from Observation 3.3.2, where, given $\mathcal{B} \cong \mathcal{A}$, one uses the Scott family to build a set $I_{\mathcal{A},\mathcal{B}} \subseteq A^{<\mathbb{N}} \times B^{<\mathbb{N}}$ with the back-and-forth property. In this case, one needs $D(\mathcal{B})'$ to enumerate $I_{\mathcal{A},\mathcal{B}}$ and then compute an isomorphism between $\mathcal{A}$ and $\mathcal{B}$.

The other direction is the interesting one. Assume (1). We claim that $\mathcal{A}'$ is relatively computably categorical. For this, let $\mathcal{B}'$ be a copy of $\mathcal{A}'$ computable from an oracle X; We need to show there is an X-computable isomorphism between them. By the second jump inversion Theorem 9.1.6, there is an oracle Y with $Y' \equiv_T X$ which computes a copy $\mathcal{C}$ of $\mathcal{B}$. Recall from the remark after the proof of the theorem that X can compute an isomorphism between $\mathcal{B}$ and $\mathcal{C}$. Since $\mathcal{A}$ is relatively Δ_2^0-categorical, there is a Y'-computable isomorphism between $\mathcal{C}$ and $\mathcal{A}$. Composing these isomorphisms, we get an X-computable isomorphism between $\mathcal{B}$ and $\mathcal{A}$, and in particular between $\mathcal{B}'$ and $\mathcal{A}'$. This proves the claim that $\mathcal{A}'$ is relatively computably categorical.

Now, by Theorem 8.2.2, $\mathcal{A}'$ has a c.e. Scott family of $\exists$-formulas over a finite tuple of parameters. It follows that $\mathcal{A}$ has a c.e. Scott family of Σ_2^c formula with parameters. □

9.2. The jump jumps—or does it?

If we are going to call this operation a jump, we should ask whether it actually jumps, or whether there is a structure that is equivalent to its own jump. The answer is not straightforward and depends on the notion of equivalence we use. For the strongest of the equivalences, namely effectively bi-interpretability, the jump does jump.

LEMMA 9.2.1. *No structure is Medvedev equivalent to its own jump. In particular, no structure is effectively bi-interpretable with its own jump.*

PROOF. We know from Lemma 6.1.10 that if $\mathcal{A}'$ were Medvedev reducible to $\mathcal{A}$, we would have $\exists\text{-}Th(\mathcal{A}') \leq_e \exists\text{-}Th(\mathcal{A})$. To show that this is not the case, we claim that $\exists\text{-}Th(\mathcal{A}')$ can enumerate the enumeration jump of $\exists\text{-}Th(\mathcal{A})$. The *enumeration jump* of a set X is defined to be

$$J(X) \oplus J(X)^c, \text{ where } J(X) = \{e : e \in \Theta_e^X\}$$

and $\{\Theta_e : e \in \mathbb{N}\}$ is an effective list of the enumeration operators as in page xviii. A standard diagonalization argument shows that X cannot enumerate the set $J(X)^c$.[84]

Let us now prove the claim that $J(\exists\text{-}Th(\mathcal{A}))^c \leq_e \exists\text{-}Th(\mathcal{A}')$. For $e \in \mathbb{N}$, $e \in J(\exists\text{-}Th(\mathcal{A}))^c$ if and only if there is no finite set $D \subset \mathbb{N}$ with $\langle \ulcorner D \urcorner, e \rangle \in \Theta_e$ and $D \subseteq \exists\text{-}Th(\mathcal{A})$. That is,

$$e \in J(\exists\text{-}Th(\mathcal{A}))^c \iff \mathcal{A} \models \bigwedge_{\substack{D \subseteq \mathbb{N} \\ \langle D, e \rangle \in \Theta_e}} \neg \bigwedge_{i \in D} \varphi_i^\exists,$$

where $\varphi_i^\exists$ is the ith existential τ-sentence. The right-hand side is a Π_1^c sentence about $\mathcal{A}$, and hence decided in the quantifier-free theory of $\mathcal{A}'$. We even get that $J(\exists\text{-}Th(\mathcal{A}))^c \leq_m \exists\text{-}Th(\mathcal{A}')$. $\square$

For the weaker notion of Muchnik equivalence, the answer gets more interesting.

THEOREM 9.2.2 (Puzarenko [Puz11] and Montalbán [Mon13c]). *There is a structure that is Muchnik equivalent to its own jump.*

The following proof, which was motivated by conversations with Schweber and Turetsky, is new and different from the original proofs of Puzarenko [Puz11] and Montalbán [Mon13c]. The three proofs build similar looking structures. Puzarenko's [Puz11] and Montalbán's [Mon13c] build an ill-founded ω-model $\mathcal{A}$ of $ZF^- + "V = L"$ where, for some $\alpha \in ON^\mathcal{A}$, $(L_\alpha)^\mathcal{A} \cong \mathcal{A}$. The structure we construct here is simpler, but only in appearance, as it is bi-interpretable with theirs. It is how such structures are built that makes the proofs so different. Montalbán [Mon13c] uses the existence of $0^\sharp$, and his proof is a paragraph long once the definition of $0^\sharp$ is understood. Puzarenko's proof (cf. [Puz11]) does not need assumptions beyond ZFC, but it uses admissibility theory and is much more complicated. The proof we present here also works within ZFC, but is simpler. It uses a lemma we will not prove until [MonP2]. This lemma is a special case of results of Morley [Mor65], Lopez-Escobar [LE66], and Barwise [Bar69] on the Hanf numbers of infinitary logic. It states that if a Π_2^c sentence has a model of arbitrary large cardinality, then it has a countable model with a non-trivial automorphism. The following proof uses ordinals and transfinite recursion which we will review in depth in [MonP2].

PROOF OF THEOREM 9.2.2. Consider the relational vocabulary $\tau = \{\leq\} \cup \{R_{i,j} : i, j \in \mathbb{N}\}$, where $\leq$ is of course binary and $R_{i,j}$ is $j + 1$-ary. The symbols $R_{i,j}$ will be used to encode Kleene's predicate $\check{K}$ on the initial segments of the structures: We say that a τ-structure $\mathcal{L}$ is a *linear jump hierarchy* if $(L; \leq_L)$ is a linear ordering and, for every $i, j \in \mathbb{N}$, $\bar{b} \in L^j$,

---

[84]If $J(X)^c$ were enumeration reducible to X, we would have $J(X)^c = \Theta_e^X$ for some e. We would then have that $e \in J(X)^c \iff e \in \Theta_e^X \iff e \in J(X)$.

and $a \in L$, with $b_\ell <_L a$ for all $\ell < j$,

$$\mathcal{L} \models R_{i,j}(\bar{b}, a) \iff \mathcal{L} \upharpoonright a \models \varphi_{i,j}^{\Sigma_1^c}(\bar{b}),$$

where $\varphi_{i,j}^{\Sigma_1^c}$ is the i-th Σ_1^c τ-formula with j free variables, and $\mathcal{L} \upharpoonright a$ is the restriction of $\mathcal{L}$ to the domain $\{b \in L : b <_L a\}$. The key property of a linear jump hierarchy is that, for every $a \in L$, the jump of $\mathcal{L} \upharpoonright a$ is effectively interpretable in $\mathcal{L}$ using a as a parameter:[85]

$$\vec{K}^{\mathcal{L} \upharpoonright a} = \{\langle i, \bar{b} \rangle \in \mathbb{N} \times (L \upharpoonright a)^{<\mathbb{N}} : \mathcal{L} \models R_{i,|\bar{b}|}(\bar{b}, a)\}.$$

The next step in the proof is to prove there is a linear jump hierarchy $\mathcal{L}$ which has an element a such that $\mathcal{L} \cong \mathcal{L} \upharpoonright a$. This will imply that the jump of $\mathcal{L}$ is effectively interpretable in $\mathcal{L}$ with parameter a, and in particular that $\mathcal{L}'$ is Muchnik reducible to $\mathcal{L}$.

It will be enough to show that there is a linear jump hierarchy $\mathcal{L}$ with a non-trivial automorphism. For if there is an automorphism mapping a to b with $a <_L b$, then $\mathcal{L} \upharpoonright a \cong \mathcal{L} \upharpoonright b$ and we can then use $\mathcal{L} \upharpoonright b$ instead of $\mathcal{L}$. The lemma we mentioned above, the one we will prove in [MonP2], states that if a Π_2^c sentence has models of arbitrary large size, then it has a countable model with a non-trivial automorphism. The class of linear jump hierarchies can be axiomatized by a Π_2^c sentence. To see that there are linear jump hierarchies of arbitrary size, let $(L; \leq_L)$ be any ordinal you want and then define the relations $R_{i,j}(\cdots, a)$ by transfinite recursion on $a \in L$. □

Even more surprising than the theorem itself is the complexity necessary to prove it. We show below that a construction of a structure that is Muchnik equivalent to its own jump must use an uncountable object. To show this we prove that Theorem 9.2.2 is not provable in second-order arithmetic. It is known that second-order arithmetic proves exactly the same Π_4^1 sentences as ZFC without the power set axiom (see [MS12] for a proof). Thus, Theorem 9.2.2 would also not be provable in ZFC without using the power set axiom.

Let us quickly describe what *second-order arithmetic* is. It has two sorts of elements, the *first-order sort* for numbers and the *second-order sort* for sets of numbers. Lower case letters are used for number variables and upper case roman letters for second-order variables. The vocabulary is $\{0, 1, \leq, +, \times, \in\}$. The axioms are those of Peano arithmetic plus comprehension for all formulas. That is, for each formula $\varphi(x)$ of second-order arithmetic with one free number variable x, plus maybe other free variables that we view as parameters, we have the axiom

$$CA(\varphi) \equiv \exists X \, \forall n \, \left(n \in X \leftrightarrow \varphi(n)\right).$$

---

[85]Furthermore, its not hard to show that $(\mathcal{L} \upharpoonright a)'$ is effectively bi-interpretable with $(\mathcal{L} \upharpoonright a, a)$.

The axiom of induction is stated as a single axiom:

$$\mathsf{IND} \equiv \forall X \left(\left(0 \in X \wedge \forall n(n \in X \rightarrow n+1 \in X) \right) \rightarrow \forall n \, (n \in X) \right).$$

THEOREM 9.2.3 (Montalbán [Mon13c]).   *Second-order arithmetic cannot prove that there exists a structure $\mathcal{A}$ which is Muchnik equivalent to its own jump.*

PROOF. We show, within second-order arithmetic, that using such a structure $\mathcal{A}$ we can build a model of second-order arithmetic. This would prove the consistency of second-order arithmetic, which, by Gödel's incompleteness theorem, cannot be proved within second-order arithmetic.

The model we build is an ω-model, that is, a model where the first-order part is the standard $(\mathbb{N}; 0, 1, \leq, +, \times)$. We define the second-order part of our model to be

$$\mathcal{M} = \{ X \subseteq \mathbb{N} : X \text{ is r.i.c.e. in } \mathcal{A} \}.$$

Let us remark that, by Corollary 2.1.24, every $X \in \mathcal{M}$ is of the form $\Theta_e^{\exists\text{-}tp_{\mathcal{A}}(\bar{p})}$ for some c.e. operator Θ_e and tuple $\bar{p} \in A^{<\mathbb{N}}$. Thus, we can use the pair $\langle e, \bar{p} \rangle \in \mathbb{N} \times A^{<\mathbb{N}}$ to name this element X of $\mathcal{M}$. We need to prove that $(\mathbb{N}, \mathcal{M})$ satisfies the axioms of second-order arithmetic. Since the first-order part of the model is standard, we immediately get that the axioms of Peano arithmetic, including induction, hold in $(\mathbb{N}, \mathcal{M})$. What is left to show is that the comprehension axioms hold in $(\mathbb{N}; \mathcal{M})$. That is, for each formula $\varphi(x)$ of second-order arithmetic with parameters from $\mathcal{M}$, we need to show that the set

$$C_\varphi = \{ n \in \mathbb{N} : (\mathbb{N}; \mathcal{M}) \models \varphi(\boldsymbol{n}) \}$$

is r.i.c.e. in $\mathcal{A}$, where $\boldsymbol{n}$ is the term $1 + 1 + \cdots + 1$ added n times. The idea is to translate φ to the language of $\mathcal{A}$. What we want is, for some $k \in \mathbb{N}$, a computable sequence of Σ_k^c sentences χ_n, for $n \in \mathbb{N}$, such that

$$\mathcal{A} \models \chi_n \iff (\mathbb{N}; \mathcal{M}) \models \varphi(\boldsymbol{n}).$$

This would then imply that C_φ is r.i. computable in $\mathcal{A}^{(k)}$. We claim that $\mathcal{A}$ is not only Muchnik equivalent to $\mathcal{A}'$, but also to $\mathcal{A}^{(k)}$ for all $k \in \mathbb{N}$. This would imply that the set C_φ is r.i. computable in $\mathcal{A}$ and hence belongs to $\mathcal{M}$. To prove the claim, let X_0 be a real that computes a copy of $\mathcal{A}$. We can recursively build a sequence of reals X_i for $i \leq k$ such that $X_i' \equiv_T X_{i-1}$ and X_i computes a copy of $\mathcal{A}$. To do this, once we have X_{i-1} computing a copy of $\mathcal{A}$, we know it must also compute a copy of $\mathcal{A}'$, and we can get X_i from the second jump-inversion theorem (Theorem 9.1.6). Once we have that X_k computes a copy of $\mathcal{A}$ and $X_k^{(k)} \equiv_T X_0$, we have that X_0 computes the kth jump of a copy of $\mathcal{A}$. This proves the claim.

The last step is to define the sequence of χ_n's from the formula $\varphi(x)$. Fix $n \in \mathbb{N}$, and replace the occurrences of x in $\varphi(x)$ by $\boldsymbol{n}$. Replace each second-order parameter $Z \in \mathcal{M}$ that shows up in φ with $\Theta_{e_Z}^{\exists\text{-}tp_{\mathcal{A}}(\bar{p}_Z)}$, where e_Z and $\bar{p}_Z$ are such that $Z = \Theta_{e_Z}^{\exists\text{-}tp_{\mathcal{A}}(\bar{p}_Z)}$. Replace each second-order variable X by a name $\langle e_X, \bar{p}_X \rangle \in \mathbb{N} \times A^{<\mathbb{N}}$: That is, replace $\forall X\ \psi(X)$ with $\forall e_X \in \mathbb{N}\ \forall \bar{p}_X \in A^{<\mathbb{N}}\ \psi(\Theta_{e_X}^{\exists\text{-}tp_{\mathcal{A}}(\bar{p}_X)})$. Do the same for existential second-order quantification. We are now left with no second-order variables in our formula. Replace each first-order quantification $\forall e \in \mathbb{N}\ \psi(e)$ with $\bigwedge_{m \in \mathbb{N}} \psi(\boldsymbol{m})$. Replace each first-order quantification $\exists e \in \mathbb{N}\ \psi(e)$ with $\bigvee_{m \in \mathbb{N}} \psi(\boldsymbol{m})$. We are now left with no variables at all in our formula. Replace each atomic arithmetic sub-formula, which by now looks something like $\boldsymbol{n} + \boldsymbol{m} \le \boldsymbol{k}$, with its truth value $\top$ or $\bot$. Finally, replace each sub-formula of the form $\boldsymbol{n} \in \Theta_e^{\exists\text{-}tp_{\mathcal{A}}(\bar{p})}$ by the equivalent Σ_1^c formula

$$\bigvee_{\substack{D \subset_{fin} \mathbb{N} \\ \langle \ulcorner D \urcorner, n \rangle \in \Theta_e}} \bigwedge_{k \in D} \varphi_{k, |\bar{p}|}^{\exists}(\bar{p})$$

(Recall that enumerator operators were defined so that $n \in \Theta^Y \iff \exists D \subseteq_{fin} \mathbb{N}\ (\langle \ulcorner D \urcorner, n \rangle \in \Theta \wedge D \subseteq Y)$, and that $\varphi_{k,j}^{\exists}$ is the k-th $\exists$-formula with arity j.) We have eliminated all traces of arithmetic and ended up with a τ-formula χ_n that is equivalent to the original $\varphi(\boldsymbol{n})$. It is not hard to see that the formula χ_n has finite depth in terms of alternations of quantifiers. This depth k depends only on the quantifiers that show up in φ and not on n. □

Generalizing this to higher orders, Montalbán [Mon13c] proved that the ω-jump of any presentation of $\mathcal{A}$ computes a countably coded ω-model of higher-order arithmetic, i.e., with a sort for n-th order sets for each $n \in \mathbb{N}$. This implies that at least ω many iterations of the power-set axiom are needed to prove such a structure $\mathcal{A}$ exists. Both, our proof and Puzarenko's proof, use ω_1^{CK} iterations of the power-set axiom. It is still unknown exactly how many iterates of the power-set axiom are needed to prove Theorem 9.2.2.

Chapter 10

Σ-SMALL CLASSES

An $\exists$-type $p(\bar{x})$ is said to be *sharply realized* in a class $\mathbb{K}$ of structures if there exists a tuple $\bar{a}$ in some structure $\mathcal{A} \in \mathbb{K}$ such that $p(\bar{x}) = \exists\text{-}tp_{\mathcal{A}}(\bar{a})$.[86]

As we saw in previous chapters, existential types capture important computability theoretic information. There are continuum many existential types realized among all τ-structures. If a class of τ-structures $\mathbb{K}$ is rich enough, there will also be continuum many existential types sharply realized in $\mathbb{K}$. If, on the contrary, the number of $\exists$-types is less than continuum, then something interesting must be going on. Descriptive set theoretic facts (which we will see in [MonP2]) imply that on definable classes of structures, the number of $\exists$-types that are sharply realized is either countable or continuum, but never in between.[87] Classes for which this number is countable have particularly nice computability theoretic properties.

DEFINITION 10.0.4 (Montalbán [Mon14a]). A class $\mathbb{K}$ of structures is Σ-*small* if there are only countably many $\exists$-types sharply realized among all the structures in $\mathbb{K}$.

A simple observation is that Σ-small classes can also be defined as the ones sharply realizing countably many $\forall$-types.[88]

The first one to analyze Σ-small classes was, indirectly, Richter with her analysis of the computable extendability condition (Definition 5.1.1).

Observation 10.0.5. A class is Σ-small if and only if there is an $X \in 2^{\mathbb{N}}$ relative to which all structures in the class have the computable embeddability condition.[89]

---

[86] A type $p(\bar{x})$ is *realized* by a tuple $\bar{a}$ in a structure $\mathcal{A}$ if $\bar{a}$ satisfies all its formulas. Thus, if $p(\bar{x})$ is an $\exists$-type, $p(\bar{x})$ is realized by $\bar{a}$ if and only if $p(x) \subseteq \exists\text{-}tp_{\mathcal{A}}(\bar{a})$.

[87] If a class is defined as the set of models of an infinitary sentence, the set of $\exists$-types sharply realized in it is Σ_1^1, and hence its size is either countable or continuum by Suslin's theorem (cf. [Sus17]).

[88] This is because tuples with the same $\exists$-type have the same $\forall$-type: $\forall\text{-}tp_{\mathcal{A}}(\bar{a}) = \{\neg\varphi : \varphi$ is an $\exists$-formula, $\varphi \notin \exists\text{-}tp_{\mathcal{A}}(\bar{a})\}$.

[89] If all structures in $\mathbb{K}$ have the computable embeddability condition relative to X, it is Σ-small because there are only countably many X-computable sets. Conversely, if $\mathbb{K}$ is Σ-small, let X be an oracle that computes all $\exists$-types sharply realized in $\mathbb{K}$.

If $\mathbb{K}$ is a natural example of a Σ-small class, this oracle X always turns out to be 0, and all structures in $\mathbb{K}$ have the computable embeddability condition. Thus, all structures in $\mathbb{K}$ would then have all the properties we proved in Section 5.1, for instance that they never have non-trivial enumeration degree, and that their degree spectra contain c.e.-minimal pairs. Richter [Ric81] showed that all linear orderings and all trees viewed as orderings have the computable embeddability condition. We will show that these classes of structures are Σ-small using a general technique inspired by Richter's proofs. Our technique will also show that Boolean algebras, adjacency linear orderings (cf. Knight [Kni86]), and equivalence structures are also Σ-small. There are other techniques used to show classes are Σ-small: Trivially, if a class $\mathbb{K}$ has only countably many structures, it is Σ-small; algebraically closed fields and vectors spaces over a fixed field are thus Σ-small. Differentially closed fields of characteristic 0 are Σ-small because they are ω-stable. All the iterates of the jump of Boolean algebras are also Σ-small (cf. Harris and Montalbán [HM12]). Abelian p-groups are another interesting example (cf. Khisamiev [Khi04]).

EXAMPLE 10.0.6. Here is an example of a class that is not Σ-small: torsion free abelian groups. For each set X of prime numbers, consider the subgroup of $(\mathbb{Q}; +)$ with domain

$$\{r/q : r \in \mathbb{Z},\ q \in \mathbb{N} \text{ satisfying that all prime factors of } q \text{ are in } X\}.$$

The ∃-type of $x = 1$ contains the formulas $(\exists z)\ p \cdot z = x$ if and only if $p \in X$, where $p \cdot z$ is shorthand for z added to itself p times. Different sets or prime numbers give rise to different ∃-types. Thus, we have continuum many ∃-types among all the torsion-free abelian groups.

10.1. Infinitary Π_1 complete relations

Understanding what the r.i.c.e. relations on a structure look like often gives us key insights into the structure's computational properties. For instance, in the case of linear ordering, we know that $\bar{\text{Adj}}^c \oplus 0'$ is a r.i.c.e.-complete relation, a fact that has been extremely helpful in proving results about linear orderings. We know of other classes of structures where the r.i.c.e.-complete relations can also be easily understood. But there are also many other classes for which we know of no such nice r.i.c.e.-complete relation. We want to understand why some classes have nice r.i.c.e.-complete relations and some do not. For that, we need to find a way to formalize what we mean by 'nice.' Every structure $\mathcal{A}$ has a r.i.c.e.-complete relation, namely $\vec{K}^{\mathcal{A}}$ (Definition 2.2.3). In contrast to $\bar{\text{Adj}}^c \oplus 0'$, the relation $\vec{K}^{\mathcal{A}}$ is much harder to visualize. One problem with Kleene's relation $\vec{K}^{\mathcal{A}}$ is that it is not necessarily *structurally* r.i.c.e.

complete (Definition 2.2.14), i.e., it is not always r.i.c.e.-complete relative to all oracles. We will see below that every structure $\mathcal{A}$ has a structurally r.i.c.e. complete relation, and structurally r.i.c.e. complete relations tend to provide the structural information we are looking for. These structurally r.i.c.e. complete relations have a disadvantage over relations like $\bar{\mathrm{Adj}}^c$ or $\bar{K}^{\mathcal{A}}$: They may be defined by different Σ_1^{in} formulas on different structures. If we are working within a class, we would like to have a structurally r.i.c.e. complete relation that is defined the same way on all structures in the class. The question now becomes: On which classes of structures do we have relations that are uniformly definable and structurally r.i.c.e.-complete? The answer is—as you might have guessed—Σ-small classes.

DEFINITION 10.1.1. A sequence of Π_1^{in} formulas $\{\varphi_i(\bar{x}_i) : i \in \mathbb{N}\}$ is Π_1^{in}-*complete* on a class $\mathbb{K}$ of structures if every Π_1^{in} formula $\psi(\bar{x})$ is equivalent to a Σ_1^{in} formula over $\tau \cup \{\varphi_i(\bar{x}_i) : i \in \mathbb{N}\}$, that is, there exists a Σ_1^{in} formula $\chi(\bar{x})$ over the vocabulary $\tau' = \tau \cup \{R_i : i \in \mathbb{N}\}$ where $R_i(\bar{x}_i)$ is interpreted as $\varphi_i(\bar{x}_i)$, such that $\forall \bar{x} \ (\chi(\bar{x}) \leftrightarrow \psi(\bar{x}))$ holds on all structures in $\mathbb{K}$.

If you negate all formulas in a Π_1^{in}-complete sequence of formulas, you get a Σ_1^{in}-complete sequence of formulas. For a structure $\mathcal{A} \in \mathbb{K}$, a Σ_1^{in}-complete sequence of formulas defines a structurally u.r.i.c.e.-complete relation on $\mathcal{A}$: A relation R is *structurally u.r.i.c.e.* if it is u.r.i.c.e. in $(\mathcal{A}, X)$ for some $X \in 2^{\mathbb{N}}$; this is equivalent to R being Σ_1^{in}-definable without parameters. R is *structurally u.r.i.c.e.complete* if also, for every other structurally u.r.i.c.e. relation $Q \subseteq \mathbb{N} \times A^{<\mathbb{N}}$, Q is u.r.i. computable in $(\mathcal{A}, R, Y)$ for some $Y \in 2^{\mathbb{N}}$; this is equivalent to Q being Δ_1^{in}-definable in $(\mathcal{A}, R)$ without parameters. (C.f. Section 2.2.3, where we introduced the non-uniform versions.)

LEMMA 10.1.2. *If $\mathbb{K}$ is Σ-small, the set of formulas of the form $\bigwedge p$, where p is a $\forall$-type sharply realized in $\mathbb{K}$, is a Π_1^{in}-complete sequence of formulas.*[90]

PROOF. Enumerate the $\forall$-types sharply realized in $\mathbb{K}$ as $p_0, p_1, \ldots$. For every Π_1^{in} formula $\psi(\bar{x})$, let

$$I_\psi = \{i \in \mathbb{N} : \text{all the conjuncts of } \psi \text{ are in } p_i\}.$$

We claim that

$$\forall \bar{x} \left(\psi(\bar{x}) \iff \bigvee_{i \in I_\psi} \left(\bigwedge p_i(\bar{x}) \right) \right)$$

holds on all structures in $\mathbb{K}$. This would show that ψ is equivalent to a Σ_1^{in} formula over $\{\bigwedge p_i : i \in \mathbb{N}\}$. As for the proof of the claim: clearly, if $i \in I_\psi$, then $\bigwedge p_i \Longrightarrow \psi$. Thus $\bigvee_{i \in I_\psi} \bigwedge p_i \Longrightarrow \psi$. Conversely, if ψ holds of some tuple $\bar{a}$ in some $\mathcal{A} \in \mathbb{K}$, let p_i be the $\forall$-type of $\bar{a}$ in $\mathcal{A}$.

---

[90] $\bigwedge p$ stands for $\bigwedge_{\chi(\bar{x}) \in p} \chi(\bar{x})$.

Then all conjuncts in ψ are part of p_i, and hence $i \in I_\psi$ and $\bar{a}$ satisfies $\bigvee_{i \in I_\psi} \bigwedge p_i$. □

COROLLARY 10.1.3. *On every structure $\mathcal{A}$, there is a structurally r.i.c.e.-complete sequence of formulas.*

PROOF. Just take the class $\mathbb{K} = \{\mathcal{A}\}$. □

In all natural examples of Σ-small classes, not only are all ∃-types computable, they are uniformly computable. In that case, the structurally r.i.c.e.-complete sequence of formulas also defines a r.i.c.e.-complete relation.

COROLLARY 10.1.4. *Let $\mathbb{K}$ be a Σ-small class for which there is a computable listing $\{p_i : i \in \mathbb{N}\}$ of all the ∀-types sharply realized in $\mathbb{K}$.[91] Then, for every $\mathcal{A} \in \mathbb{K}$, the relation*

$$\{\langle i, \bar{a} \rangle : \mathcal{A} \models \neg \bigwedge p_i(\bar{a})\} \oplus 0'$$

is u.r.i.c.e. complete in $\mathcal{A}$.

PROOF. The relation $\{\langle i, \bar{a} \rangle : \mathcal{A} \models \neg \bigwedge p_i(\bar{a})\}$ is u.r.i.c.e. because the formulas $\neg \bigwedge p_i(\bar{x})$ are uniformly Σ_1^c. To see they are u.r.i.c.e. complete once we add $0'$, we refer to the proof of Lemma 10.1.2 above. There, it is shown that if ψ is a Π_1^c formula, it is equivalent to $\bigvee_{i \in I_\psi} \bigwedge p_i(\bar{x})$. Notice that this formula is $\Sigma_1^{c^{0'}}$ over $\{\bigwedge p_i : i \in \mathbb{N}\}$, as I_ψ is Π_1^0, and hence computable in $0'$. □

Let us now prove the reversal: If a Π_1^{in}-complete sequence of formulas exists, then the class is Σ-small.

DEFINITION 10.1.5. Given a formula $\psi(\bar{x})$, the *∀-type generated* by ψ over $\mathbb{K}$ is the set of all ∀-formulas $\varphi(\bar{x})$ that are implied by $\psi(\bar{x})$ on all structures in $\mathbb{K}$.

LEMMA 10.1.6. *Let $\{\varphi_i(\bar{x}_i) : i \in \mathbb{N}\}$ be a Π_1^{in}-complete sequence of formulas over a class $\mathbb{K}$. Then every ∀-type sharply realized in $\mathbb{K}$ is generated by a ∃-formula over $\tau \cup \{\varphi_i(\bar{x}_i) : i \in \mathbb{N}\}$.*

PROOF. Let p be a ∀-type sharply realized in $\mathbb{K}$. By the completeness of $\{\varphi_i(\bar{x}_i) : i \in \mathbb{N}\}$, $\bigwedge p$ must be equivalent to a Σ_1^{in} formula χ over $\tau \cup \{\varphi_i(\bar{x}_i) : i \in \mathbb{N}\}$. We claim that p is generated by one of the disjuncts of χ. Consider a tuple $\bar{a}$ with ∀-type p in some structure $\mathcal{A} \in \mathbb{K}$. Then one of the disjuncts of χ must be true of $\bar{a}$; call it ψ. On the one hand, ψ implies χ, and χ implies $\bigwedge p$. On the other hand, ψ cannot imply any other ∀-formula, as $\bar{a}$ satisfies ψ and no ∀-formula outside of p. Thus ψ generates p. □

COROLLARY 10.1.7. *A class $\mathbb{K}$ of structures is Σ-small if and only if there exists a sequence of Π_1^{in} formulas that is Π_1^{in}-complete on $\mathbb{K}$.*

---

[91] A computable listing of ∀-types is coded by a computable set $C \subseteq \mathbb{N}^2$ such that $\langle i, j \rangle \in C$ if and only if the jth ∀-formula belongs to p_i.

PROOF. If $\mathbb{K}$ is Σ-small, we showed that the formulas formed as the conjunctions of the $\forall$-types form a Π_1^{in}-complete sequence of formulas in Lemma 10.1.2. The other direction follows from the previous lemma, as there are only countably many finitary $\exists$-formulas over a countable vocabulary.   □

This corollary answers our original question about which classes of structures have *nice* r.i.c.e.-complete relations—somewhat. It definitely proves that if a class is not Σ-small, it does not have a nice r.i.c.e.-complete relation. However, in the case when the class is Σ-small and we do have a Π_1^{in}-complete sequence of formulas, it might still be a bit of a stretch to say that the formulas we get from Lemma 10.1.2 are nice. In practice, they usually are. Let us look at linear orderings. A $\forall$-type $p(\bar{x})$ is determined by a permutation of $|\bar{x}|$ that describes the order among the variables, and a tuple of numbers saying, for each pair of variables x_i, x_j, that there are no more than so many elements in between them, to their left, and to their right. These are nice enough relations. As we know, the co-adjacency relation alone is already structurally r.i.c.e. complete. This is because all these relations defined from $\forall$-types are r.i. computable from $\bar{\mathrm{Adj}}$.

EXERCISE 10.1.8. Consider the Π_1^c-definable relation

$$\{\langle i, \bar{a}\rangle : \mathcal{A} \models \neg \bigwedge\!\!\!\bigwedge p_i(\bar{a})\},$$

as in Corollary 10.1.4, for the class of linear orderings. Prove that it is r.i. computably equivalent to the adjacency relation.

EXERCISE 10.1.9. Consider the Π_1^c-definable relation

$$\{\langle i, \bar{a}\rangle : \mathcal{A} \models \neg \bigwedge\!\!\!\bigwedge p_i(\bar{a})\},$$

as in Corollary 10.1.4, for the class of Boolean algebras. Prove that it is r.i. computably equivalent to the atom relation. (An element in a Boolean algebra is an *atom* if it is non-zero and has no elements below it other than zero. Recall that in Exercise 2.3.8 we showed that the atom-relation is structurally complete.)

In these two examples, we have r.i.c.e. complete relations of arity two and one, respectively. This is quite nice, but not very common. We do not know much about which structures have structurally r.i.c.e. complete relations of bounded arity. We know this is the case for adjacency linear orders [Mon12, Lemma 7.1], Boolean algebras and their jumps (cf. [HM12]), and equivalence structures (cf. [Mon12, Section 7.4]). On the other hand, it follows from Exercise 2.3.9 that the infinite dimension $\mathbb{Q}$-vector space does not have a r.i.c.e. complete relation of bounded arity (i.e., a subset of $\mathbb{N} \times A^{\le k}$ for $k \in \mathbb{N}$).

10.2. A sufficient condition

In this section, we give a sufficient condition for a class of structures to be Σ-small. This condition has a stronger consequence than just Σ-smallness: It implies that every infinitary Π_1 formula is equivalent to a finitary $\forall$-formula. This implies that there are countably many $\forall$-types, as the conjunction of all the formulas in a $\forall$-type would then be equivalent to a single $\forall$-formula, and there are only countably many $\forall$-formulas. Despite not being necessary, this is still quite a useful condition, as it holds in many of the examples of Σ-small classes we know.

Throughout this section, let τ be a finite vocabulary and let $\mathbb{K}$ be a uniformly locally finite[92] class of τ-structures. Let $\mathbb{K}^{fin}$ be the set of all finite substructures of structures in $\mathbb{K}$.[93] Given a set A, define τ_A by augmenting the vocabulary τ with new constant symbols, one for each element of A. A τ_A-structure is thus determined by a τ-structure $\mathcal{B}$ and a map $f : A \to B$ describing the assignments of the new constants. Given a finite structure $\mathcal{A} \in \mathbb{K}^{fin}$, let $\mathbb{K}_A$ be the set of τ_A-structures that consists of a τ-structure $\mathcal{B}$ in $\mathbb{K}$ together with a τ-embedding from $\mathcal{A}$ to $\mathcal{B}$. Let $\mathbb{K}_A^{fin}$ be the set of finite τ_A-substructures of structures in $\mathbb{K}_A$. $\mathbb{K}_A^{fin}$ is essentially a set of pairs $\langle f, \mathcal{B} \rangle$, where $\mathcal{B} \in \mathbb{K}^{fin}$ and f is a τ-embedding from $\mathcal{A}$ into $\mathcal{B}$.

Before stating our theorem, we need one more definition.

DEFINITION 10.2.1. A partial ordering is *well-quasi-ordered* if it has no infinite descending sequences and no infinite anti-chains.

It is not hard to show that a partial ordering is well-quasi-ordered if and only if every subset X has a finite subset $F \subseteq X$ such that $\forall x \in X \, \exists y \in F \, (y \leq x)$. Such an F can be chosen to be an anti-chain, the anti-chain of the minimal elements of X. It is also not hard to show that the product of well-quasi-orderings is a well-quasi-ordering. Then, for instance, we get Dickson's lemma, which says that $\mathbb{N}^k$ is well-quasi-ordered, with the ordering where a tuple is below another if each coordinate of the first tuple is below the corresponding one of the second. Here are two well-known results about well-quasi-orderings we will use: One is Higman's lemma (cf. [Hig52]), which says that $\mathbb{N}^{<\mathbb{N}}$ is well-quasi-ordered under the *embeddability partial ordering*: $\langle x_0, \ldots, x_k \rangle \preccurlyeq \langle y_0, \ldots, y_\ell \rangle$ if there is an increasing map $f : \{0, \ldots, k\} \to \{0, \ldots, \ell\}$ such that, for all $i \leq k$, $x_i \leq y_{f(i)}$. The other is Kurskal's theorem (cf. [Kru60]), which says

---

[92] $\mathcal{A}$ is *locally finite* if every finitely generated substructure of a structure in $\mathcal{A}$ is finite. $\mathbb{K}$ is *uniformly locally finite* if, for every $n \in \mathbb{N}$, there exists an $m \in \mathbb{N}$ such that every substructure of a structure in $\mathbb{K}$ generated by n elements has size at most m. For a finite vocabulary, this implies that the number of quantifier-free types on a fixed tuple of variables is finite.

[93] $\mathbb{K}^{fin}$ was defined in 3.6.2 to be the set of diagrams of the finite $\tau_{|.|}$-substructures of the structures in $\mathbb{K}$. That is still the formal definition, but in this section, it is easier to visualize $\mathbb{K}^{fin}$ as a set of structures. Also, since we assume τ is finite, $\tau_{|.|}$-structures are just τ-structures.

that the set of finite trees is well-quasi-ordered by embeddability even if we require the embeddings to preserve meets (i.e., greatest lower bounds). Here, by embeddability, we mean as partial orderings, that is, a tree is below another if there is a one-to-one, order-preserving map which preserves meets.

THEOREM 10.2.2. *Let $\mathbb{K}$ be a uniformly locally finite class of structures over a finite vocabulary. Suppose that, for every finite substructure $\mathcal{A} \in \mathbb{K}^{fin}$, $\mathbb{K}_{\mathcal{A}}^{fin}$ is well-quasi-ordered under the embeddability relation. Then, in $\mathbb{K}$, every Σ_1^{in} formula is equivalent to a finitary $\exists$-formula.*

Recall that this implies that $\mathbb{K}$ is Σ-small.

PROOF. Consider a Σ_1^{in} formula $\psi(\bar{x})$. Since $\mathbb{K}$ is a uniformly locally finite class over a finite vocabulary, there are only finitely many quantifier-free types on the variables $\bar{x}$: Call them $q_0(\bar{x}), \dots, q_k(\bar{x})$. Since $\psi \equiv \bigvee_{i \le k}(\psi \land \bigwedge q_i)$, it is enough to show that each of the formulas $\psi(\bar{x}) \land \bigwedge q_i(\bar{x})$ for $i \le k$ is equivalent to a finitary $\exists$-formula. Fix $i \le k$ and let $\mathcal{A} \in \mathbb{K}^{fin}$ be the finite substructure generated by a tuple $\bar{a}$ that satisfies the type $q_i(\bar{x})$. Let $S \subseteq \mathbb{K}_{\mathcal{A}}^{fin}$ be the set of τ_A-structures $\mathcal{B}$ that satisfy $\psi(\bar{a})$. S is closed upwards under inclusion. Since $\mathbb{K}_{\mathcal{A}}^{fin}$ is a well-quasi-ordering, there is a finite subset S of minimal elements $\mathcal{F}_0, \dots, \mathcal{F}_k$. Thus, for a structure $\mathcal{B} \in \mathbb{K}_{\mathcal{A}}^{fin}$, we have that $\mathcal{B} \in S$ if and only if there is a τ_A-embedding from one of the $\mathcal{F}_i$'s into $\mathcal{B}$. Now, for a structure $\mathcal{C} \in \mathbb{K}$, a tuple $\bar{c} \in C^{|\bar{x}|}$ satisfies $\psi(\bar{x}) \land \bigwedge q_i(\bar{x})$ if and only if there is an embedding of $\mathcal{A}$ into $\mathcal{C}$ mapping $\bar{a}$ to $\bar{c}$ so that one of the $\mathcal{F}_i$'s τ_A-embeds into $\mathcal{C}_A$, where $\mathcal{C}_A$ is, of course, the τ_A-structure corresponding to $\mathcal{C}$ and the embedding from $\mathcal{A}$ to $\mathcal{C}$. This is equivalent to saying that there exists a tuple $\bar{y} \in C^{<\mathbb{N}}$ that, together with $\bar{c}$, has the atomic diagram of $\mathcal{F}_i$. Since $\mathbb{K}$ is uniformly locally finite and the vocabulary is finite, this can be expressed by a finitary $\exists$-formula:

$$\bigvee_{i \le k} \exists \bar{y} \, D(\bar{x}, \bar{y}) = D(\mathcal{F}_i).$$

This $\exists$-formula is equivalent to $\psi(\bar{x}) \land \bigwedge q_i(\bar{x})$. □

Let us remark that $\mathbb{K}_{\mathcal{A}}^{fin}$ is always well-founded: These structures are finite, so we could never have an infinite descending sequence. Thus, stating that $\mathbb{K}_{\mathcal{A}}^{fin}$ is well-quasi-ordered is equivalent to saying that it contains no anti-chains.

EXERCISE 10.2.3. Let $\mathbb{K}$ be a uniformly locally finite class of structures over a finite vocabulary. Suppose that $\mathbb{K}$ is the class of models of a Π_1^{in} sentence. Prove that the theorem above reverses. That is, that if every Π_1^{in} formula is equivalent to a finitary $\forall$-formula, then $\mathbb{K}_{\mathcal{A}}^{fin}$ is well-quasi-ordered for every finite substructure $\mathcal{A} \in \mathbb{K}^{fin}$. Hint in footnote.[94]

---

[94] Use that $\mathbb{K}^{fin} \subseteq \mathbb{K}$ and that finitary $\exists$-formulas can be expressed in terms of embedding structures from $\mathbb{K}^{fin}$.

LEMMA 10.2.4. *Let* $\mathbb{LO}$ *be the class of all linear orderings. Then, for every finite linear ordering* $\mathcal{L}$, $\mathbb{LO}_{\mathcal{L}}^{fin}$ *is well-quasi-ordered by embeddability.*

This gives a new angle on the proof that $\mathbb{LO}$ is Σ-small.

PROOF. This is essentially what was happening in Claim 2.3.4.

Fix a finite linear ordering $\mathcal{L}$. Note that each structure in $\mathbb{LO}_{\mathcal{L}}^{fin}$ can be described by a tuple in $\mathbb{N}^{|\mathcal{L}|+1}$ saying how many elements there are between each pair of consecutive elements of $\mathcal{L}$, how many there are to the left of the first element of $\mathcal{L}$, and how many there are to the right of the last. It is not hard to see that a structure from $\mathbb{LO}_{\mathcal{L}}^{fin}$ embeds in another if and only if the tuple corresponding to the first structure is below the tuple for the latter. When we say 'below,' we mean coordinate-wise; That is, for each $i \leq |\mathcal{L}|$, the i-th entry of the first tuple is less than or equal to the i-th element of the second. Dickson's lemma, which follows from the closure of well-quasi-orders under products, states that $\mathbb{N}^{k+1}$ is well-quasi-ordered under this coordinate-wise ordering.   □

Let $\mathbb{ALO}$ be the class of adjacency linear orderings, which consists of linear orderings together with a relation $\bar{\text{Adj}}$ for adjacency. (Recall that $\bar{\text{Adj}}$ consists not only of the usual binary adjacency relation Adj, but also includes the relations for first and last elements; see Definition 2.3.6.)

LEMMA 10.2.5. *For every finite* $\mathcal{L} \in \mathbb{ALO}^{fin}$, $\mathbb{ALO}_{\mathcal{L}}^{fin}$ *is well-quasi-ordered by embeddability.*

This shows that $\mathbb{ALO}$ is Σ-small (cf. [Kni86]). Recall that $\mathbb{ALO} \oplus 0'$ is effectively bi-interpretable with the jump of $\mathbb{LO}$. Notice that the finite structures $\mathcal{L} \in \mathbb{ALO}^{fin}$ need not be adjacency linear orderings themselves: There might be consecutive elements $a <_{\mathcal{L}} b \in L$ for which $\bar{\text{Adj}}^{\mathcal{L}}$ does not hold because $\mathcal{L}$ is a subset of a larger linear ordering where a and b are not adjacent. The structures in $\mathbb{ALO}^{fin}$ satisfy that if $\bar{\text{Adj}}^{\mathcal{L}}(a, b)$, then a and b are consecutive, but not vice versa.

PROOF. For this proof, we need to use Higman's theorem that $\mathbb{N}^{<\mathbb{N}}$ is well-quasi-ordered under the following partial ordering:

$$\langle x_0, \ldots, x_k \rangle \preccurlyeq \langle y_0, \ldots, y_\ell \rangle \iff$$
$$\exists f : \{0, \ldots, k\} \to \{0, \ldots, \ell\} \text{ increasing, } \forall i \leq k \; (x_i \leq y_{f(i)}).$$

We will show that the well-quasi-orderness of $\mathbb{ALO}_{\mathcal{L}}^{fin}$ can be reduced to the well-quasi-orderness of a product of orderings of the form $\mathbb{N}$ and $\mathbb{N}^{<\mathbb{N}}$. Recall that well-quasi-orderness is preserved under products.

A set of elements of a finite adjacency linear ordering is said to be an *adjacency chain* if it is a maximal sequence of $\bar{\text{Adj}}$-adjacent elements. Every $\mathcal{A} \in \mathbb{ALO}^{fin}$ can be partitioned into adjacency chains. Let $\langle l_{\mathcal{A}}, \bar{a}_{\mathcal{A}}, r_{\mathcal{A}} \rangle \in \mathbb{N} \times \mathbb{N}^{<\mathbb{N}} \times \mathbb{N}$ be the tuple of sizes of the adjacency chains in $\mathcal{A}$, ordered from left to right, where $l_{\mathcal{A}}$ is the size of the adjacency chain of $-\infty$, $r_{\mathcal{A}}$ is the size of the adjacency chain of $+\infty$, and $\bar{a}_A$ is the tuple of non-zero sizes of the adjacency chains in between. That is, l_A is the size of the adjacency chain containing the element f satisfying $\bar{\mathsf{Adj}}(-\infty, f)$ if there is any, and $l_A = 0$ if there is no such element. Same with r_A and $+\infty$. It is not hard to see that if we have $l_A \leq l_B$, $r_A \leq r_B$, and $\bar{a}_A \preccurlyeq \bar{a}_B$ (the latter as in the Higman's ordering), then A embeds into B preserving $\bar{\mathsf{Adj}}$. This is not an if-and-only-if equivalence. But this implication is enough to conclude that ALO^{fin} is a well-quasi-ordering: Because if $\{A_i : i \in \mathbb{N}\}$ were an anti-chain of finite adjacency linear orderings, then $\{\langle l_{A_i}, \bar{a}_{A_i}, r_{A_i}\rangle : i \in \mathbb{N}\}$ would be an anti-chain in the well-quasi-ordering $\mathbb{N} \times \mathbb{N}^{<\mathbb{N}} \times \mathbb{N}$ (where $\mathbb{N} \times \mathbb{N}^{<\mathbb{N}} \times \mathbb{N}$ is ordered by the product ordering, using Higman's ordering on $\mathbb{N}^{<\mathbb{N}}$). We already know that ALO^{fin} has no infinite descending sequences, as its structures are all finite.

Now, fix a finite linear ordering $\mathcal{L} = \{\ell_1 <_L \cdots <_L \ell_k\}$. A structure in $\mathrm{ALO}^{fin}_\mathcal{L}$ is determined by the intervals between consecutive elements of $\mathcal{L}$, by the interval to the left of ℓ_1, and by the interval to the right of ℓ_k. Thus, $\mathrm{ALO}^{fin}_\mathcal{L}$ is, in a sense, isomorphic to the $k + 1$-cartesian power of ALO^{fin}. Assign to each $A \in \mathrm{ALO}^{fin}_\mathcal{L}$ a tuple

$$\langle l_0, \bar{a}_0, r_0, \ l_1, \bar{a}_1, r_1, \ldots, l_k, \bar{a}_k, r_k\rangle \in (\mathbb{N} \times \mathbb{N}^{<\mathbb{N}} \times \mathbb{N})^{k+1},$$

where $\langle l_i, \bar{a}_i, r_i\rangle$ is the tuple of sizes of adjacency chains corresponding to the interval in A between ℓ_i and ℓ_{i+1} as in the first part of the proof. Of course, $\langle l_0, \bar{a}_0, r_0\rangle$ corresponds to the interval to the left of ℓ_1 and $\langle l_k, \bar{a}_k, r_k\rangle$ corresponds to the interval to the right of ℓ_k. Now, given $A, B \in \mathrm{ALO}^{fin}_\mathcal{L}$, if the tuple corresponding to A is below that of B in the product ordering, where the $\mathbb{N}^{<\mathbb{N}}$'s are ordered according to Higman's ordering, then A embeds into B. As in the argument for ALO^{fin}, we can then deduce that $\mathrm{ALO}^{fin}_\mathcal{L}$ is well-quasi-ordered using that well-quasi-orderness is preserved by products. $\square$

EXERCISE 10.2.6. Let $\mathbb{BA}$ be the class of Boolean algebras. Prove that for every finite Boolean algebra B, $\mathbb{BA}^{fin}_B$ is well-quasi-ordered by embeddability.

Harris and Montalbán [HM12] proved that all the finite jumps $\mathbb{BA}^{(n)}$ of the class of Boolean algebras are also Σ-small.

DEFINITION 10.2.7. Let $\mathbb{T}$ be the class trees viewed as partial orderings. That is, $\mathbb{T}$ is the class of partial orderings $(T; \leq, r)$ which have a least element denoted r and satisfy that the set of predecessors of every $t \in T$, namely $\{s \in t : s \leq t\}$, is finite and linearly ordered.

LEMMA 10.2.8. For every finite tree T, $\mathbb{T}^{fin}_T$ is well-quasi-ordered by embeddability.

It follows that $\mathbb{T}$ is Σ-small (cf. [Ric81]).

PROOF. Kruskal's lemma (cf. [Kru60]) states that $\mathbb{T}^{fin}$ is well-quasi-ordered by embeddability even if we require the embeddings to preserve meets (i.e., greatest lower bounds). We now need to prove that $\mathbb{T}_{\mathcal{T}}^{fin}$ is well-quasi-ordered by embeddability for any given finite tree $\mathcal{T}$. We prove this by induction on the size of $\mathcal{T}$. In the case when $\mathcal{T}$ contains just one element, namely the root, $\mathbb{T}_{\mathcal{T}}^{fin}$ is exactly the same as $\mathbb{T}^{fin}$. Suppose that now $\mathcal{T}$ contains more than just the root. If $\mathcal{T}$ has subtrees $T_1, \ldots, T_k$ coming out of the root, then $\mathbb{T}_{\mathcal{T}}^{fin}$ is isomorphic to $\mathbb{T}^{fin} \times \mathbb{T}_{T_1}^{fin} \times \cdots \times \mathbb{T}_{T_k}^{fin}$: To see this, notice that every tree $\mathcal{A}$ in $\mathbb{T}_{\mathcal{T}}^{fin}$ can be split into $k+1$ trees $\mathcal{A}_0, \ldots, \mathcal{A}_k$ as follows. Let $\mathcal{A}_0$ consist of all the nodes in $\mathcal{A}$ whose only predecessor in $\mathcal{T}$ is the root, and let $\mathcal{A}_i$ be the set of nodes in $\mathcal{A}$ which have a predecessor in T_i other than the root. It is not hard to see that for $\mathcal{A}, \mathcal{B} \in \mathbb{T}_{\mathcal{T}}^{fin}$, $\mathcal{A} \preccurlyeq \mathcal{B}$ if and only if $\mathcal{A}_j \preccurlyeq \mathcal{B}_j$ for each $j \leq k$. Since well-quasi-orderness is preserved by products, and we are assuming by the induction hypothesis that each $\mathbb{T}_{T_i}^{fin}$ is well-quasi-ordered, we get that $\mathbb{T}_{\mathcal{T}}^{fin}$ is well-quasi-ordered too.   □

EXERCISE 10.2.9. Prove that Lemma 10.2.8 also holds for trees as graphs of finite height. Hint in footnote.[95]

EXERCISE 10.2.10. Prove that the class of all trees viewed as graphs is not Σ-small. Hint in footnote.[96]

10.3. The canonical structural jump

As we argued in Section 10.1, a class of structures has a nice r.i.c.e.-complete relation if and only if it is Σ-small. What we actually proved is that a class has a Π_1^{in}-complete sequence of formulas if and only if it is Σ-small, and that these formulas are given by the conjunctions of the ∀-types sharply realized in the class. That these formulas are nice is arguable. Nice or not, what we do get is a canonical way to define a *structural jump*. Structurally r.i.c.e. complete relations are unique up to structurally r.i. computability, but not up to plain r.i. computability. So, in principle, it is unclear what the canonical structural jump of a structure should be. For instance, we want the structural jump of a linear ordering to be the linear ordering together with the adjacency relation, and the structural jump of a Boolean algebra to be the Boolean algebra together with the atom relation, without having to add a relation for $0'$ or anything else.

DEFINITION 10.3.1. Let $\mathbb{K}$ be a Σ-small class of τ-structures and $\{p_i : i \in \mathbb{N}\}$ a computable listing of the ∀-types sharply realized in $\mathbb{K}$. We define the *canonical structural jump* of a structure $\mathcal{A} \in \mathbb{K}$ by adding to $\mathcal{A}$ the relations $\{\bar{a} \in A^{<\mathbb{N}} : \bigwedge p_i(\bar{a})\}$ for $i \in \mathbb{N}$. We denote this new structure $\mathcal{A}_{(1)}$, and we use $\mathbb{K}_{(1)}$ to denote $\{\mathcal{A}_{(1)} : \mathcal{A} \in \mathbb{K}\}$. We use $\tau_{(1)}$ to denote the new

---

[95]Prove it for trees of height bounded by a fixed k.
[96]Think of Y-shaped finite trees.

vocabulary, defined by adding to τ relations symbols $\bigwedge p_i(\cdot)$ for $i \in \mathbb{N}$. The finite approximations to $\tau_{(1)}$ are defined by $\tau_{(1)_s} = \tau_s \cup \{\bigwedge p_i(\cdot) : i < s\}$.

If we add $0'$ as a relation to $\mathcal{A}_{(1)}$, we get a structure effectively bi-interpretable with $\mathcal{A}'$. It follows from Exercise 10.1.8 that, for a linear ordering $\mathcal{L}$, the class $\mathbb{LO}_{(1)}$ is effectively bi-interpretable with $\mathbb{ALO}$, the class of adjacency linear orderings. One can also show that, for Boolean algebras, the class $\mathbb{BA}_{(1)}$ is effectively bi-interpretable with $\mathbb{ABA}$, the class of Boolean algebras with an added relation that distinguishes atoms. We think of the canonical structural jump as a structure up to effective bi-interpretability. Thus, we think of adjacency linear orderings as the canonical structural jumps of linear orderings, and atom Boolean algebras as the canonical structural jumps of Boolean algebras.

Notice that the definition of $\mathcal{A}_{(1)}$ depends on the enumeration of the $\forall$-types sharply realized in $\mathbb{K}$. In most natural Σ-small classes, there is a natural such enumeration that is unique up to a computable re-ordering, so this is not usually an issue. For such natural classes, another effectiveness property we always get is that $\mathbb{K}_{(1)}^{fin}$ is computable. Recall that $\mathbb{K}_{(1)}^{fin}$ is the set of diagrams of the finite $\tau_{(1)|\cdot|}$-substructures of the structures in $\mathbb{K}$, and recall that a $\tau_{(1)|\cdot|}$-structure is a $\tau_{(1)_s}$-structure where s is the size of the structure.

DEFINITION 10.3.2. A Σ-small class $\mathbb{K}$ is *effectively Σ-small* if there is a computable listing of the $\forall$-types sharply realized in $\mathbb{K}$, and $\mathbb{K}_{(1)}^{fin}$ is computable.

Of course, all the natural Σ-small classes we know are effectively Σ-small. Proving that this is so sometimes takes a bit of work, as it requires understanding the space of $\forall$-types and the compatibilities between the different types. For instance, the class of differentially closed fields of characteristic zero, denoted $\mathbb{DCF}_0$, is Σ-small just because there are countably many first-order types. Understanding the structure of the $\forall$-types requires some model theory, as, for instance, it requires proving that $\mathbb{DCF}_0$ has quantifier elimination. It can then be shown that the Π_1^c-relations $R_{m,n}(x, y_1, \ldots, y_m)$, which say that x is not a root of any differential polynomial of degree n over $\mathbb{Q}\langle y_1, \ldots, y_m \rangle$ for $m \in \mathbb{N}$ and $n \in \mathbb{N} \cup \{\infty\}$, are enough to define the canonical structural jump of $\mathbb{DCF}_0$.

Let us observe that if $\mathbb{K}$ is effectively Σ-small and Π_2^c, then $\mathbb{K}_{(1)}$ is also Π_2^c. This is because the definitions of the new symbols in $\tau_{(1)}$ are Π_1^c.

10.4. The low property

Downey and Jockusch [DJ94] proved that Boolean algebras have the *low property*, that is, that every low Boolean algebra has a computable copy (Theorem 10.6.11). Jockusch and Soare [JS91] showed this is not the case for linear orderings. This property interests computability theorists for two reasons. On the one hand, understanding when structures have computable copies is a general theme of computable structure theory, and these results give us useful information about it. On the other hand, as we will see below in Lemma 10.4.2, the low property implies that the degree spectrum of the structure is determined by the degree spectrum of its jump. Thus, in a sense, in terms of the information encoded in the isomorphism type of a structure, the low property says that no information is lost when taking a jump.

DEFINITION 10.4.1. A class $\mathbb{K}$ has the *low property* if, for every $X \in 2^{\mathbb{N}}$, every structure from $\mathbb{K}$ that has a copy that is low over X also has a copy that is computable in X.

This is an interesting property of the degree spectra of all structures in $\mathbb{K}$. It implies that the degree spectra of the structures in $\mathbb{K}$ are determined by the spectra of their jumps:

LEMMA 10.4.2. *A class $\mathbb{K}$ has the low property if and only if, for every structure $\mathcal{A} \in \mathbb{K}$,*

$$DgSp(\mathcal{A}) = \{X \in 2^{\mathbb{N}} : X' \in DgSp(\mathcal{A}')\}.$$

PROOF. For the left-to-right direction, assume that $\mathbb{K}$ has the low property. First, let us observe that the inclusion

$$DgSp(\mathcal{A}) \subseteq \{X \in 2^{\mathbb{N}} : X' \in DgSp(\mathcal{A}')\}$$

always holds, as whenever X computes a copy of $\mathcal{A}$, X' computes a copy of $\mathcal{A}'$. For the other inclusion, we need to use the low property. If X' computes a copy of $\mathcal{A}'$, by the Second Jump Inversion Theorem 9.1.6, $\mathcal{A}$ has a copy that is low over X. Then, by the low property, X computes a copy of $\mathcal{A}$.

For the right-to-left direction, suppose that $\mathcal{A}$ has a copy that is low over X. Then $\mathcal{A}'$ has a copy computable in X', and by our assumption regarding spectra, $X \in DgSp(\mathcal{A})$. Thus X computes a copy of $\mathcal{A}$.   □

One of the most interesting examples of a class with the low property is differentially closed fields of characteristic zero. This was recently proved by Marker and Miller [MM17]. They also showed that the jump of $\mathbb{DCF}_0$ is *universal for degree spectrum*, that is, every degree spectrum of a structure that computably codes $0'$ is equal to the degree spectrum of the jump of a differentially closed field. This gives a full description of the degree spectra of differentially closed fields of characteristic zero: They are the jump inversions of all the possible degree spectra.

As we mentioned above, the first example of a class with the low property was the class of Boolean algebras (cf. [DJ94]). The following year, Thurber [Thu95] showed that Boolean algebras have the low$_2$ property, that is, that every low$_2$ Boolean algebra has a computable copy. This is equivalent to stating that $\mathbb{BA}'$ has the low property. A few years later, Knight and Stob showed that Boolean algebras have the low$_4$ property, that is, that every low$_4$ Boolean algebra has a computable copy, or, equivalently, that $\mathbb{BA}'''$ has the low property. It is not known if Boolean algebras have the low$_n$ property for all $n \in \mathbb{N}$, or even if they have the low$_5$ property. Harris and Montalbán [HM14] showed that the low$_5$ problem for Boolean algebras is qualitatively more difficult that the previous ones: While for $n = 1, 2, 3, 4$, every low$_n$ Boolean algebra is $0^{(n+2)}$-isomorphic to a computable one, they built a low$_5$ Boolean algebra not $0^{(7)}$-isomorphic to any computable one.

As we will see in [MonP2], the class of ordinals satisfies much more than the low$_n$ property for all $n \in \mathbb{N}$: Every arithmetic (even hyperarithmetic) ordinal has a computable copy. We will see, also in [MonP2], that the same behavior would occur on counterexamples to Vaught's conjecture—if there are any.

A sharper example is the class of linear orderings with finitely many descending sequences.[97] Kach and Montalbán [KM] showed that they have the low$_n$ property for all $n \in \mathbb{N}$, but not much more: They built a Δ_2^0 intermediate linear ordering with exactly one descending cut and with no computable copy.

The next theorem shows that classes with the low property are necessarily Σ-small. Not all Σ-small classes have the low property though. We will try to characterize the ones that do later. For now, we need the following technical lemma, whose proof uses the techniques from Chapter 7.

LEMMA 10.4.3. *If $A \subseteq \mathbb{N}$ is not c.e., there is a $G \in 2^{\mathbb{N}}$ such that $G' \geq_T A'$, but A is still not c.e. in G.*

PROOF. To get $A' \leq_T G'$, we will build $G \in 2^{\mathbb{N} \times \mathbb{N}}$ so that $\lim_t G(n, t) = A'(n)$ for all $n \in \mathbb{N}$. At each stage s, we define an approximation $G[s] \in 2^{k_s \times \mathbb{N}}$ for some $k_s \in \mathbb{N}$. The approximations $G[s]$ will be compatible throughout the construction: That is, $G[s] = G[s+1] \restriction k_s \times \mathbb{N}$ for all $s \in \mathbb{N}$. This will allow us to define G at the end of the construction as the union of the $G[s]$'s so that $G[s] = G \restriction k_s \times \mathbb{N}$. Note that, even if $G[s]$ is an infinite binary string in $2^{k_s \times \mathbb{N}}$, it can be described using only finitely much information, as it must satisfy that $\lim_t G[s](n, t) = A'(n)$ for all $n < k_s$.

At stage 0, let $k_0 = 0$ and $G[0] = \emptyset$. At each stage $s + 1 = e$, we define $G[s+1]$ as to ensure the satisfaction of one more requirement:

Requirement R_e: $A \neq W_e^G$.

Ask if there exists a finite string $\bar{q} \in 2^{<\mathbb{N} \times <\mathbb{N}}$ and an $n \in \mathbb{N}$ such that $\bar{q}$ is compatible with $G[s]$, $n \in W_e^{\bar{q}}$, but $n \notin A$. If there is one, let $G[s+1]$ be an extension of $G[s]$ compatible with $\bar{q}$ satisfying $\lim_t G[s+1](n, t) = A'(n)$

---

[97] By linear orderings with finitely many descending sequences, we mean linear orderings with finitely many cuts which are limits from the right.

for all $n < k_{s+1}$. This ensures that $n \in W_e^G \smallsetminus A$, and in particular that $A \neq W_e^G$. If there are no such $\bar{q}$'s, we define $G[s+1] = G[s]$ and we claim that R_e is automatically satisfied. This is because the set of n such that there exists a $\bar{q}$ compatible with $G[s]$ for which $n \in W_e^{\bar{q}}$ is a c.e. set, and hence it is different from A. If there is no such n outside A, it means that that set must be properly included in A. It follows that W_e^G is properly included in A for any extension G of $G[s]$, and in particular that $A \neq W_e^G$.   □

THEOREM 10.4.4. *Let* $\mathbb{K}$ *be a* Π_2^{in} *class. If* $\mathbb{K}$ *has the low property, then* $\mathbb{K}$ *is* Σ-*small.*

PROOF. Suppose $\mathbb{K}$ is not Σ-small. Let Z be such that $\mathbb{K}^{fin}$ is Z-computable and the Π_2^{in} sentence defining $\mathbb{K}$ is $\Pi^{\mathscr{C}_Z}$. Since only countably many sets are c.e. in Z, but uncountably many ∃-types are sharply realized in $\mathbb{K}$, one of those ∃-types must be not c.e. in Z; call it q. Let $G \in 2^\mathbb{N}$, $G \geq_T Z$, be such that $G' \geq_T (q \oplus Z)'$ but q is still not c.e. in G. Such a G is given by the previous lemma relativized to Z. Now, using Corollary 3.6.7, we get a structure $\mathcal{A}$ in $\mathbb{K}$ computable in $q \oplus Z$ and with a tuple having type q.[98] That structure has no copies computable in G, as q is c.e.-coded in the structure and q is not c.e. in G. Since $(q \oplus Z)' \leq_T G'$, $\mathcal{A}$ has a copy computable form G'. Thus, by Lemma 10.4.2, we get that $\mathbb{K}$ does not have the low property.   □

10.5. Listable classes

DEFINITION 10.5.1. A class of infinite structures $\mathbb{K}$ is *listable* if there is an operator Φ such that, for every $X \in 2^\mathbb{N}$, Φ^X is a sequence of ω-presentations of structures in $\mathbb{K}$ listing all X-computable structures in the class. Repetitions are allowed.

Again, even if this definition is new as is, it is not a new idea. Nurtazin [Nur74], four decades ago, considered a similar notion, but it allowed the lists to have $(\subseteq \omega)$-presentations of finite structures too. She gave a sufficient condition for a class of structures to be listable in her sense that includes the classes of linear orderings, Boolean algebras, equivalence structures, Abelian p-groups, and algebraic fields of characteristic p. Allowing for finite structures makes a huge difference though. Even if we were to allow finite structures in our definition, for our purposes we would have to use $(\sqsubseteq \omega)$-presentations instead of $(\subseteq \omega)$-presentations, the difference not being minor at all: In the case of $(\sqsubseteq \omega)$-presentations, one is forced to eventually state that there are no more elements in the domain, while with $(\subseteq \omega)$-presentations, one can always extend the domain later.

---

[98] Corollary 3.6.7 uses ∃-theories instead of ∃-types, so we have to add constants to the language and turn the type into a theory.

Goncharov and Knight [GK02, Section 5] considered a similar idea as their "third approach" to defining what it means to have a computable characterization for a class. Their notion is not the same as ours, as they allow their listing of computable structures to be hyperarithmetic. An interesting variation we should mention is described in [MonP2]: A class is *hyperarithmetically listable* if there is a hyperarithmetic listing of all its hyperarithmetic structures. We will see that, on a cone, this is equivalent to being a counterexample to Vaught's conjecture.

The objective of the rest of the section is to show that, under some effectiveness conditions, a Σ-small class has the low property if and only if its jump is listable.

THEOREM 10.5.2. *Let* $\mathbb{K}$ *be an effectively* Σ-*small,* Π_2^c *class of infinite structures. If* $\mathbb{K}$ *has the low property,* $\mathbb{K}_{(1)}$ *is listable relative to* $0'$.

PROOF. Let $X \in 2^{\mathbb{N}}$ be given. We need to build a list of all $(X \oplus 0')$-computable structures in $\mathbb{K}_{(1)}$ in an $(X \oplus 0')$-computably uniform way. Let Y be obtained from the Friedberg jump inversion theorem (Theorem 4.1.6) so that $Y' \equiv_T X \oplus 0'$. Since $\mathbb{K}$ has the low property, we have that for a structure $\mathcal{A} \in \mathbb{K}$, $\mathcal{A}'$ has an X-computable copy if and only if $\mathcal{A}$ has a Y computable copy. Thus, what we need is a listing of all the structures $\mathcal{A}'$ for $\mathcal{A} \in \mathbb{K}$ with a Y-computable copy.

For every Y-computable ω-presentation $\mathcal{A} \in \mathbb{K}$, we can build a Y-computable approximation $\mathcal{A}[0] \subseteq \mathcal{A}[1] \subseteq \cdots$ of $\mathcal{A}$ by finite structures in $\mathbb{K}^{fin}$ in a way that, at each step, we satisfy more and more of the Π_2^c sentence defining $\mathbb{K}$, as we did in Lemma 3.6.6. Let us call such sequences *witnessed approximations*. Such a procedure is uniform: If we are given a Y-partial computable function Φ_e^Y that outputs the diagram of a structure in $\mathbb{K}$, we can uniformly build such a witnessed approximation $\mathcal{A}_e[0] \subseteq \mathcal{A}_e[1] \subseteq \cdots$ to a structure $\mathcal{A}_e$. We do it in a way that if Φ_e^Y turns out to be either not total or not the diagram of a structure in $\mathbb{K}$, then there will be a step s in the sequence at which $\mathcal{A}_e[s]$ is undefined, while at all the previous stages t, $\mathcal{A}_e[t]$ is defined and satisfies ($\star$) from Lemma 3.6.6.

For each e, we will build a structure $\mathcal{B}_e \in \mathbb{K}_{(1)}$ so that if Φ_e^Y outputs the diagram of a structure $\mathcal{A}_e \in \mathbb{K}$, then $\mathcal{B}_e = \mathcal{A}_e'$. If the eth partial computable function with domain Y is either not total or not the diagram of a structure in $\mathbb{K}$, we allow $\mathcal{B}_e$ to be any structure in $\mathbb{K}_{(1)}$. Fix e; the rest of this construction is uniform in e. At stage s, we define $\mathcal{B}_e[s]$ to be a finite structure in $\mathbb{K}_{(1)}^{fin}$ which properly extends $\mathcal{B}_e[s-1]$ and is a $\tau_{(1)_s}$-expansion of $\mathcal{A}_e[s]$ as follows. We need to add to $\mathcal{A}_e[s]$ interpretations for the new symbols of $\tau_{(1)_s}$ that are not in τ. These new symbols are given by the conjunctions of $\forall$-types. Given such a type $p_i(\bar{x})$, deciding if it holds in $\mathcal{A}_e$ on a tuple $\bar{a}$ is clearly Π_1^0 in $D(\mathcal{A}_e) = \Phi_e^Y$ and hence computable in $Y' \equiv_T X \oplus 0'$. However, we need to be a bit careful, as Φ_e^Y might not be fully defined. Instead, given such a tuple $\bar{a} \in A_e[s]^{<\mathbb{N}}$, we ask if, for every

$t \geq s$ for which $\mathcal{A}_e[t]$ is defined, we have $\mathcal{A}_e[t] \models p_i(\bar{a})$. This question is still Π^0_1 in Y and gives the correct answer when $\mathcal{A}_e[t]$ is indeed defined for all t. For the finitely many new relation symbols added to the vocabulary of $\mathcal{B}[s]$, we can then decide their truth values to get $\mathcal{B}_e[s] \in \mathbb{K}^{fin}_{(1)}$. If $\mathcal{A}_e[t]$ is not defined for all t, then Y' will eventually find out. It could also happen that before Y' finds this out, the structure we would like to define as $\mathcal{B}_e[s]$ is not in $\mathbb{K}^{fin}_{(1)}$. This could only be because $\mathcal{A}_e[t]$ was not defined for all t and we were getting non-compatible answers to which of the ∀-types hold. In this case, or in the case when we find out that some $\mathcal{A}_e[t]$ is undefined, which Y' can detect, we need to build $\mathcal{B}_e$ in a different way. All we do is define $\mathcal{B}_e$ to be any structure in $\mathbb{K}_{(1)}$ extending $\mathcal{B}[s-1]$. We can do this because $\mathbb{K}^{fin}_{(1)}$ is computable and $\mathbb{K}_{(1)}$ is Π^c_2 (Lemma 3.6.6).   □

The following theorem is the key combinatorial core in the proof that if $\mathbb{K}_{(1)}$ is listable, $\mathbb{K}$ has the low property.

THEOREM 10.5.3. *Let $\mathbb{K}$ be an effectively Σ-small class of τ-structures. Suppose we have a computable operator $\mathcal{C}_{(1)}$ that, given an oracle $X \in 2^\omega$, outputs the diagram of a structure $\mathcal{C}^X_{(1)}$ in $\mathbb{K}_{(1)}$ in a way that if $X \equiv_T Y$, then $\mathcal{C}^X_{(1)} \cong \mathcal{C}^Y_{(1)}$. Then for every $X \in 2^\mathbb{N}$, $\mathcal{C}^{X'}$ has an X-computable copy.*

A small clarification is in order. The structure $\mathcal{C}^{X'}$ is the τ-structure corresponding to X'. This is not $\mathcal{C}^X_{(1)}$, which would be the canonical structural jump of $\mathcal{C}^X$. $\mathcal{C}_{(1)}$ is an operator that builds $\tau_{(1)}$-structures in $\mathbb{K}_{(1)}$, while $\mathcal{C}$ outputs the τ-restrictions that belong to $\mathbb{K}$.

PROOF. Since $\mathcal{C}_{(1)} \colon 2^\mathbb{N} \to \mathbb{K}_{(1)}$ is a computable operator, we can assume we have a computable map $\mathcal{C}_{(1)} \colon 2^{<\mathbb{N}} \to \mathbb{K}^{fin}_{(1)}$ such that

$$\delta \subseteq \gamma \in 2^{<\mathbb{N}} \implies \mathcal{C}^\delta_{(1)} \subseteq \mathcal{C}^\gamma_{(1)}$$

and that, on every path $Y \in 2^\mathbb{N}$, it produces a structure $\mathcal{C}^Y_{(1)} = \bigcup_s \mathcal{C}^{Y \restriction s}_{(1)}$ in $\mathbb{K}_{(1)}$.

Assume $X = \emptyset$. The general case is a straightforward relativization. We need to produce a computable copy of $\mathcal{C}^{0'}$. Instead of using $0'$, we will define an oracle $Y \in 2^\mathbb{N}$ that is Turing equivalent to $0'$ and produce a computable copy of $\mathcal{C}^Y$, which by assumption is isomorphic to $\mathcal{C}^{0'}$. We will define Y as the pointwise limit of a computable sequence of finite strings $\{\pi(\sigma_s) : s \in \mathbb{N}\}$. To get Y to compute $0'$, we will make sure the 1s in Y are so far apart that the function that lists the positions of the 1s dominates the settling-time function ∇ for $0'$ (see Definition 7.2.1).

To simplify the notation, we will work with strings in $\mathbb{N}^{<\mathbb{N}}$ instead of $2^{<\mathbb{N}}$: Given $\sigma \in \mathbb{N}^{\leq\mathbb{N}}$, let

$$\pi(\sigma) = 0^{\sigma(0)} {}^\frown 1 {}^\frown 0^{\sigma(1)} {}^\frown \cdots {}^\frown 0^{\sigma(|\sigma|-1)} \in 2^{\leq\mathbb{N}},$$

and

$$\pi^*(\sigma) = \pi(\sigma)^\frown 1^\frown 0^{\mathbb{N}} \in 2^{\mathbb{N}}.$$

Notice that, as k grows to infinity, the finite structures $C_{(1)}^{\pi(\sigma)^\frown k}$ form a nested increasing chain converging to $C_{(1)}^{\pi^*(\sigma)}$.

The core of the proof is the computable construction of $\{\sigma_s \in \mathbb{N}^{<\mathbb{N}} : s \in \mathbb{N}\}$. We will then define Y as the pointwise limit of $\pi(\sigma_s)$ as $s \to \infty$, i.e, $Y(i) = \lim_s \sigma_s(i)$ for all $i \in \mathbb{N}$. This sequence must satisfy the following properties:

(S1) For each s and $i < |\sigma_s|$, $\sigma_s(i) \geq \nabla_s(i)$ (see Definition 7.3.1).

(S2) There are infinitely many stages t satisfying that $(\forall s \geq t) \, \sigma_s \supseteq \sigma_t$. We call these the *true stages* of the sequence of σ_s's.

(S3) For every s, there is a τembedding $f_{s,s+1} : C^{\pi(\sigma_s)} \hookrightarrow C^{\pi(\sigma_{s+1})}$ that keeps $C^{\pi(\delta)}$ fixed, where δ is the largest common initial segment of σ_s and σ_{s+1}.

The first two conditions make the σ_s's behave similarly to the ∇_s's. If $t_0 < t_1 < t_2 < \cdots$ is the sequence of all the true stages, then $\sigma_{t_0} \subseteq \sigma_{t_1} \subseteq \sigma_{t_2} \subseteq \cdots$ is an increasing sequence. The union $Z = \bigcup_j \sigma_{t_j} \in \mathbb{N}^{\mathbb{N}}$ is the pointwise limit of the σ_s's, and hence it is Δ_2^0. By (S1), Z dominates ∇ and hence computes $0'$. The sequence $Y = \pi(Z) \in 2^{\mathbb{N}}$ is then Turing equivalent to $0'$.

(S3) allows us to build a computable copy of C^Y: For $s < r$, we can define embeddings $f_{s,r} : C^{\pi(\sigma_s)} \hookrightarrow C^{\pi(\sigma_r)}$ by composing the intermediate embeddings. Consider the direct limit of this sequence. By Exercise 1.1.13, the direct limit of a computable sequence of embeddings is computable. Note that these are $\tau_{|\cdot|}$-embeddings and not $\tau_{(1)_{|\cdot|}}$-embeddings, and therefore we end up with a computable τ-structure in the limit whose canonical structural jump might not be computable. If t is a true stage, then $f_{s,s+1}$ preserves $C^{\pi(\sigma_t)}$ for all $s \geq t$. Taking compositions, we can see that $C^{\pi(\sigma_t)}$ is also preserved by $f_{s,r}$ for all $r > s \geq t$. It follows that, along the true stages, the embeddings $f_{t_0,t_1}, f_{t_1,t_2}, \ldots$ are the inclusion embeddings. Thus the direct limit coincides with the limit of the increasing sequence $C^{\pi(\sigma_{t_0})} \subseteq C^{\pi(\sigma_{t_1})} \subseteq \cdots$, namely $C^{\pi(Z)}$. This shows that C^Y has a computable copy.

Before we move on to building the sequence of σ_s's, there is an issue we need to examine. Suppose that after defining $\sigma_0, \ldots, \sigma_s$, we find out that ∇_{s+1} has changed its value at some i, and we need σ_{s+1} to update the value of $\sigma_s(i)$. What we have to do is define σ_{s+1} to be $(\sigma_s \restriction i)^\frown k$ for some large enough k. But we need to do it in a way that $C^{\pi(\sigma_s)}$ embeds into $C^{\pi(\sigma_{s+1})}$ fixing $C^{\pi(\sigma_s \restriction i)}$. There is no reason why such an embedding would exist unless we take precautions ahead of time.

We say that γ is a *good extension* of σ if $\gamma \supseteq \sigma$ and there is an embedding from $C^{\pi(\gamma)}$ to $C^{\pi^*(\sigma)}$ fixing $C^{\pi(\sigma)}$. We call a string σ *good* if it is a good extension of $\sigma \restriction i$ for every $i < |\sigma|$. To resolve the issue mentioned above, we need to make sure we only work with good strings. Notice that being good is a Σ_1^0 property, so we can always wait for verification that a string is good before we use it. The next claim shows that there are plenty of good strings.

CLAIM 10.5.4. *For every $\sigma \in \mathbb{N}^{<\mathbb{N}}$, there is a $k_\sigma \in \mathbb{N}$ such that every $\gamma \supset \sigma$ with $\gamma(|\sigma|) \geq k_\sigma$ is a good extension of σ.*

Consider the $\forall$-type of the elements of $C^{\pi(\sigma)}$ within the structure $C^{\pi^*(\sigma)}$. That is, let $\bar{c}^\sigma$ be the tuple that consists of all the elements of $C^{\pi(\sigma)}$ and let

$$p_\sigma(\bar{x}) = \forall\text{-}tp_{C^{\pi^*(\sigma)}}(\bar{c}^\sigma).$$

Recall that the vocabulary of $C^{\pi(\sigma)}_{(1)}$ is $\tau_{(1)_s}$, the step-s approximation to $\tau_{(1)}$, where $s = |C^{\pi(\sigma)}|$. This finite vocabulary might not have a symbol for $\bigwedge p_\sigma$ yet. Let k_σ be large enough so that the relation symbol for $\bigwedge p_\sigma$ appears in the vocabulary of $C^{\pi(\sigma^\frown k_\sigma)}_{(1)}$. Since $C^{\pi(\sigma^\frown k_\sigma)}_{(1)} \subseteq C^{\pi^*(\sigma)}_{(1)}$, we have that

$$C^{\pi(\sigma^\frown k_\sigma)}_{(1)} \models \bigwedge p_\sigma(\bar{c}^\sigma).$$

Let us now show that k_σ is as wanted. Consider $\gamma \supset \sigma$ with $\gamma(|\sigma|) \geq k_\sigma$.

$$
\begin{array}{ccccccc}
& & & & C^{\pi(\gamma)} & \!\!\!\!C & \\
& & & \hookrightarrow & & & \searrow \\
C^{\pi(\sigma)}_{(1)} & \subseteq & C^{\pi(\sigma^\frown k_\sigma)}_{(1)} & & \subseteq & & C^{\pi^*(\sigma)}_{(1)}
\end{array}
$$

Since $C^{\pi(\gamma)}_{(1)}$ extends $C^{\pi(\sigma^\frown k_s)}_{(1)}$, we have that $C^{\pi(\gamma)} \models \bigwedge p_\sigma(\bar{b}^\sigma)$ too. In particular, the $\exists$-formula $\psi_{\sigma,\gamma}(\bar{x})$ that says that $\bar{c}^\sigma$ has an extension that looks like $C^{\pi(\gamma)}$, namely

$$\psi_{\sigma,\gamma}(\bar{x}) \equiv \exists \bar{y} \, (D(\bar{x},\bar{y}) = D(C^{\pi(\gamma)})),$$

(where $|\bar{x}| = |\bar{c}_\sigma|$) is obviously true in $C^{\pi(\gamma)}$. Hence its negation is not part of $p_\sigma(\bar{x})$. The formula $\psi_{\sigma,\gamma}(\bar{x})$ is then true of $\bar{c}^\sigma$ in $C^{\pi^*(\sigma)}$ too. This implies that $C^{\pi(\gamma)}$ embeds into $C^{\pi^*(\sigma)}$ preserving $\bar{c}^\sigma$, and finishes the proof of the claim.

We are now ready to define the sequence of σ_s's. Suppose we have defined $\sigma_0, \ldots, \sigma_s$ already. We want to find appropriate $i \leq |\sigma_s|$ and $k > \sigma_s(i)$ to define $\sigma_{s+1} = (\sigma_s \restriction i)^\frown k$. (We could have $i = |\sigma_s|$ and $\sigma_{s+1} = \sigma_s^\frown k$.) We say that a pair $\langle i, k \rangle$ is *appropriate* for σ_{s+1} if:

(A1) $k \geq \nabla_{s+1}(i)$ and $\sigma_s(j) \geq \nabla_{s+1}(j)$ for all $j < i$.

(A2) $\sigma_s \restriction i^\frown k$ is good.

(A3) There is an embedding $C^{\pi(\sigma_s)} \hookrightarrow C^{\pi(\sigma_s \restriction i^\frown k)}$ keeping $C^{\pi(\sigma_s \restriction i)}$ fixed.

(A4) If $i < |\sigma_s|$, then either $\sigma_s(i) < \nabla_{s+1}(i)$ or $\sigma_s(i) < k_{\sigma \restriction i}$.

As soon as we find such i and k for which we have verification that they are appropriate, we go ahead and define $\sigma_{s+1} = (\sigma_s \restriction i)^\frown k$. For most of the items above, it is clear what we mean by "having verification." The only item we should comment on is $\sigma_s(i) < k_{\sigma \restriction i}$: This means that the symbol for the $\forall$-type of $\bar{c}^{\sigma \restriction i}$ in $C^{\pi^*(\sigma \restriction i)}$ has not yet appeared in the vocabulary of $C_{(1)}^{\pi(\sigma \restriction i)}$ (recall that $\sigma \parallel i = \sigma \restriction i + 1$). A verification for this would be to find a symbol on $\tau_{(1)}$ for a $\forall$-type p true of $\bar{c}^{\sigma \restriction i}$ in $C^{\pi^*(\sigma \restriction i)}$, but which is not implied by any of the $\forall$-types q which are true of $\bar{c}^{\sigma \restriction i}$ and whose relation symbols $\bigwedge q$ appear in the vocabulary of $C_{(1)}^{\pi(\sigma \restriction i)}$. A verification that a symbol for a $\forall$-type does not imply another would be a finite structure in $\mathbb{K}_{(1)}^{fin}$ which has a tuple satisfying the former but not the later.

If we manage to define such a sequence, it clearly satisfies conditions (S1) and (S3). To see why (S2) holds, first observe that, for fixed i, $\sigma_s(i)$ is non-decreasing in s and grows at most once beyond $\nabla(i)$ or $k_{\sigma \restriction i}$. It thus eventually stabilizes. Let t be the last stage at which the value of $\sigma_t(i)$ changes. Then $|\sigma_t| = i + 1$. From then on, σ_t is an initial segment of all σ_s, and hence t is a true stage.

Finally, we need to show that appropriate i and k exist. Once we know they exist, we know we will find them. Let $i_0 \leq |\sigma_s|$ be the greatest i such that for all $j < i$, $\sigma_s(j) \geq \nabla_{s+1}(j)$. Notice that if $i_0 < |\sigma_s|$, then $\sigma_s(i_0) < \nabla_{s+1}(i_0)$. Let $i_1 \leq i_0$ be the greatest i such that for all $j < i$, there exists no verification that $\sigma_s(j) < k_{\sigma_s \restriction j}$. Notice that if $i_1 < i_0$, then such a verification exists for i_1. Thus i_1 satisfies (A4) one way or the other. Let $k \geq \nabla_{s+1}(i_1)$ be such that there is an embedding $C^{\pi(\sigma_s)} \hookrightarrow C^{\pi(\sigma_s \restriction i_1 ^\frown k)}$ keeping $C^{\pi(\sigma_s \restriction i_1)}$: We know such a k exists because σ_s is good, and hence it is a good extension of $\sigma_s \restriction i_1$. The pair $\langle i_1, k \rangle$ now satisfies (A1), (A3), and (A4). We need to show that $\gamma = \sigma_s \restriction i_1 ^\frown k$ is good. Suppose it is not, and that γ is not a good extension of $\gamma \restriction j$ for some $j < i_0$. That means that the $\forall$-formula $\neg \psi_{\gamma \restriction j, \gamma}(\bar{x})$ saying that $\bar{c}^{\gamma \restriction j}$ has no extension that looks like $C^{\pi(\gamma)}$ is true in $C^{\pi^*(\gamma \restriction j)}$, and hence belongs to the $\forall$-type $p_{\gamma \restriction j}(\bar{x})$. Since this formula is not true in $C^{\pi(\gamma)}$, it must be that the relation symbol for $\bigwedge p_{\gamma \restriction j}$ is not in the vocabulary of $C_{(1)}^{\pi(\gamma \parallel j)}$ yet. Remember we defined $k_{\gamma \restriction j}$ so that the symbol for $\bigwedge p_{\gamma \restriction j}$ would be in the vocabulary of $C_{(1)}^{\pi(\gamma \restriction j ^\frown k_{\gamma \restriction j})}$. Thus, we have found verification that $\gamma(j) < k_{\gamma \restriction j}$. Recall that we have chosen i_1 so that there are no verifications that $\gamma(j) < k_{\gamma \restriction j}$ for $j < i_1$. It follows that $\sigma_s \restriction i_1 ^\frown k$ is good, and (A2) holds. $\square$

COROLLARY 10.5.5. *Let $\mathbb{K}$ be an effectively Σ-small Π_2^c class of structures. If $\mathbb{K}_{(1)}$ is listable, $\mathbb{K}$ has the low property.*

PROOF. Let $\mathbb{S}$ be the class of structures that consist of infinitely many disjoint copies of structures in $\mathbb{K}$. That is, the vocabulary of $\mathbb{S}$ consists of the vocabulary of $\mathbb{K}$ plus a binary symbol E which defines an equivalence relation so that, on each equivalence class, we have a structure from $\mathbb{K}$.

Since $\mathbb{K}_{(1)}$ is listable, we can build an operator $\mathcal{S}_{(1)}^X$ that outputs a structure in $\mathbb{S}_{(1)}$ that contains copies of all X-computable structures in $\mathbb{K}_{(1)}$, each one repeated infinitely often. (We will see below that the canonical jumps of structures in $\mathbb{S}$ are essentially given by the canonical jumps of their components in $\mathbb{K}$.) This map is Turing-to-isomorphism invariant, that is, if $X \equiv_T Y$, then $\mathcal{S}^X \cong \mathcal{S}^Y$, as the lists of X-computable structures and Y-computable structures in $\mathbb{K}_{(1)}$ are the same. We want to apply the previous theorem and get that for every X, $\mathcal{S}^{X'}$ has an X computable copy. This would imply that $\mathbb{K}$ has the low property as follows: Consider a structure $\mathcal{A}$ that has a presentation that is low over X. Then X' computes a copy of $\mathcal{A}_{(1)}$, and hence $\mathcal{A}_{(1)}$ is one of the structures that appears within one of the equivalence classes in $\mathcal{S}_{(1)}^{X'}$. Since $\mathcal{S}^{X'}$ has an X-computable copy, so does $\mathcal{A}$.

To be able to apply the previous theorem we need to verify that $\mathbb{S}$ is effectively Σ-small. This requires us to modify the definition of $\mathbb{S}$ slightly. For a structure to be in $\mathbb{S}$ we impose the additional condition that for each ∃-theory that is sharply realized in $\mathbb{K}$, there are infinitely many equivalence classes which have that theory. Remember that an ∃-theory is an ∃-0-type, and since $\mathbb{K}$ is effectively Σ-small, we have a computable list of all ∃-theories. Since $\mathbb{K}_{(1)}$ is Π_2^c and $\mathbb{K}_{(1)}^{fin}$ computable, we can use Lemma 3.6.6 to build a list of computable structures in $\mathbb{K}_{(1)}$ with all possible ∃-theories. Adding these structures to the ones in $\mathcal{S}_{(1)}^X$, we get an operator that outputs a structure in the new $\mathbb{S}_{(1)}$. We claim that, with this modification, $\mathbb{S}$ is effectively Σ-small. The ∃-type of a tuple $\bar{x}$ is determined by the following information:

1. a partition of the variables $\bar{x}$ into E-equivalence classes $\bar{x}_1, \bar{x}_2, \ldots,$ $\bar{x}_k$ (i.e., the variables within each sub-tuple $\bar{x}_i$ are E-equivalent to each other and E-inequivalent to the rest);

2. the ∃-type of each tuple $\bar{x}_i$ within its equivalence class, which is one of the ∃-types of $\mathbb{K}$;

3. the ∃-theory of the rest of the structure, the part that consists of the equivalence classes that do not intersect $\bar{x}$. Since each ∃-theory repeats infinitely often, this ∃-theory is independent of the tuple $\bar{x}$, and is the same as the ∃-theory of all the structures in $\mathbb{S}$. It is not hard to see that this ∃-theory is computable.

It is then not hard to analyze these types and show that if $\mathbb{K}$ is effectively Σ-small, then so is $\mathbb{S}$. Such analysis would also yield that the canonical structural jump of a structure in $\mathbb{S}$ is determined by the canonical structural jumps of its components.   □

EXERCISE 10.5.6. Show that being listable is preserved by effective bi-interpretability of classes.

10.6. The copy-vs-diagonalize game

The copy-vs-diagonalize game provides a structural way of speaking about the listability property without having to refer to Turing operators or lists of X-computable structures. This game is the combinatorial core behind any proof that a class is listable or not. It was introduced in [Mon13b], where different variants of the game were analyzed. We only introduce the plain version and the ∞-version. The latter is the one that provides a notion equivalent to listability.

Fix a class $\mathbb{K}$ of infinite structures. Let us define the game $G(\mathbb{K})$. Two players, C and D, play alternatively. It does not matter who starts. On the sth move, D plays a finite $\tau_{|\cdot|}$-structure $\mathcal{D}[s] \in \mathbb{K}^{fin}$, and C plays a finite $\tau_{|\cdot|}$-structure $\mathcal{C}[s] \in \mathbb{K}^{fin}$. Structures must be nested, i.e., $\mathcal{D}[s] \subseteq \mathcal{D}[s+1]$ and $\mathcal{C}[s] \subseteq \mathcal{C}[s+1]$. If a player does not follow this rule, he or she loses. At the end of stages, we end up with two structures: $\mathcal{C} = \bigcup_s \mathcal{C}[s]$ and $\mathcal{D} = \bigcup_s \mathcal{D}[s]$.

Player D	$\mathcal{D}[0]$	$\subseteq$	$\mathcal{D}[1]$	$\subseteq$	$\mathcal{D}[2]$	$\cdots$	$\mathcal{D} = \bigcup_s \mathcal{D}[s]$
Player C	$\mathcal{C}[0]$	$\subseteq$	$\mathcal{C}[1]$	$\subseteq$	$\mathcal{C}[2]$	$\subseteq \cdots$	$\mathcal{C} = \bigcup_s \mathcal{C}[s]$

The winner of $G(\mathbb{K})$ is decided as follows:
1. If $\mathcal{C} \notin \mathbb{K}$, then D wins.
2. If $\mathcal{C} \in \mathbb{K}$, but $\mathcal{D} \notin \mathbb{K}$, then C wins.
3. If $\mathcal{D}, \mathcal{C} \in \mathbb{K}$ and $\mathcal{D} \cong \mathcal{C}$, then C wins.
4. If $\mathcal{D}, \mathcal{C} \in \mathbb{K}$ and $\mathcal{D} \not\cong \mathcal{C}$, then D wins.

DEFINITION 10.6.1. We say that $\mathbb{K}$ is *copyable* if C has a winning strategy in the game $G(\mathbb{K})$, and that $\mathbb{K}$ is *diagonalizable* if D has a winning strategy.[99]

Notice from the winning conditions that if neither player builds a structure in $\mathbb{K}$, then D wins. This seemingly minor point is actually what creates the tension between the players. It allows D to "pass" (i.e., play $\mathcal{D}[s+1] = \mathcal{D}[s]$) and wait for C to make a move she can take advantage of. On the contrary, C does not have the luxury of "passing," as if both C and D pass forever, they will end up with structures outside $\mathbb{K}$, and D will win. Then, for instance, it would be safe for D to pass whenever C passes. Thus, we may very well assume C is never allowed to pass, while D is. This forces C to build ahead of D, making his job of copying $\mathcal{D}$ harder.

As an example, let us show that the class of infinite linear orderings is diagonalizable. This proof is the combinatorial core of Jockusch and Soare's proof that there is a low linear ordering without a computable copy (cf. [JS91]). We will show later that the class of atom Boolean algebras

---

[99] Assuming enough determinacy, one of the two players must have a winning strategy. $\mathbb{K}$ is *effectively copyable* if C has a computable winning strategy.

with infinitely many atoms is effectively copyable and deduce that $\mathbb{BA}$ has the low property (cf. [DJ94]).

LEMMA 10.6.2 (Kach & Montalbán [Mon13b, Lemma 7.3]). *The class of infinite linear orderings is diagonalizable.*

PROOF. The strategy for player D is the following: Before we even start, we pledge that the structure $\mathcal{D}$ will be isomorphic to either ω or $m + \omega^*$ for some $m \in \mathbb{N}$. Every time C passes, we, namely player D, pass too. So we may assume C starts and then never passes. Pick an element c in $\mathcal{C}[0]$, which we fix for the rest of the construction. At each stage s, let $n[s]$ be the number of predecessors of c in $\mathcal{C}[s]$. Throughout the construction, we keep track of an auxiliary *restraint-function* $m[s]$; we will never add elements to $\mathcal{D}[s]$ below its $m[s]$'th element. Start by setting $m[s] = 0$. Throughout the game, we will make sure that $m[s] \leq n[s]$.

Here is what we do at stage $s + 1$:

(1) If C enumerates an element to the left of c, we pass, unless $n[s]$ becomes greater than $|\mathcal{D}[s]|$. In that case, we take this opportunity to move one step towards building $\mathcal{D} = \omega$: For this, we set $m[s] = |\mathcal{D}_s|$, ensuring that every element enumerated in the future is to the right of all the elements of $\mathcal{D}[s]$. Notice that if this happens infinitely often, we will end up with $\mathcal{D} = \omega$, while c will end up having infinitely many elements to its left in $\mathcal{C}$.

(2) If C enumerates an element to the right of c at stage $s + 1$, we define $\mathcal{D}[s + 1]$ by adding one more element to the right of the $m[s]$'th element of $\mathcal{D}[s]$. That is, we take one step towards enumerating a copy of $m[s] + \omega^*$. Note that if this happens from some point on without ever changing the value of $m[s]$, then c in $\mathcal{C}$ will have infinitely many elements to its right and $n[s]$ many to its left, while no element in $m[s] + \omega^*$ will have this property.

If C is actually building an infinite linear ordering, then either (2) will occur infinitely often or (1) will occur from some point on without $m[s]$ changing. As we already argued, in either case, $\mathcal{D} \not\cong \mathcal{C}$.   □

EXERCISE 10.6.3. Show that linear orderings with no maximal elements are copyable.

EXERCISE 10.6.4. Show that the properties of being copyable or diagonalizable are preserved by effective bi-interpretability of classes.

The notion of listability, which we already know is connected to the low property, is connected to a modification of this game. Fix a class of structures $\mathbb{K}$. The new game is called $G^\infty(\mathbb{K})$. Two players, C and D, play alternatively, and it does not matter who starts. Here comes the new feature: On the sth move, D plays a finite $\tau_{|\cdot|}$-structure $\mathcal{D}[s] \in \mathbb{K}^{fin}$, and C plays $s + 1$ many finite $\tau_{|\cdot|}$-structures $\mathcal{C}^j[s - j] \in \mathbb{K}^{fin}$ for $j = 0, \ldots, s$. Structures must be nested, i.e., $\mathcal{D}[s] \subseteq \mathcal{D}[s + 1]$, and $\mathcal{C}^j[s] \subseteq \mathcal{C}^j[s + 1]$. If a player does not follow this rule, he or she loses. At the end of stages, we end up with structures $C^j = \bigcup_s C^j[s]$ and $\mathcal{D} = \bigcup_s \mathcal{D}[s]$.

Player D	$\mathcal{D}[0]$	$\subseteq$	$\mathcal{D}[1]$	$\subseteq$	$\mathcal{D}[2]$	$\cdots$	$\mathcal{D} = \bigcup_s \mathcal{D}[s]$
	$C^0[0]$	$\subseteq$	$C^0[1]$	$\subseteq$	$C^0[2]$	$\subseteq \cdots$	$C^0 = \bigcup_s C^0[s]$
Player C			$C^1[0]$	$\subseteq$	$C^1[1]$	$\subseteq \cdots$	$C^1 = \bigcup_s C^1[s]$
					$C^2[0]$	$\subseteq \cdots$	$C^2 = \bigcup_s C^1[s]$
						$\ddots \quad \vdots$	$\vdots$

The winner of $\mathsf{G}^\infty(\mathbb{K})$ is decided as follows:

1. If for some j, $C^j \notin \mathbb{K}$, then D wins.
2. If for all j, $C^j \in \mathbb{K}$, but $\mathcal{D} \notin \mathbb{K}$, then C wins.
3. If $\mathcal{D}, C^0, C^1, \cdots \in \mathbb{K}$ and, for some j, $\mathcal{D} \cong C^j$, then C wins.
4. If $\mathcal{D}, C^0, C^1, \cdots \in \mathbb{K}$ and, for all j, $\mathcal{D} \not\cong C^j$, then D wins.

DEFINITION 10.6.5. We say that $\mathbb{K}$ is ∞-*copyable* if C has a winning strategy in the game $\mathsf{G}^\infty(\mathbb{K})$, and that $\mathbb{K}$ is ∞-*diagonalizable* if D has a winning strategy. $\mathbb{K}$ is *effectively* ∞-*copyable* if C has a computable winning strategy in the game $\mathsf{G}^\infty(\mathbb{K})$.

If a class is copyable, it is clearly ∞-copyable. The reverse direction does not hold. For instance, the class of infinite linear orderings is ∞-copyable (see exercise below), though diagonalizable. However, the few proofs of ∞-copyability we know usually proceed by splitting the class $\mathbb{K}$ (maybe with possible added constants) into countably many copyable classes $\mathbb{K}_n$, $n \in \mathbb{N}$, and building C_n by following the strategy for $\mathsf{G}(\mathbb{K}_n)$.

EXERCISE 10.6.6 (Montalbán [Mon13b, Lemma 7.2]). Show that infinite linear orderings are ∞-copyable by, first adding two constant symbols, a and b, and then partitioning the class of infinite linear orderings according to the following properties, and treating each case separately:

- no maximal elements
- no minimal elements
- a is a limit from the right
- b is a limit from the left
- the interval between a and b looks like $\omega + \omega^*$
- the whole linear ordering looks like $\omega + \mathbb{Z} \cdot \mathbb{Q} + \omega^*$

The connection with the low property comes from the following theorem. The connection is actually quite straightforward, and it seems that one is not adding much by considering ∞-games instead of just working with listability. However, the proofs of listability can usually be understood as a copy-vs-diagonalize game. We feel that visualizing these proofs as game proofs helps to see what is really going on.

THEOREM 10.6.7. *A class $\mathbb{K}$ of infinite structures is listable if and only if it is effectively ∞-copyable.*

PROOF. Suppose first that $\mathbb{K}$ is ∞-copyable. For each $e \in \mathbb{N}$, we build an infinite sequence of structures $\{C_e^j : j \in \omega\}$ in $\mathbb{K}$ such that if Φ_e^Y is the diagram of a structure in $\mathbb{K}$, then one of the C_e^j's is isomorphic to it. To do this, we let D play the diagram of Φ_e^Y so long at it converges, and we let D pass while we wait for convergence. Independently of whether Φ_e^Y is total or in $\mathbb{K}$, player C is forced to play a structure in $\mathbb{K}$. The answer by player C given from the effective strategy gives us the desired list. Putting together all these lists over all $e \in \mathbb{N}$, we get a list of structures in $\mathbb{K}$ that includes all the structures which have diagrams of the form Φ_e^Y for some e.

Suppose now that $\mathbb{K}$ is listable; we need to define a strategy for C. Let X be the sequence of indices of the finite structures played by D. We let C play the X-computable list of all X-computable structures in $\mathbb{K}$ in response. Since $\mathcal{D}$ is computable in X, it will be isomorphic to one of the structures played by C.   □

Jockusch and Soare's proof that linear orderings do not have the low property is, in essence, a proof that adjacency linear orderings are ∞-diagonalizable. About that proof, let us just say that it requires an adaptation of the proof of Lemma 10.6.2 to show that adjacency linear orderings are $0''$-diagonalizable (cf. [Mon13b, Lemma 7.4]), and then uses $0''$-separators to split an ordering into infinitely many disjoint intervals.

DEFINITION 10.6.8. $\mathbb{K}$ has the *low property on a cone* if there is $Y \in 2^{\mathbb{N}}$ such that, for every $X \geq_T Y$, if some structure from $\mathbb{K}$ has a copy that is low over X, it also has an X-computable copy.

THEOREM 10.6.9 (Montalbán [Mon13b]). *Let $\mathbb{K}$ be a Π_2^{in} class of structures. The following are equivalent:*

1. $\mathbb{K}$ *has the low property on a cone.*
2. $\mathbb{K}$ *is Σ-small and $\mathbb{K}'$ is listable on a cone.*
3. $\mathbb{K}$ *is Σ-small and $\mathbb{K}'$ is ∞-copyable.*

10.6.1. Low Boolean algebras. In this section, we prove Downey and Jockusch's result that every low Boolean algebra has a computable copy. The original proof uses interval algebras and is very hands on. We give a different presentation using the machinery we just developed. This will allow us to recognize the different parts of the proof and isolate its combinatorial core, namely the proof that $\mathbb{ABA}$ is effectively ∞-copyable. Recall that $\mathbb{ABA}$ is the class of Boolean algebras with an added relation that recognizes atoms, which we call *atom Boolean algebras*. From Exercise 10.1.9, we get that $\mathbb{BA}_{(1)}$ is effectively bi-interpretable with $\mathbb{ABA}$. By Exercise 10.6.4, we get that it is enough to show that $\mathbb{ABA}$ is effectively ∞-copyable.

The following technical lemma will be quite useful.

LEMMA 10.6.10 (Remmel [Rem81b]). *Suppose $\mathcal{A} \subseteq \mathcal{B}$ are Boolean algebras with infinitely many atoms satisfying that*

- *every atom of $\mathcal{A}$ is a finite sum of atoms of $\mathcal{B}$;*
- *$\mathcal{B}$ is generated by $\mathcal{A}$ together with the atoms of $\mathcal{B}$ that are below the atoms of $\mathcal{A}$.[100]*

Then $\mathcal{A}$ and $\mathcal{B}$ are isomorphic.

PROOF. Let I be the set of finite partial embeddings p from a finite sub-algebra of $\mathcal{A}$ to $\mathcal{B}$ that satisfy that, for every $a \in \mathrm{dom}(p)$,

- a and $p(a)$ are finitely apart in $\mathcal{B}$ (i.e., $a \triangle p(a)$ is a finite sum of atoms in $\mathcal{B}$), and
- the number of atoms below a in $\mathcal{A}$ is the same as the number of atoms below $p(a)$ in $\mathcal{B}$ (this number could be infinite).

Observe that the trivial partial embedding that maps 0_A to 0_B and 1_A to 1_B belongs to I because of our assumption that both $\mathcal{A}$ and $\mathcal{B}$ have infinitely many atoms. We claim that I has the back-and-forth property.[101] Consider $p \in I$ with domain A_0 and $c \in A$. We want to build $q \in I$ which extends p and whose domain includes c. Let $A_0[c]$ be the finite sub-Boolean algebra of $\mathcal{A}$ generated by A_0 and c. We want q to have domain $A_0[c]$. If $a_1, \ldots, a_k$ are the minimal non-zero elements of A_0, then the minimal elements of $A_0[c]$ are $a_1 \wedge c, a_1 \wedge c^c, \ldots, a_k \wedge c, a_k \wedge c^c$. Let us divide the proof into k steps and work first below a_1, then below a_2, etc. For the first step, let us just assume that c is below the minimal element a_1 of A_0. The other steps work the same way. To define an extension q of p with domain $A_0[c]$, all we need to do is to find an appropriate element $d \in B$ to define $d = p(c)$. Such a d needs to satisfy that $d < p(a_1)$, that $c \triangle d$ is a finite sum of atoms, that the number of atoms below c and d are the same, and that the number of atoms below $a_i \wedge c^c$ and $p(a_1) \wedge d^c$ are also the same.

The first candidate for d is $p(a_i) \wedge c$, which is finitely apart from c since a_i is finitely apart from $p(a_i)$. However, $p(a_i) \wedge c$ and c might have a different number of atoms below. Since a_i and $p(a_i)$ have the same number of atoms below, by adding or removing a few atoms to $p(a_i) \wedge c$, we can find a $d < p(a_i)$ which has the same number of atoms below as c, and so that the number of atoms below $a_i \wedge c^c$ and $p(a_i) \wedge d^c$ are also the same.

---

[100]We say that a subset X of a Boolean algebra $\mathcal{B}$ *generates* $\mathcal{B}$ if every element of B can be written as a Boolean combination of elements of X.

[101]This is not exactly the same back-and-forth property from Definition 3.3.1, but almost. Of course, one can think of a finite partial embedding p as the pair of tuples $\langle \bar{a}, \bar{b} \rangle \in A^{<\mathbb{N}} \times B^{<\mathbb{N}}$, where $\bar{a}$ is the domain of p and $\bar{b} = p(\bar{a})$. That they satisfy the same atomic formulas follows from p being an embedding. The versions of the back-and-forth properties we need are as follows: For every $p \in I$ and $c \in A$, there exists a q extending p whose domain includes c; and for every $p \in I$ and $d \in B$, there exists a q extending p whose image includes d. That every p in a back-and-forth set extends to an isomorphism is proved exactly as in Lemma 3.3.3.

The proof of the back condition of the back-and-forth property is the same. We conclude that I has the back-and-forth property and that $\mathcal{A}$ and $\mathcal{B}$ are isomorphic.   □

Before going into the proof that atom Boolean algebras are ∞-copyable, we need to look more closely at how we approximate Boolean algebras. In Definition 1.1.6, we defined finite approximations for structures only for relational vocabularies. We could make the vocabulary of Boolean algebras relational, for instance by considering only $\leq$, but we would then have to deal with partial Boolean algebras, which causes unnecessary complications. However, since Boolean algebras are locally finite,[102] we can assume every step in a finite approximation to a Boolean algebra is a finite Boolean algebra. That is, $\mathcal{B}$ has a computable copy if and only if there is a computable sequence of finite Boolean algebras $\mathcal{B}_0 \subseteq \mathcal{B}_1 \subseteq \cdots$ whose union is $\mathcal{B}$. The same is true for atom Boolean algebras. In the case of atom Boolean algebras, the finite approximations are not atom Boolean algebras themselves: There might be *minimal elements* (i.e., non-zero elements with only zero below them) which are not labeled atoms because they will not remain minimal in the later steps of the approximation. Even though minimal elements and atoms are the same thing, when we are working with $\mathbb{ABA}^{fin}$, we use the word *atom* to refer to the elements that are labeled as atoms, and use the word *minimal* for the ones that are minimal in the given finite approximation, even if they do not stay minimal later.

A useful way to visualize this dynamic approximation process is via trees. Suppose we are given a finite approximation $\mathcal{B}_1 \subseteq \mathcal{B}_2 \subseteq \cdots$ to a Boolean algebra $\mathcal{B}$. Assume that each $\mathcal{B}_{s+1}$ is generated by $\mathcal{B}_s$ and at most one extra element c_{s+1}. We define a finite approximation $T_0 \subseteq T_1 \subseteq \cdots$ to a binary tree T so that, for every $s \in \mathbb{N}$, the leaves of T_s are in one-to-one correspondence with the minimal elements of $\mathcal{B}_s$. Think of these trees as growing downwards. Let T_0 be the tree that contains just a root node. Suppose we have defined T_s. In $\mathcal{B}_{s+1}$, some of the minimal elements from $\mathcal{B}_s$ get split in two, and some stay minimal: That is, if a is a minimal element of $\mathcal{B}_s$, it splits into $a \wedge c_{s+1}$ and $a \wedge c^c_{s+1}$ unless one of these is 0, in which case a stays minimal in $\mathcal{B}_{s+1}$. For the minimal elements that split in two, we add in T_{s+1} two extensions to the corresponding leaf in T_s. For the ones that stay minimal, we leave the corresponding leaf in T_s a leaf in T_{s+1}. Notice that the new elements of T_{s+1} always come in pairs as children of some leaf of T_s.

When we approximate atom Boolean algebras, we also use an atom relation symbol on the trees which does not change throughout the T_s's. Notice that only leaves can be labeled atom. When we have a leaf with an atom label, that leaf will remain a leaf in all subsequent trees. If a leaf does

---

[102]Finitely generated Boolean algebras are finite: n elements generate a Boolean algebra of size at most 2^{2^n}.

not have an atom label, it must eventually split. By slightly modifying the trees, we may assume that if a leaf of T_s is not labeled atom, it must always split into two leaves in T_{s+1}. We may thus assume that a leaf of a tree T_s is labeled atom if and only it does not split in T_{s+1}.

THEOREM 10.6.11 (Downey and Jockusch [DJ94]). *Atom Boolean algebras are effectively ∞-copyable.*

It follows from Theorem 10.6.7 and Corollary 10.5.5 that Boolean algebras have the low property.

PROOF. We need to describe an effective strategy for C in the game $G^\infty(\mathbb{ABA})$. For each $n \in \mathbb{N}$, there is a unique Boolean algebra with exactly n atoms, denoted by $Int(n + 1 + \mathbb{Q})$,[103] which can be built computably uniformly in n. For $n > 0$, player C builds C_n to be that algebra. We only need to concentrate on building C_0, trying to copy $\mathcal{D}$ under the assumption that $\mathcal{D}$ has infinitely many atoms. Thus it is enough to show that the class of atom Boolean algebras with infinitely many atoms is copyable. We describe a strategy for C in that game.

As we mentioned above, we can assume the players play *atom finite trees* $T_0^C \subseteq T_1^C \subseteq T_2^C \subseteq \cdots$ and $T_0^D \subseteq T_1^D \subseteq T_2^D \subseteq \cdots$ as in the paragraph before the theorem. These trees satisfy that the minimal elements of $C[s]$ are in one-to-one correspondence with the leaves of T_s^C and some of those leaves are labeled atom. If a leaf is not labeled atom, it must split in T_{s+1}^C. The same happens with T_s^D.

We are now ready to define player C's strategy when responding to a sequence $T_0^D \subseteq T_1^D \subseteq T_2^D \subseteq \cdots$ played by D. At each stage s of the game, C will build a tree T_s^C and an embedding $f_s \colon T_s^D \to T_s^C$. This is a tree-as-a-graph embedding; that is, the root is mapped to the root, and the two children of a node t are mapped to the children of $f_s(t)$. Furthermore, we require that if a leaf ℓ in T_s^D is labeled atom, then all the leaves below $f_s(\ell)$ are also labeled atoms, and if ℓ is not labeled atom, then none of the leaves below $f_s(\ell)$ are labeled atom. (This way we get that the image of an atom ℓ in $\mathcal{D}[s]$ is mapped to a finite sum of atoms in $C[s]$.) Building such a sequence of trees is rather straightforward: Whether D passes (i.e., $T_{s+1}^D = T_s^D$) or not, we, namely player C, always extend the non-atom leaves of T_s^C by adding two non-atom children. If D does extend his tree, we need to extend f_s to the new domain. If D adds leaves ℓ_1 and ℓ_2 below a leaf t of T_s^D, we know $f_s(t)$ was not labeled atom, and hence it splits in T_{s+1}^C into two nodes which are not labeled atom either. If ℓ_1 or ℓ_2 are not labeled atom, we can just map them to children of $f_s(t)$ in T_{s+1}^C. If ℓ_1 is labeled atom, let $f_{s+1}(\ell_s)$ be one of those two children, add new leaves in

---

[103] $Int(\mathcal{L})$ denotes the *interval Boolean algebra* of $\mathcal{L}$ whose elements are the subsets of $\mathcal{L}$ which are finite unions of left-closed, right-open intervals. An easy back-and-forth argument shows that the interval algebra of $1 + \mathbb{Q}$ is the unique countable Boolean algebra without atoms.

T^C_{s+1} below all the leaves in T^C_s below $f_{s+1}(\ell)$ and label them atoms. Same for ℓ_2.

At the end of the game, we end up with an embedding $f = \bigcup_s f_s$ from $T^D = \bigcup_s T^D_s$ to $T^C = \bigcup_s T^D_s$. This can be viewed as an embedding between the corresponding Boolean algebras $\mathcal{C}$ and $\mathcal{D}$. By construction, it is easy to see that f maps atoms to finite sums of atoms. To see that it is almost onto, one needs to prove that every element c of T^C either is eventually in the image of f_s, or there is a stage where a predecessor of c is assigned to an atom of T^D and all successors of c that are leaves are labeled atom. We then get that every element in T^C that is not in the image of f is a finite sum of atoms that are below the image of an atom from T^D. We thus get that the Boolean algebras $f(\mathcal{C}) \subseteq \mathcal{D}$ satisfy the hypothesis of Remmel's Lemma 10.6.10, and thus $\mathcal{C}$ and $\mathcal{D}$ are isomorphic. □

The proofs of the low$_2$, low$_3$, and low$_4$ Boolean algebra theorems by Thurber [Thu95], and Knight and Stob [KS00] essentially give an understanding of $\mathbb{BA}_{(n)}$ for $n = 2, 3, 4$ even if the concept was not defined yet. For instance, for $n = 2$, Thurber worked with $\mathbb{BA}$ with added relations for atom, atomless (i.e., no atoms below), and infinite (infinitely many elements below). It turned out that with these three extra relations, we get a class that is effectively bi-interpretable with $\mathbb{BA}_{(2)}$ (cf. [HM12]). For $n = 3$, Knight and Stob added relations for atomic (not atomless elements below), 1-atomic (infinite, but without splitting into two infinite elements), and atominf (infinitely many atoms below). It turned out that with these three extra relations, we get a class that is effectively bi-interpretable with $\mathbb{BA}_{(3)}$ (cf. [HM12]). They then added five more relations for $\mathbb{BA}_{(4)}$. The copy-vs-diagonalize game was not known by then, but Thurber's and Knight and Stob's proofs become clearer when seen in terms of games. Harris and Montalbán [HM12] gave an in-depth analysis of $\mathbb{BA}_{(n)}$ and $\mathbb{BA}^{fin}_{(n)}$ for all $n \in \mathbb{N}$.

Whether $\mathbb{BA}_{(5)}$, of $\mathbb{BA}_{(n)}$ for $n > 5$, is ∞-copyable remains open.

BIBLIOGRAPHY

[ACK⁺16] URI ANDREWS, MINGZHONG CAI, ISKANDER SH. KALIMULLIN, STEFFEN LEMPP, JOSEPH S. MILLER, and ANTONIO MONTALBÁN, *The complements of lower cones of degrees and the degree spectra of structures*, **The Journal of Symbolic Logic**, vol. 81 (2016), no. 3, pp. 997–1006.

[AGK⁺] URI ANDREWS, HRISTO GANCHEV, RUITGER KUYPER, STEFFEN LEMPP, JOSEPH S. MILLER, ALEXANDRA A. SOSKOVA, and MARIYA I. SOSKOVA, *On cototality and the skip operator in the enumeration degrees*, submitted for publication.

[AM15] URI ANDREWS and JOSEPH S. MILLER, *Spectra of theories and structures*, **Proceedings of the American Mathematical Society**, vol. 143 (2015), no. 3, pp. 1283–1298.

[AKMS89] CHRIS ASH, JULIA KNIGHT, MARK MANASSE, and THEODORE SLAMAN, *Generic copies of countable structures*, **Annals of Pure and Applied Logic**, vol. 42 (1989), no. 3, pp. 195–205.

[AK00] CHRIS J. ASH and JULIA KNIGHT, **Computable Structures and the Hyperarithmetical Hierarchy**, Elsevier Science, 2000.

[Bad77] SERIKZHAN A. BADAEV, *Computable enumerations of families of general recursive functions*, **Algebra i Logika**, vol. 16 (1977), no. 2, pp. 129–148, 249.

[Bal06] VESSELA BALEVA, *The jump operation for structure degrees*, **Archive for Mathematical Logic**, vol. 45 (2006), no. 3, pp. 249–265.

[Bar69] JON BARWISE, *Infinitary logic and admissible sets*, **The Journal of Symbolic Logic**, vol. 34 (1969), no. 2, pp. 226–252.

[Bar75] ———, **Admissible Sets and Structures**, Springer-Verlag, Berlin, 1975, An approach to definability theory, Perspectives in Mathematical Logic.

[BT79] V. JA. BELJAEV and MIKHAIL ABRAMOVICH TAĬCLIN, *Elementary properties of existentially closed systems*, **Uspekhi Mat. Nauk**, vol. 34 (1979), no. 2(206), pp. 39–94.

[CCKM04] WESLEY CALVERT, DESMOND CUMMINS, JULIA F. KNIGHT, and SARA MILLER, *Comparison of classes of finite structures*, **Algebra Logika**, vol. 43 (2004), no. 6, pp. 666–701, 759.

[CHS07] WESLEY CALVERT, VALENTINA HARIZANOV, and ALEXANDRA SHLAPENTOKH, *Turing degrees of isomorphism types of algebraic objects*, **Journal of the London Mathematical Society (2)**, vol. 75 (2007), no. 2, pp. 273–286.

[Chi90] JOHN CHISHOLM, *Effective model theory vs. recursive model theory*, **The Journal of Symbolic Logic**, vol. 55 (1990), no. 3, pp. 1168–1191.

[Coo04] S. BARRY COOPER, **Computability Theory**, Chapman & Hall/CRC, Boca Raton, FL, 2004.

[Cut80] NIGEL CUTLAND, **Computability**, Cambridge University Press, Cambridge-New York, 1980.

[DHK⁺07] RODNEY G. DOWNEY, DENIS R. HIRSCHFELDT, ASHER M. KACH, STEFFEN LEMPP, JOSEPH R. MILETI, and ANTONIO MONTALBÁN, *Subspaces of computable vector spaces*, **Journal of Algebra**, vol. 314 (2007), no. 2, pp. 888–894.

[DHK03] RODNEY G. DOWNEY, DENIS R. HIRSCHFELDT, and BAKHADYR KHOUSSAINOV, *Uniformity in the theory of computable structures*, **Algebra Logika**, vol. 42 (2003), no. 5, pp. 566–593, 637.

[DJ94] RODNEY G. DOWNEY and CARL G. JOCKUSCH, *Every low Boolean algebra is isomorphic to a recursive one*, **Proceedings of the American Mathematical Society**, vol. 122 (1994), no. 3, pp. 871–880.

[DKL⁺15] RODNEY G. DOWNEY, ASHER M. KACH, STEFFEN LEMPP, ANDREW E. M. LEWIS-PYE, ANTONIO MONTALBÁN, and DANIEL D. TURETSKY, *The complexity of computable categoricity*, **Advances in Mathematics**, vol. 268 (2015), pp. 423–466.

[DKLT13] RODNEY G. DOWNEY, ASHER M. KACH, STEFFEN LEMPP, and DANIEL D. TURETSKY, *Computable categoricity versus relative computable categoricity*, **Fundamenta Mathematicae**, vol. 221 (2013), no. 2, pp. 129–159.

[DK92] RODNEY G. DOWNEY and JULIA F. KNIGHT, *Orderings with αth jump degree $0^{(\alpha)}$*, **Proceedings of the American Mathematical Society**, vol. 114 (1992), no. 2, pp. 545–552.

[End11] HERBERT B. ENDERTON, **Computability Theory**, Elsevier, Academic Press, Amsterdam, 2011.

[Ers77] YURI L. ERSHOV, *Theorie der Numerierungen. III*, **Z. Math. Logik Grundlagen Math.**, vol. 23 (1977), no. 4, pp. 289–371, Translated from the Russian and edited by G. Asser and H.-D. Hecker.

[Ers96] ———, **Definability and Computability**, Siberian School of Algebra and Logic, Consultants Bureau, New York, 1996.

[FK15] MARAT FAIZRAHMANOV and ISKANDER KALIMULLIN, *Limitwise monotonic sets of reals*, submitted for publication.

[Fel76] STEPHEN MARTIN FELLNER, **Recursiveness and Finite Axiomatizability of Linear Orderings**, Ph.D. thesis, Rutgers, The State University of New Jersey, New Brunswick, 1976.

[FF09] EKATERINA B. FOKINA and SY-DAVID FRIEDMAN, *Equivalence relations on classes of computable structures*, **Mathematical Theory and Computational Practice**, Lecture Notes in Computer Science, vol. 5635, Springer, Berlin, 2009, pp. 198–207.

[FFH⁺12] EKATERINA B. FOKINA, SY-DAVID FRIEDMAN, VALENTINA HARIZANOV, JULIA F. KNIGHT, CHARLES MCCOY, and ANTONIO MONTALBÁN, *Isomorphism relations on computable structures*, **The Journal of Symbolic Logic**, vol. 77 (2012), no. 1, pp. 122–132.

[Fri57a] RICHARD M. FRIEDBERG, *A criterion for completeness of degrees of unsolvability*, **The Journal of Symbolic Logic**, vol. 22 (1957), pp. 159–160.

[Fri57b] ———, *Two recursively enumerable sets of incomparable degrees of unsolvability* (*solution of Post's problem*, 1944), **Proc. Nat. Acad. Sci. U.S.A.**, vol. 43 (1957), pp. 236–238.

[FS89] HARVEY FRIEDMAN and LEE STANLEY, *A Borel reducibility theory for classes of countable structures*, **The Journal of Symbolic Logic**, vol. 54 (1989), no. 3, pp. 894–914.

[FSS83] HARVEY M. FRIEDMAN, STEPHEN G. SIMPSON, and RICK L. SMITH, *Countable algebra and set existence axioms*, **Annals of Pure and Applied Logic**, vol. 25 (1983), no. 2, pp. 141–181.

[FKM09] ANDREY FROLOV, ISKANDER KALIMULLIN, and RUSSELL MILLER, *Spectra of algebraic fields and subfields*, **Mathematical Theory and Computational Practice**, Lecture Notes in Comput. Sci., vol. 5635, Springer, Berlin, 2009, pp. 232–241.

[Gon75a] SERGEY S. GONCHAROV, *Selfstability, and computable families of constructivizations*, **Algebra i Logika**, vol. 14 (1975), no. 6, pp. 647–680, 727.

[Gon75b] ———, *Some properties of the constructivization of boolean algebras*, **Sibirskii Matematicheskii Zhurnal**, vol. 16 (1975), no. 2, pp. 264–278.

[Gon77] ———, *The number of nonautoequivalent constructivizations*, **Algebra i Logika**, vol. 16 (1977), no. 3, pp. 257–282, 377.

[Gon80] ———, *Autostability of models and abelian groups*, **Algebra i Logika**, vol. 19 (1980), no. 1, pp. 23–44, 132.

[DG80] SERGEY S. GONCHAROV and V. D. DZGOEV, *Autostability of models*, **Algebra i Logika**, vol. 19 (1980), no. 1, pp. 45–58, 132.

[GHK⁺05] SERGEY S. GONCHAROV, VALENTINA HARIZANOV, JULIA KNIGHT, CHARLES MCCOY, RUSSELL MILLER, and REED SOLOMON, *Enumerations in computable structure theory*, **Annals of Pure and Applied Logic**, vol. 136 (2005), no. 3, pp. 219–246.

[GLS03] SERGEY S. GONCHAROV, STEFFEN LEMPP, and REED SOLOMON, *The computable dimension of ordered abelian groups*, **Advances in Mathematics**, vol. 175 (2003), no. 1, pp. 102–143.

[GK02] SERGEY S. GONCHAROV and DZH. NAĬT, *Computable structure and antistructure theorems*, **Algebra Logika**, vol. 41 (2002), no. 6, pp. 639–681, 757.

[Gor70] CARL E. GORDON, *Comparisons between some generalizations of recursion theory*, **Compositio Math.**, vol. 22 (1970), pp. 333–346.

[Har78] LEO HARRINGTON, *Analytic determinacy and $0^\sharp$*, **The Journal of Symbolic Logic**, vol. 43 (1978), no. 4, pp. 685–693.

[HM12] KENNETH HARRIS and ANTONIO MONTALBÁN, *On the n-back-and-forth types of Boolean algebras*, **Transactions of the American Mathematical Society**, vol. 364 (2012), no. 2, pp. 827–866.

[HM14] ———, *Boolean algebra approximations*, **Transactions of the American Mathematical Society**, vol. 366 (2014), no. 10, pp. 5223–5256.

[HTMMM] MATTHEW HARRISON-TRAINOR, ALEXANDER MELNIKOV, RUSSELL MILLER, and ANTONIO MONTALBÁN, *Computable functors and effective interpretability*, submitted for publication.

[HTMM] MATTHEW HARRISON-TRAINOR, RUSSELL MILLER, and ANTONIO MONTALBÁN, *Generic functors and infinitary interpretations*, in preparation.

[HTM] MATTHEW HARRISON-TRAINOR and ANTONIO MONTALBÁN, *The tree of tuples of a structure*, submitted for publication.

[HLZ99] BERNHARD HERWIG, STEFFEN LEMPP, and MARTIN ZIEGLER, *Constructive models of uncountably categorical theories*, **Proceedings of the American Mathematical Society**, vol. 127 (1999), no. 12, pp. 3711–3719.

[Hig52] GRAHAM HIGMAN, *Ordering by divisibility in abstract algebras*, **Proceedings of the London Mathematical Society (3)**, vol. 2 (1952), pp. 326–336.

[Hir06] DENIS R. HIRSCHFELDT, *Computable trees, prime models, and relative decidability*, **Proceedings of the American Mathematical Society**, vol. 134 (2006), no. 5, pp. 1495–1498.

[HKSS02] DENIS R. HIRSCHFELDT, BAKHADYR KHOUSSAINOV, RICHARD A. SHORE, and ARKADII M. SLINKO, *Degree spectra and computable dimensions in algebraic structures*, **Annals of Pure and Applied Logic**, vol. 115 (2002), no. 1-3, pp. 71–113.

[HS07] DENIS R. HIRSCHFELDT and RICHARD A. SHORE, *Combinatorial principles weaker than Ramsey's theorem for pairs*, **The Journal of Symbolic Logic**, vol. 72 (2007), no. 1, pp. 171–206.

[Joc68] CARL G. JOCKUSCH, JR., *Semirecursive sets and positive reducibility*, **Transactions of the American Mathematical Society**, vol. 131 (1968), pp. 420–436.

[Joc80] ———, *Degrees of generic sets*, **Recursion Theory: Its Generalisation and Applications (Proc. Logic Colloq., Univ. Leeds, Leeds, 1979)**, London Mathematical Society Lecture Note Series, vol. 45, Cambridge University Press, Cambridge – New York, 1980, pp. 110–139.

[JS91] CARL G. JOCKUSCH, JR. and ROBERT I. SOARE, *Degrees of orderings not isomorphic to recursive linear orderings*, **Annals of Pure and Applied Logic**, vol. 52 (1991), no. 1-2, pp. 39–64, International Symposium on Mathematical Logic and its Applications (Nagoya, 1988).

[KM] ASHER KACH and ANTONIO MONTALBÁN, *Linear orders with finitely many descending cuts*, in preparation.

[Kal08] ISKANDER SH. KALIMULLIN, *Almost computably enumerable families of sets*, **Mat. Sb.**, vol. 199 (2008), no. 10, pp. 33–40.

[Kal09] ———, *Uniform reducibility of representability problems for algebraic structures.*, **Sibirskii Matematicheskii Zhurnal**, vol. 50 (2009), no. 2, pp. 334–343.

[Khi04] A. N. KHISAMIEV, *On the Ershov upper semilattice L_E*, **Sibirsk. Mat. Zh.**, vol. 45 (2004), no. 1, pp. 211–228.

[KSS07] BAKHADYR KHOUSSAINOV, PAVEL SEMUKHIN, and FRANK STEPHAN, *Applications of Kolmogorov complexity to computable model theory*, **The Journal of Symbolic Logic**, vol. 72 (2007), no. 3, pp. 1041–1054.

[KS98] BAKHADYR KHOUSSAINOV and RICHARD A. SHORE, *Computable isomorphisms, degree spectra of relations, and Scott families*, **Annals of Pure and Applied Logic**, vol. 93 (1998), no. 1-3, pp. 153–193.

[KP54] STEPHEN C. KLEENE and EMIL L. POST, *The upper semi-lattice of the degrees of recursive unsolvability*, **Annals of Mathematics**, vol. 59 (1954), pp. 379–407.

[Kni86] JULIA F. KNIGHT, *Degrees coded in jumps of orderings*, **The Journal of Symbolic Logic**, vol. 51 (1986), no. 4, pp. 1034–1042.

[Kni98] ———, *Degrees of models*, **Handbook of Recursive Mathematics, Vol. 1**, Stud. Logic Found. Math., vol. 138, North-Holland, Amsterdam, 1998, pp. 289–309.

[KMVB07] JULIA F. KNIGHT, SARA MILLER, and M. VANDEN BOOM, *Turing computable embeddings*, **The Journal of Symbolic Logic**, vol. 72 (2007), no. 3, pp. 901–918.

[KS00] JULIA F. KNIGHT and MICHAEL STOB, *Computable Boolean algebras*, **The Journal of Symbolic Logic**, vol. 65 (2000), no. 4, pp. 1605–1623.

[Kru60] JOSEPH B. KRUSKAL, *Well-quasi-ordering, the Tree Theorem, and Vazsonyi's conjecture*, **Transactions of the American Mathematical Society**, vol. 95 (1960), pp. 210–225.

[Kud96a] OLEG V. KUDINOV, *An autostable 1-decidable model without a computable Scott family of $\exists$-formulas*, **Algebra i Logika**, vol. 35 (1996), no. 4, pp. 458–467, 498.

[Kud96b] ———, *Some properties of autostable models*, **Algebra i Logika**, vol. 35 (1996), no. 6, pp. 685–698, 752.

[Kud97] ——— , *The problem of describing autostable models*, **Algebra i Logika**, vol. 36 (1997), no. 1, pp. 26–36, 117.

[LR78] PETER E. LA ROCHE, **Contributions to Recursive Algebra**, Ph.D. thesis, Cornell University, 1978, p. 33.

[Lac73] ALISTAIR H. LACHLAN, *The priority method for the construction of recursively enumerable sets*, **Cambridge Summer School in Mathematical Logic (Cambridge, 1971)**, Springer, Berlin, 1973, pp. 299–310. Lecture Notes in Math., Vol. 337.

[Lav63] IGOR A. LAVROV, *The effective non-separability of the set of identically true formulae and the set of finitely refutable formulae for certain elementary theories*, **Algebra i Logika Sem.**, vol. 2 (1963), no. 1, pp. 5–18.

[LMMS05] STEFFEN LEMPP, CHARLES MCCOY, RUSSELL MILLER, and REED SOLOMON, *Computable categoricity of trees of finite height*, **The Journal of Symbolic Logic**, vol. 70 (2005), no. 1, pp. 151–215.

[Ler83] MANUEL LERMAN, **Degrees of Unsolvability**, Perspectives in Mathematical Logic, Springer-Verlag, Berlin, 1983, Local and global theory.

[LE66] EDGAR G. K. LOPEZ-ESCOBAR, *On defining well-orderings*, **Fundamenta Mathematicae**, vol. 59 (1966), pp. 13–21.

[Mal62] ANATOLII I. MAL'CEV, *On recursive Abelian groups*, **Dokl. Akad. Nauk SSSR**, vol. 146 (1962), pp. 1009–1012.

[Mar82] DAVID MARKER, *Degrees of models of true arithmetic*, **Proceedings of the Herbrand symposium (Marseilles, 1981)**, Stud. Logic Found. Math., vol. 107, North-Holland, Amsterdam, 1982, pp. 233–242.

[MM17] DAVID MARKER and RUSSELL MILLER, *Turing degree spectra of differentially closed fields*, **The Journal of Symbolic Logic**, vol. 82 (2017), no. 1, pp. 1–25.

[Mar68] DONALD A. MARTIN, *The axiom of determinateness and reduction principles in the analytical hierarchy*, **Bulletin of the American Mathematical Society**, vol. 74 (1968), pp. 687–689.

[Mar75] ——— , *Borel determinacy*, **Annals of Mathematics (2)**, vol. 102 (1975), no. 2, pp. 363–371.

[McC] ETHAN MCCARTHY, *Cototal enumeration degrees and the Turing degree spectra of minimal subshifts*, to appear in the Proceedings of the American Mathematical Society DOI: 10.1090/proc/13783.

[McC03] CHARLES F. D. MCCOY, Δ_2^0-*categoricity in Boolean algebras and linear orderings*, **Annals of Pure and Applied Logic**, vol. 119 (2003), no. 1-3, pp. 85–120.

[Med55] YURI T. MEDVEDEV, *Degrees of difficulty of the mass problem*, **Dokl. Akad. Nauk SSSR (N.S.)**, vol. 104 (1955), pp. 501–504.

[MM18] ALEXANDER MELNIKOV and ANTONIO MONTALBÁN, *Computable polish group actions*, **The Journal of Symbolic Logic**, vol. 83 (2018), no. 2, pp. 443–460.

[MN79] GEORGE METAKIDES and ANIL NERODE, *Effective content of field theory*, **Annals of Mathematical Logic**, vol. 17 (1979), no. 3, pp. 289–320.

[Mil83] TERRENCE MILLAR, *Omitting types, type spectrums, and decidability*, **The Journal of Symbolic Logic**, vol. 48 (1983), no. 1, pp. 171–181.

[Mil04] JOSEPH S. MILLER, *Degrees of unsolvability of continuous functions*, **The Journal of Symbolic Logic**, vol. 69 (2004), no. 2, pp. 555–584.

[MPSS] RUSSELL MILLER, BJORN POONEN, HANS SCHOUTENS, and ALEXANDRA SHLAPENTOKH, *A computable functor from graphs to fields*, to appear.

[Mon09] ANTONIO MONTALBÁN, *Notes on the jump of a structure*, **Mathematical Theory and Computational Practice**, (2009), pp. 372–378.

[Mon10] ——, *Counting the back-and-forth types*, **Journal of Logic and Computability**, (2010), pp. 857–876, doi: 10.1093/logcom/exq048.

[Mon12] ——, *Rice sequences of relations*, **Philosophical Transactions of the Royal Society A**, vol. 370 (2012), pp. 3464–3487.

[Mon13a] ——, *A computability theoretic equivalent to Vaught's conjecture*, **Advances in Mathematics**, vol. 235 (2013), pp. 56–73.

[Mon13b] ——, *Copyable structures*, **The Journal of Symbolic Logic**, vol. 78 (2013), no. 4, pp. 1025–1346.

[Mon13c] ——, *A fixed point for the jump operator on structures*, **The Journal of Symbolic Logic**, vol. 78 (2013), no. 2, pp. 425–438.

[Mon 14a] ——, *Computability theoretic classifications for classes of structures*, **Proceedings of ICM 2014**, vol. 2 (2014), pp. 79–101.

[Mon14b] ——, *Priority arguments via true stages*, **The Journal of Symbolic Logic**, vol. 79 (2014), no. 4, pp. 1315–1335.

[Mon15] ——, *Analytic equivalence relations satisfying hyperarithmetic-is-recursive*, **Forum Math. Sigma**, vol. 3 (2015), pp. e8, 11.

[Mon16b] ——, *Classes of structures with no intermediate isomorphism problems*, **The Journal of Symbolic Logic**, vol. 81 (2016), no. 1, pp. 127–150.

[Mon16a] ——, *Effectively existentially-atomic structures*, **Computability and Complexity**, Lecture Notes in Computer Science, vol. 10010, Springer, 2016, pp. 221–237.

[MonP2] ——, **Computable Structure Theory: Beyond the arithmetic**, Cambridge University Press, P2, in preparation.

[MS12] ANTONIO MONTALBÁN and RICHARD A. SHORE, *The limits of determinacy in second order arithmetic*, **Proceedings of the London Mathematical Society**, vol. 104 (2012), no. 2, pp. 223–252.

[Mor65] MICHAEL MORLEY, *Omitting classes of elements*, **Theory of Models (Proc. 1963 Internat. Sympos. Berkeley)**, North-Holland, Amsterdam, 1965, pp. 265–273.

[Mor04] ANDREI S. MOROZOV, *On the relation of Σ-reducibility between admissible sets*, **Sibirsk. Mat. Zh.**, vol. 45 (2004), no. 3, pp. 634–652.

[Mos69] YIANNIS N. MOSCHOVAKIS, *Abstract first order computability. I, II*, **Transactions of the American Mathematical Society**, vol. 138 (1969), pp. 427–464.

[Muc56] ALBERT A. MUCHNIK, *On the unsolvability of the problem of reducibility in the theory of algorithms*, **Dokl. Akad. Nauk SSSR, N.S.**, vol. 108 (1956), pp. 194–197.

[Muc63] ———, *On strong and weak reducibility of algorithmic problems*, **Sibirsk. Mat. Ž.**, vol. 4 (1963), pp. 1328–1341.

[Nie09] ANDRÉ NIES, **Computability and Randomness**, Oxford Logic Guides, vol. 51, Oxford University Press, Oxford, 2009.

[Nur74] ABYZ T. NURTAZIN, *Strong and weak constructivizations, and enumerable families*, **Algebra i Logika**, vol. 13 (1974), pp. 311–323, 364.

[Pos44] EMIL L. POST, *Recursively enumerable sets of positive integers and their decision problems*, **Bulletin of the American Mathematical Society**, vol. 50 (1944), pp. 284–316.

[Pou72] MAURICE POUZET, *Modèle universel d'une théorie n-complète: Modèle uniformément préhomogène*, **Comptes Rendus Hebdomadaires des Séances de l'Academie des Sciences, Série A-B**, vol. 274 (1972), pp. A695–A698.

[Puz09] VADIM PUZARENKO, *On a certain reducibility on admissible sets*, **Sibirsk. Mat. Zh.**, vol. 50 (2009), no. 2, pp. 415–429.

[Puz11] ———, *Fixed points of the jump operator*, **Algebra and Logic**, vol. 5 (2011), pp. 418–438.

[RS] MICHAEL O. RABIN and DANA SCOTT, *The undecidability of some simple theories*, unpublished notes.

[Rem81b] JEFFREY B. REMMEL, *Recursive Boolean algebras with recursive atoms*, **The Journal of Symbolic Logic**, vol. 46 (1981), no. 3, pp. 595–616.

[Rem81] ———, *Recursively categorical linear orderings*, **Proceedings of the American Mathematical Society**, vol. 83 (1981), no. 2, pp. 387–391.

[Ric77] LINDA RICHTER, **Degrees of Unsolvability of Models**, Ph.D. thesis, University of Illinois at Urbana-Champaign, 1977.

[Ric81] ———, *Degrees of structures*, **The Journal of Symbolic Logic**, vol. 46 (1981), no. 4, pp. 723–731.

[Sch16] NOAH SCHWEBER, **Interactions Between Computability Theory and Set Theory**, Ph.D. thesis, University of California, Berkeley, 2016, p. 137.

[Sel71] ALAN L. SELMAN, *Arithmetical reducibilities. I*, **Z. Math. Logik Grundlagen Math.**, vol. 17 (1971), pp. 335–350.

[Sho78] RICHARD A. SHORE, *Controlling the dependence degree of a recursively enumerable vector space*, **The Journal of Symbolic Logic**, vol. 43 (1978), no. 1, pp. 13–22.

[Sim76] HAROLD SIMMONS, *Large and small existentially closed structures*, **The Journal of Symbolic Logic**, vol. 41 (1976), no. 2, pp. 379–390.

[Sla98] THEODORE A. SLAMAN, *Relative to any nonrecursive set*, **Proceedings of the American Mathematical Society**, vol. 126 (1998), no. 7, pp. 2117–2122.

[Smi81] RICK L. SMITH, *Two theorems on autostability in p-groups*, **Logic Year 1979–80 (Proc. Seminars and Conf. Math. Logic, Univ. Connecticut, Storrs, Conn., 1979/80)**, Lecture Notes in Math., vol. 859, Springer, Berlin, 1981, pp. 302–311.

[Soa16] ROBERT I. SOARE, *Turing Computability*, Theory and Applications of Computability, Springer-Verlag, Berlin, 2016, Theory and applications.

[Sos04] IVAN N. SOSKOV, *Degree spectra and co-spectra of structures*, **Annuaire Univ. Sofia Fac. Math. Inform.**, vol. 96 (2004), pp. 45–68.

[Sos07] ALEXANDRA A. SOSKOVA, *A jump inversion theorem for the degree spectra*, **Proceedings of CiE 2007**, Lecture Notes in Computer Science, vol. 4497, Springer-Verlag, 2007, pp. 716–726.

[SS09] ALEXANDRA A. SOSKOVA and IVAN N. SOSKOV, *A jump inversion theorem for the degree spectra*, **Journal of Logic and Computation**, vol. 19 (2009), no. 1, pp. 199–215.

[Ste13] REBECCA M. STEINER, *Effective algebraicity*, **Archive for Mathematical Logic**, vol. 52 (2013), no. 1-2, pp. 91–112.

[Stu07] ALEKSEY I. STUKACHEV, *Degrees of presentability of models. I*, **Algebra Logika**, vol. 46 (2007), no. 6, pp. 763–788, 793–794.

[Stu09] ——— , *A jump inversion theorem for semilattices of Σ-degrees*, **Sib. Èlektron. Mat. Izv.**, vol. 6 (2009), pp. 182–190.

[Stu10] ——— , *A jump inversion theorem for the semilattices of Sigma-degrees* [*translation of mr*2586684], **Siberian Advances in Mathematics**, vol. 20 (2010), no. 1, pp. 68–74.

[Stu] ——— , *Effective model theory: approach via σ-definability*, **Effective Mathematics of the Uncountable** (Noam Greenberg, Denis Hirschfeldt, Joel David Hamkins, and Russell Miller, editors), Lecture Notes in Logic, vol. 41, Cambridge University Press, 2013, pp. 164–197.

[Sus17] MIKHAIL YA. SUSLIN, *Sur un définition des ensembles measurables B sans nombres transfinis*, **Comptes Rendus de l'Academie des Sciences Paris**, vol. 164 (1917), pp. 88–91.

[Thu95] JOHN J. THURBER, *Every low$_2$ Boolean algebra has a recursive copy*, **Proceedings of the American Mathematical Society**, vol. 123 (1995), no. 12, pp. 3859–3866.

[Vai89] RIMANTAS VAĬTSENAVICHYUS, *Inner-resolvent feasible sets*, **Mat. Logika Primenen.**, vol. 1 (1989), no. 6, pp. 9–20.

[Vat11] STEFAN VATEV, *Conservative extensions of abstract structures*, **CiE** (Benedikt Löwe, Dag Normann, Ivan N. Soskov, and Alexandra A. Soskova, editors), Lecture Notes in Computer Science, vol. 6735, Springer, 2011, pp. 300–309.

[Ven92] YURI G. VENTSOV, *The effective choice problem for relations and reducibilities in classes of constructive and positive models*, **Algebra i Logika**, vol. 31 (1992), no. 2, pp. 101–118, 220.

[Weh98] STEPHAN WEHNER, *Enumerations, countable structures and Turing degrees*, **Proceedings of the American Mathematical Society**, vol. 126 (1998), no. 7, pp. 2131–2139.

INDEX

Printed in the United States
by Baker & Taylor Publisher Services